Sustainable Metal Extraction from Waste Streams

Sustainable Metal Extraction from Waste Streams

Garima Chauhan
Perminder Jit Kaur
K.K. Pant
K.D.P. Nigam

Authors

Dr. Garima Chauhan
University of Alberta
Department of Chemical & Materials Engineering
9211 - 116 Street NW
Edmonton AB, T6G 1H9
Canada

Dr. Perminder Jit Kaur
Indian Institute of Technology
Center for Rural Development and Technology
Hauz Khas
New Delhi 110016
India

Prof. K.K. Pant
Indian Institute of Technology
Department of Chemical Engineering
Hauz Khas
New Delhi 110016
India

Prof. K.D.P. Nigam
Indian Institute of Technology
Department of Chemical Engineering
Hauz Khas
New Delhi 110016
India

Cover
Aerial top view Water treatment plant, shutterstock 1239611791 / Avigator Fortuner.

All books published by **Wiley-VCH** are carefully produced. Nevertheless, authors, editors, and publisher do not warrant the information contained in these books, including this book, to be free of errors. Readers are advised to keep in mind that statements, data, illustrations, procedural details or other items may inadvertently be inaccurate.

Library of Congress Card No.:
applied for

British Library Cataloguing-in-Publication Data
A catalogue record for this book is available from the British Library.

Bibliographic information published by the Deutsche Nationalbibliothek
The Deutsche Nationalbibliothek lists this publication in the Deutsche Nationalbibliografie; detailed bibliographic data are available on the Internet at <http://dnb.d-nb.de>.

© 2020 Wiley-VCH Verlag GmbH & Co. KGaA, Boschstr. 12, 69469 Weinheim, Germany

All rights reserved (including those of translation into other languages). No part of this book may be reproduced in any form – by photoprinting, microfilm, or any other means – nor transmitted or translated into a machine language without written permission from the publishers. Registered names, trademarks, etc. used in this book, even when not specifically marked as such, are not to be considered unprotected by law.

Print ISBN: 978-3-527-34755-1
ePDF ISBN: 978-3-527-82672-8
ePub ISBN: 978-3-527-82671-1
oBook ISBN: 978-3-527-82670-4

Typesetting SPi Global, Chennai, India
Printing and Binding CPI books GmbH, Leck

Printed on acid-free paper

Contents

Graphical Abstract *xi*
Preface *xiii*

1 Introduction to Sustainability and Green Chemistry *1*
1.1 Introduction *1*
1.2 Defining "Sustainability" *2*
1.3 Dimensions of Sustainability *3*
1.4 New Conceptual Frameworks to Define Sustainability *5*
1.4.1 Five Dimension Framework *5*
1.4.2 Four Force Model *5*
1.4.3 Corporate Sustainable Management *7*
1.5 Green Value Stream Mapping (GVSM) *7*
1.6 "Greening the Waste" *8*
1.7 Green Chemistry Terminology *10*
1.8 Green Ways of Metal Extraction: Core of the Book *11*
Questions *13*

2 Waste Handling and Pre-treatment *15*
2.1 Introduction *15*
2.2 Waste Categorization *17*
2.2.1 Waste Electrical and Electronic Equipment (WEEE) *17*
2.2.2 Agro-Residue Waste *20*
2.2.3 Industrial Waste *21*
2.3 Legislations and Regulations for Hazardous Wastes *27*
2.4 Handling/Management of Hazardous Waste *28*
2.4.1 Secured Landfilling *29*
2.4.2 Incineration *30*
2.4.3 Recycling of Hazardous Waste *31*
2.5 A Call for Metal Recovery from Waste *31*
2.5.1 Threat to Human Health and Environment *31*
2.5.2 Waste: An Artificial Ore *32*
2.5.3 "Waste" to Wealth *33*

2.6	Pretreatment of Waste	*34*
2.6.1	Disassembling the Waste	*34*
2.6.2	Size Reduction (Comminution)	*34*
2.6.3	Screening/Sieving	*35*
2.6.4	Classification	*36*
2.6.5	Segregation	*36*
2.6.6	Calcination and Chemical Pretreatment	*37*
2.7	Summary and Outlook	*37*
	Questions	*38*

3 Conventional Technologies for Metal Extraction from Waste *39*

3.1	Introduction	*39*
3.2	Pyrometallurgical Operations	*40*
3.2.1	Pyrometallurgical Treatment of Industrial Waste	*40*
3.2.2	Pyrometallurgical Treatment of WEEE	*45*
3.2.3	Major Challenges Associated with Pyrometallurgical Operations	*49*
3.3	Hydrometallurgical Treatment of Waste	*50*
3.3.1	Leaching of Metals in Acidic Medium	*50*
3.3.2	Leaching of Metals in Alkali Medium	*57*
3.3.3	Leaching with Lixiviants (Cyanide, Thiourea, Thiosulfate)	*60*
3.3.4	Halide Leaching	*66*
3.4	Summary and Outlook	*69*
	Questions	*70*

4 Emerging Technology for Metal Extraction from Waste: I. Green Adsorption *71*

4.1	Introduction	*71*
4.2	Adsorption	*71*
4.2.1	Hydrophilic Compounds	*72*
4.2.2	Hydrophobic Compounds	*72*
4.2.3	Polymer Matrix	*73*
4.3	Green Adsorption	*74*
4.4	Parameters Affecting the Adsorption Capacity of Green Adsorbents	*75*
4.4.1	Influence of pH	*75*
4.4.2	Influence of Temperature	*76*
4.4.3	Effect of Initial Concentration	*76*
4.4.4	Effect of Adsorbent Dosage	*76*
4.4.5	Effect of Co-ions	*77*
4.5	Adsorption Kinetic Models	*77*
4.6	Mechanism of Metal Uptake	*78*
4.7	Green Adsorbents: Relevant Literature	*79*
4.7.1	Agricultural Resources	*79*
4.7.2	Zeolites	*81*
4.7.3	Clay	*84*

4.7.4	Industrial Waste	*85*
4.7.5	Modified Biopolymers	*88*
4.8	Innovative Applications of Adsorption	*88*
4.9	Case Study	*89*
4.10	Summary and Outlook	*90*
	Questions	*91*

5 Emerging Technologies for Extraction of Metals from Waste II. Bioleaching *93*

5.1	Introduction	*93*
5.2	Bioleaching Process Description	*94*
5.3	Factors Affecting the Process Efficiency	*95*
5.3.1	Types of Microorganisms	*95*
5.3.1.1	Mesophiles	*95*
5.3.1.2	Thermophiles	*96*
5.3.1.3	Heterotrophic Microbes	*97*
5.3.2	Affinity Between Microorganisms and Metal Surfaces	*97*
5.3.3	Physicochemical Factors	*98*
5.3.3.1	Surface Properties	*98*
5.3.3.2	Oxygen and Carbon Dioxide Content	*98*
5.3.3.3	pH Value of Solution	*99*
5.3.3.4	Temperature	*99*
5.3.3.5	Mineral Substrate	*99*
5.3.3.6	Surface Chemistry of Metals	*99*
5.3.3.7	Surfactant and Organic Extractants	*100*
5.3.4	Reactor Design	*100*
5.4	Mechanism of Bioleaching Process	*101*
5.4.1	Biochemical Reaction (Direct vs. Indirect) Mechanism	*102*
5.4.2	Mechanism of Metal Sulfide Dissolution (Polysulfide Pathway)	*103*
5.5	Engineering Practices in Bioleaching Process	*104*
5.5.1	Batch Process	*105*
5.5.2	Continuous Process	*106*
5.5.3	Hybrid Processes	*110*
5.6	Application of Bioleaching in Extracting Metals from Waste	*110*
5.6.1	Extraction of Metals from WEEE	*111*
5.6.2	Extraction of Metals from Industrial Waste	*115*
5.6.3	Extraction of Metals from Mineral Waste	*118*
5.6.4	Extraction of Metals from Municipal Sewage Sludge	*119*
5.7	Technoeconomic Opportunities and Challenges	*119*
5.8	Summary and Outlook	*121*
	Questions	*122*

6 Future Technology for Metal Extraction from Waste: I. Chelation Technology *123*

	Abbreviations	*123*
6.1	Introduction	*123*

6.2	Defining "Chelation" *124*	
6.3	Classification of Ligands *124*	
6.4	Chemistry Associated with Chelation *127*	
6.4.1	Theories Derived for Metal–Ligand Complexation *127*	
6.4.2	Attributes of Metal Ions for Complexation *129*	
6.4.3	Metal–Chelate Complex Formation *130*	
6.4.4	The Chelate Effect *132*	
6.5	Chelation Process for Extraction of Metals *133*	
6.5.1	Framework for Chelating Agent Assisted Metal Extraction from Solid Waste *133*	
6.5.2	Process Parameters Affecting the Metal Extraction Process *135*	
6.5.2.1	Effect of Reaction pH *135*	
6.5.2.2	Effect of Molar Concentration of Chelating Agent *138*	
6.5.2.3	Effect of Reaction Temperature *140*	
6.5.2.4	Presence of Competing Ions in Reaction Zone *141*	
6.5.3	Factors Affecting Stability of Metal–Ligand Complex *142*	
6.6	Novel Applications of Chelating Agents *143*	
6.6.1	Chelating Agents Used for Metal Extraction from Metal-Contaminated Soil *144*	
6.6.1.1	Hydrometallurgical Route of Chelation Process (Direct Use) *144*	
6.6.1.2	Phyto-remediation of Soils in Presence of Chelating Agents *145*	
6.6.2	Chelating Agents Used for Metal Extraction from Industrial Waste *147*	
6.6.3	Chelating Agents Used for Metal Extraction from WEEE *149*	
6.7	Ecotoxicological Concerns and Biodegradability *151*	
6.8	Summary and Outlook *155*	
	Questions *155*	
7	**Future Technology for Metal Extraction from Waste: II. Ionic Liquids** *157*	
	Abbreviation *157*	
7.1	Introduction *158*	
7.2	What Are Ionic Liquids? *158*	
7.3	Characteristic Properties of Ionic Liquids *161*	
7.3.1	Melting Point *162*	
7.3.2	Vapor Pressure and Nonflammability *162*	
7.3.3	Thermal Stability *163*	
7.3.4	Density *164*	
7.3.5	Viscosity *164*	
7.3.6	Polarity *166*	
7.3.7	Coordination Ability *166*	
7.3.8	Conductivity *167*	
7.3.9	Solubility *167*	
7.4	Classification of Ionic Liquids *169*	
7.5	Environmental Scrutiny of Ionic Liquids *171*	
7.6	Applications of Ionic Liquids *173*	

7.6.1	Extraction of Metals from Aqueous Media	173
7.6.2	Extraction of Metals from Industrial Solid Waste/Ores	176
7.6.3	Extraction of Metals from WEEE	177
7.7	Summary and Outlook	179
	Questions	179

8   Scale-up Process for Metal Extraction from Solid Waste   181
Nomenclature   181
8.1   Introduction   182
8.2   Process Intensification   183
8.3   Intensification of Metal Extraction Processes   185
8.3.1   Centrifugation   185
8.3.2   Liquid–Liquid Extraction   185
8.3.3   Mixing   186
8.3.4   Reactors   188
8.3.5   Comminution   188
8.3.6   Drying   189
8.4   Scaling Up from Batch to Continuous Process   189
8.4.1   Process Design Fabrication   190
8.4.2   Designing of Pilot Plant   191
8.4.2.1   Material Balance   191
8.4.2.2   Development of Comminution Circuit   193
8.4.3   Reactor Sizing and Agitator Selection   197
8.4.4   Design of Filtration System   199
8.4.5   Design of Heat Exchanger   201
8.4.6   Design of Precipitator Unit   202
8.4.7   Batch Scheduling   203
8.5   Summary and Outlook   204
Questions   205

9   Process Intensification for Micro-flow Extraction: Batch to Continuous Process   207
Jogender Singh, Loveleen Sharma, and Jamal Chaouki
Abbreviations   207
9.1   Introduction   208
9.2   Miniaturized Extraction Devices   208
9.2.1   Intensification in Miniaturized Extraction Devices   209
9.2.2   Application of Miniaturized Extraction Devices   211
9.3   CFI for Continuous Micro-flow Extraction   212
9.3.1   Designing CFI as an Extractor   216
9.3.2   Extraction Parameters   218
9.3.3   Methodology and Setup for Micro-flow Extraction   218
9.3.4   Liquid–Liquid Micro-flow Extraction   220
9.3.4.1   Typical Flow Patterns   220

9.3.4.2　Extraction Efficiency *222*
9.3.4.3　Effect of Aqueous Phase Volume Fractions on Extraction Efficiency *223*
9.3.5　Micro-flow Extraction of Co and Ni *225*
9.3.5.1　Effect of pH *225*
9.3.5.2　Effect of Residence Time *225*
9.3.5.3　Effect of Extractant Concentration *229*
9.4　Summary and Future Challenges *229*
　　Questions *229*

Bibliography *231*

Index *273*

Graphical Abstract

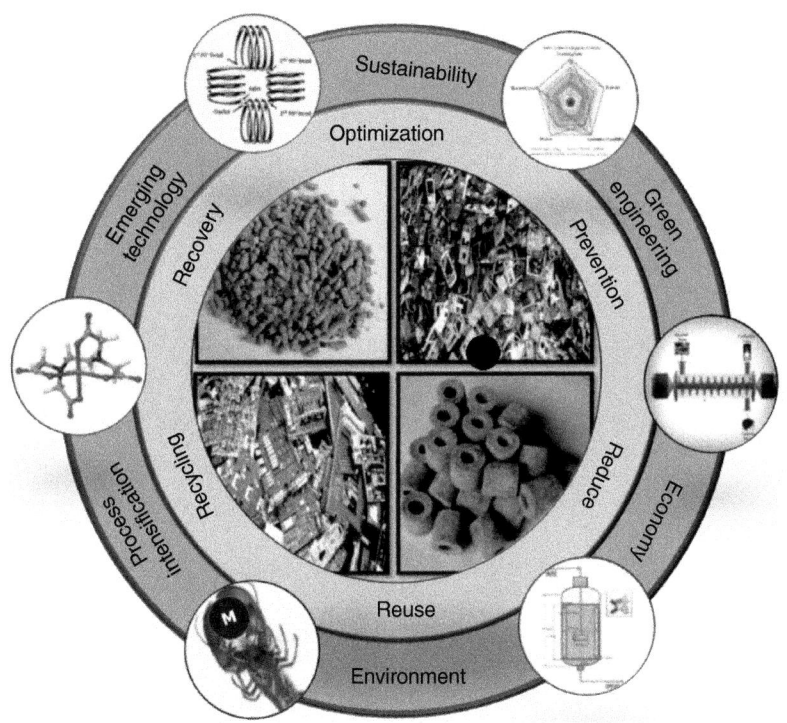

Preface

Environment friendly technologies and improved resource intensity is the call of the present day where various types of wastes pose many hazards for the eco-system. The key concept of writing this book is to demonstrate transdisciplinary research as an executable direction to achieve sustainable governance of natural resources and solid waste management. Different aspects associated with solid waste management including types of solid waste, conventional waste disposal methodologies, rules and regulations for waste disposal, and the global scenario for solid waste generation are discussed in brief. Recovery and recycling of heavy metals from solid waste can always be a good option to reduce fresh inputs and hazardous waste output. This book primarily focuses on showcasing various recent technological innovations that have the potential to extract metals from solid waste and thus to offer answers to challenges in the field of resource optimization.

While many books and research articles have already been published highlighting the developments in the field of solid waste management, information about the emerging processes of waste treatment that would be in the forefront in near future is scanty. The concept of developing sustainable practices for waste minimization defines the theme of our book. This book provides a comprehensive collection of the conventional, emerging and future technologies for metal extraction from industrial waste and waste electrical and electronic equipment(s) in a sustainable way. Various aspects of the novel processes ranging from basic concepts, benchmark performance of these technologies on laboratory scale, recent research trends in metal extraction, challenges to the implementation of these at large scale, and the future perspectives have been summarized in this book. Process intensification is another interesting research area that has not been dealt with extensively in the literature. This book presents a dedicated section on scale-up and process intensification of metallurgical processes. Covering a variety of interdisciplinary topics on resource optimization and waste minimization makes it one book for all that serves as an excellent reading material for engineers, science students, entrepreneurs, and organizations who are working in the field of waste management and wish to gain information on upcoming sustainable processes.

The chapters of this book present various novel technologies that show the potential of metal extraction in a greener way and offer several new opportunities to attain sustainability. It is important to understand how crucial it is to learn about solid waste management in the present context. Chapter 1 provides an overview of sustainability-related constructs and illustrates the prerequisites for sustainable management of waste streams. Chapter 2 covers the basics of solid waste handling ranging from an analysis of the relevance, categories of wastes, consequences of untreated waste disposal into the environment, government initiatives, management strategies, and unit operations for pretreatment of wastes.

Some of the widely accepted conventional technologies for metal extraction such as hydro- and pyrometallurgical methods are discussed in Chapter 3. Next, the book brings out the possibility of sustainable green processes for metal extraction. Chapters 4 and 5 discuss the approaches (adsorption, bioleaching) that have been investigated already at industrial/pilot scale for several applications and that are now being investigated (slow to be adapted) to explore their novel applications in the field of waste management (especially the extraction of metals from waste) and are included in the category of "emerging technology." Chelation technology and ionic liquids (Chapters 6 and 7) have been included in the list of future technologies. These approaches are still in their infancy and have been tested only at the laboratory scale (only academic). Chapter 8 addresses the extent of commercialization of the processes discussed through scaling-up of the bench scale processes. A case study is illustrated to process a large amount of industrial/electronic waste while scaling up various unit operations used in the processing. A brief insight has been provided on the technical complexities and economic feasibility of the aforementioned eco-friendly processes of metal extraction from waste. Chapter 9 introduces coiled flow inverter, a recent example of process intensification that has been used earlier for heat transfer/mass transfer applications and that is now being explored for recovery of metals from the aqueous and solid waste stream. If these processes are deployed on a large scale, certainly it will deepen their impact.

Each new book gives recognition to those whose name appears on the title page, yet a book is, in fact, a collective effort of many more individuals who have contributed either directly or indirectly to its successful completion. The authors would like to extend their gratitude to every individual for their support and motivation throughout the journey. The most obvious debt is certainly to the researcher(s) who are constantly working to find out new sustainable approaches for the betterment of human lives and future generations. The authors would also like to acknowledge Dr. Jogender Singh, Dr. Loveleen Sharma, and Prof. Jamal Chaouki for their help in completing Chapter 9.

New Delhi, India *Garima Chauhan, Perminder jit Kaur,*
June 01, 2019 *K.K. Pant, and K.D.P. Nigam*

1

Introduction to Sustainability and Green Chemistry

> *"We need a new system of values, a system of the organic unity between humankind and nature and the ethic of global responsibility."*
> — Mikhail Gorbachev

1.1 Introduction

The growing necessity in the past few decades has initiated a mechanical thinking. Rapid industrialization has the impetus to increase the consumption of natural resources and consequently to degrade the environment by disposing off industrial effluents into the soil–water–air continuum in a cavalier manner. These activities result in change in weather patterns, global warming, foodborne contamination, and increasing rate of health risk, which have led to significant concern of researchers toward the concept of sustainable development (SD). A call for sustainability is required for preserving nonrenewable natural resources and improving industrial economy.

The World Commission on Environment and Development (WCED) launched the concept of SD to illustrate new schemes orientated to economic, social, and ecological benefits. A major step toward SD was initiated in 1987 by a commission under the Chair of Gro Harlem Brundtland (WCED 1987), convened by the United Nations in 1983. The report was commissioned to address issues such as the accelerating rate of deterioration of the environment and natural resources and its consequences on economic and social development. In the Brundtland report, SD was defined as "The development that meets the needs of present generation without compromising the ability of future generations to meet their own needs." This widely popular definition of SD posed several challenges to researchers and industrialists to equilibrate the ecological, economic, and technical development. Eventually, in order to respond to this challenge, the concept of Green Chemistry was nurtured.

Green chemistry is defined as "The process of manufacturing of chemical product in which raw materials are environment friendly, without use of any hazardous or toxic chemicals, energy efficient, inherently safer processes and involves minimum production of waste."

Sustainable Metal Extraction from Waste Streams, First Edition.
Garima Chauhan, Perminder Jit Kaur, K.K. Pant, and K.D.P. Nigam.
© 2020 Wiley-VCH Verlag GmbH & Co. KGaA. Published 2020 by Wiley-VCH Verlag GmbH & Co. KGaA.

Although the terms "Green chemistry" and "Sustainability" are similar constructs that are used interchangeably in general, it is believed that the goal of sustainability is achieved using Green chemistry approaches. An overview of sustainability-related constructs is provided in this chapter to encourage readers, paying particular attention to the use of these terms in their studies. Dimensions of sustainability and new frameworks to define SD are also outlined. Green chemistry terminology and value stream mapping (VSM) are discussed in brief to make readers aware of the basic concepts of the green processes. A section at the end introduces the readers to the core theme of the book, which is centered around the recovery of metals from waste in a sustainable manner using green chemistry approaches.

1.2 Defining "Sustainability"

Sustainability can be defined as "The expectations of improving the social and environmental performance of the present generation without comprising the ability of future generations to meet their social and environmental needs" (Hart and Milsten 2003). Sustainability was cited in various ways in the literature based on the source; however, the core concept always refers to the society and the environment. Aras and Crowther (2009) referred to it as a fundamental and complex construct that mandates the balance of several factors on the planet to continually exist. This construct is also delineated as a belief of the preservation of natural resources.

Sustainability also considers the financial advantages in the present along with the assurance of benefits for the future generations. Elkington (2005) suggested "economics" as an integral factor along with social and environmental factors to attain sustainability and coined the term "triple bottom line (TBL)." This term was cited as the practical framework of sustainability by Rogers and Hudson (2011). The economic line of TBL framework concerns the capability of economy to survive and evolve into the future in order to support future generations. Industrial practices that are beneficial and fair to the human capital and to the community are included in the social line (Elkington 2005). Environmental line refers to the efficient use of energy resources and reducing the environmental pollution. A complex balance among the economy, environment, and society is the construct that leads to SD. These aspects of SD are essential to be considered by individuals and industries.

However, the lack of a rigid framework and poor interpretation of the term "sustainability" confronted inconsistency in its usage. Some sustainability studies are centered on one line only whereas some authors tried to include two or more lines in their studies. Yan et al. (2009) employed the term "sustainability" to refer to the environmental line, whereas the social line was emphasized by Bibri (2008). Some of the literature studies included the importance of the economic line into the term "sustainability" (Collins et al. 2007). Although very few studies managed to use this term to refer to all three bottom lines together (Marcus and Fremeth 2009), efforts are being made across the world to emphasize SD at national and international levels. Several organizations/industries are trying to incorporate these three lines of sustainability in their policies and cultures. For example, the International Organization for Standardization (ISO) provides Ecolabels to market products under various categories as follows:

Type 1 (ISO 14024: Environment labels and declaration): This type of products are in accordance with environmental criteria released by a third party organization. For instance, European Ecolabels are awarded on the basis of life cycle assessment (LCA) of a product under the supervision of the Ecolabel Committee. These labels are considered as a clear sign of environmental excellence. The products labeled under the European Ecolabel can become a product of choice for the consumers who are concerned for sustainability.

Type II (ISO 14021: Self-declared environmental claims): These labels are self-declared by companies or producers, based on environmental performances of their products, for example, the recyclability at end of the product life.

Type III (ISO 14025: Environment labels and declaration): These labels require a quantified declaration of the life cycle of the product, which can be verified by a third party organization. This helps to provide transparent information to the consumer for comparison purposes.

These labels consider the source of raw material, processing involved in the preparation of the final product, use, and end of life of the product. Moreover, these international environment sustainability labels are also based on the following aspects of a product:

i. Amount of energy and natural resources consumed during the production process
ii. Emission of toxic substances to the environment
iii. Impact of process on habitat and natural resources such as water, air, and soil
iv. Waste management practices followed by the organization.

Therefore, here it has become necessary to understand the dimensions of sustainability in order to maximize profits as well as to handle all the social and environmental responsibilities.

1.3 Dimensions of Sustainability

Three fundamental factors, i.e. People, Planet, and Profit (the three Ps) associated with the TBL construct, can be accounted as the dimensions of sustainability and outlined here as the traditional sustainable governance. Sustainability of any process thus banks upon the economic, social, and environmental pillars as shown in Figure 1.1. If any of these pillars is weak then the process cannot be considered a sustainable practice. The interlocking circles indicate that these three dimensions should be integrated in such a way as to be able to maintain balance between all the factors. It is observed that even most of the distinguished organizations focus on only one dimension at a time. The United Nations Environment Programme (UNEP), the Environmental Protection Agencies (EPAs) of

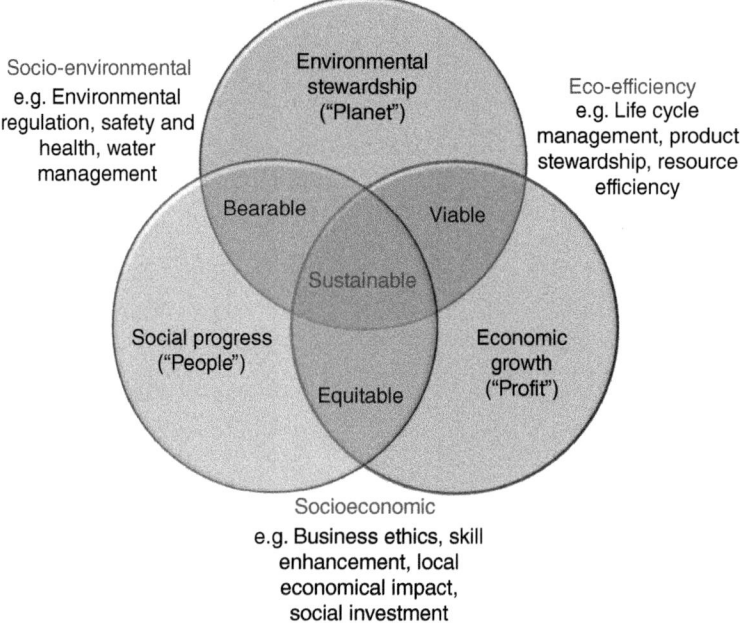

Figure 1.1 Dimensions of sustainability (three Ps).

many nations, and environmental nongovernmental organizations (NGOs) focus only on the environmental pillar. On the other hand, the primary focus area of the World Trade Organization (WTO) and the Organization for Economic Cooperation and Development (OECD) is centered on economic growth. It is difficult to include all the three pillars in one process; still efforts should be made in this direction for the continuance of natural resources indefinitely in future.

Economic pillar (the "profit" factor) of sustainability is referred to as the condition of necessary economic benefits projected indefinitely into the future. Economically sustainable industry should focus on the maximization of profit per unit of product, maintenance of a strong competitive position in the industry, and high level of operational efficiency along with being consistently profitable. Carroll (1991) presented a corporate socio-economic model that suggests that "initiating a strategy on socio-economic responsibility is the foundation on which other three dimensions are based on." Any new concept for sustainability should consider the costs of implementing sustainability and identify potential benefits to ensure long-term financial benefits. Thus, economic feasibility and process productivity are the most desired characteristics to define sustainable practices.

Social dimension of sustainability (the "people" factor) refers to a shared sense of social purpose to foster social integration and cohesion. An organization with social responsibility must ensure that it contributes the resources to the community and improves quality of life. Social responsibilities are also connected to the environmental pillar in order to provide a healthy ecosystem to the society. In this regard, safety, health care, environmental regulations, access to potable water, crisis management, and many other environmental practices become a part of the social pillar to achieve sustainability.

The contribution of the environment (the "planet" factor) to the economy and society is perceived to be through the operation of a wide range of environmental functions. De Groot (1992), defines "environmental functions" as "the capacity of natural processes and resources to provide services in order to satisfy human needs." Also, a natural process is based on complex interactions between living organisms, i.e. biotic components and chemical and physical components of ecosystems through the universal driving forces of matter and energy.

In support of the aforementioned definition, preservation of natural resources is one of the major responsibilities of the scientific community, which is further extended to the three Rs (Reduce, Recycle, Reuse) scheme to achieve sustainability. Several research efforts are being made in order to develop novel environment-friendly technologies, minimize waste generation, use open- and closed-loop recycling processes, use alternative energy sources, effectively utilize renewable/nonrenewable resources, and foster harmony between supply chains and nature.

1.4 New Conceptual Frameworks to Define Sustainability

1.4.1 Five Dimension Framework

The aforementioned WCED definition (WCED 1987) is widely accepted to delineate the term "sustainability"; however, some authors contended the anthropocentrism of the WCED definition. Anthropocentrism indicates that the gratification of human needs inherently contravenes with environmental constraints and, therefore, society and environment are represented as separate "pillars" to define the dimensions of sustainability.

Seghezzo (2009) considered this limitation of the WCED sustainability definition and proposed a new conceptual framework that includes five dimensions of sustainability. Territorial (place), temporal (permanence), and personal (persons) aspects of development were covered in this new definition of sustainability as shown in Figure 1.2. Place contains the three dimensions of space, Permanence is the fourth dimension of time, and the Persons category represents a fifth, human dimension. The importance of economy is also considered to be overestimated in the WCED definition, which has also been addressed in the Seghezzo framework. Spatial and temporal boundaries must also be taken into account to assess sustainability, which has been ignored in the WCED definition of sustainability. The importance of time in the complexities associated with problem solving is also acknowledged.

1.4.2 Four Force Model

A conceptual framework, the "Four Force Model," assists in the transformation of unsustainable development (USD) to SD. It is a structured approach to identify

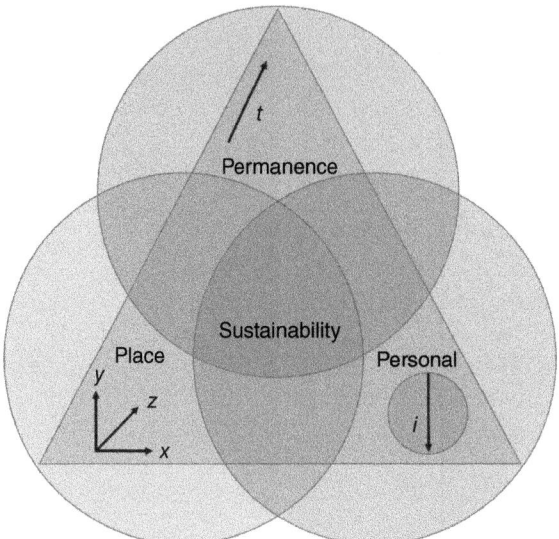

Figure 1.2 The new five-dimensional sustainability triangle. Source: Seghezzo (2009). Reproduced with permission of Taylor & Francis.

the need, cause, and process to accomplish SD. Five issues, i.e. current USD, environmental degradation, future SD, greening force, and greening process, have been emphasized in this model (Gandhi et al. 2006). The model suggests that the move of an enterprise lower end solution to a higher end solution leads to a green process. By enhancing the resource utilization, recycling, and extraction of maximum valuable components from waste streams, an organization can easily meet the target of minimum waste and green process. With increasing international cooperation and technology transfer, it is possible to minimize waste without causing any harm to the environment.

Figure 1.3 illustrates the link between current growth with USD and the most requisite SD in future. The back-casting approach utilizes a particular future scenario to be considered initially. Backward direction was defined by fulfilling the needed outcomes in order to link the future with the present. The four force model defines the four indicators of environment as the driving force, state, reactive response, and proactive response (Banerjee 2002). The model shows the present state as environment degradation. The future state is targeted as SD using either proactive or reactive responses. It is also known as a process of determining the steps to be taken or events to be realized for the needed future.

In the present context, environmental degradation is the major outcome of the current USD. Industrial growth occurs by extracting and utilizing natural resources and therefore nature shrinks with the expansion in industrial activities and economics. Certain factors such as rapid industrialization, increase in per capita income, subsequent changes in resource consumption pattern, exponential increase in population, and continuous depletion of nonrenewable resources are responsible for the increasing rate of environmental degradation. The greening process may be either reactive or proactive responses to the four driving forces. These four driving forces include three external forces (regulatory, community, and consumer force) and one internal force (financial advantage). These

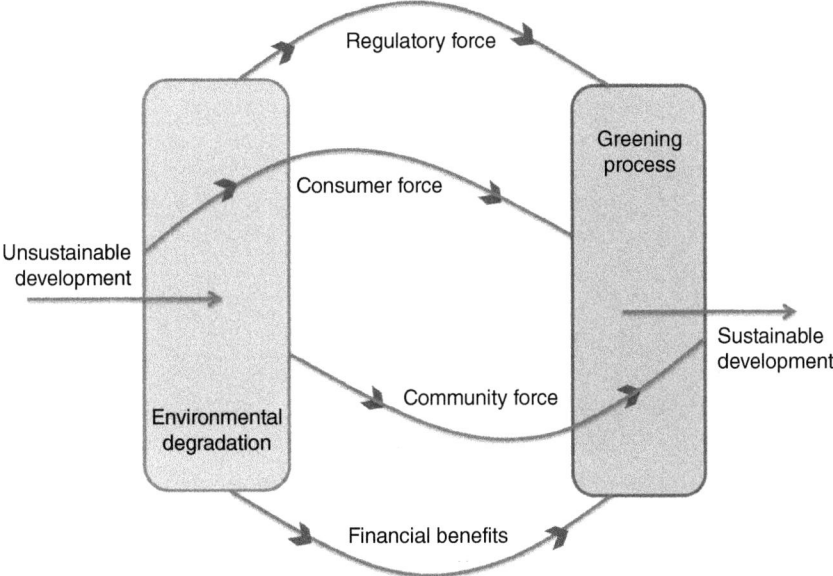

Figure 1.3 Four force model. Source: Gandhi et al. (2006). Reproduced with permission of Emerald Group Publishing Limited.

forces exert extreme pressure on the industries to adopt green processes, which may lead from USD to novel sustainable technologies and products.

1.4.3 Corporate Sustainable Management

Corporate sustainable management is a very new concept that is adopted by several industries to involve their own framework to manage sustainability issues. Companies are constantly making efforts to differentiate their products on the basis of sustainability practices. Some companies have developed the "carbon neutral" programs, whereas other industries are working to minimize hazardous gas emissions and waste generation from the source. Corporate sustainability index is also an emerging idea to measure sustainability performance. The index provides vital and timely information to customers and shareholders on the environmental health and sustainability of companies. Industries are coerced to pay sincere attention to minimize environmental impacts such as waste generation, greenhouse gas emissions, deforestation, and other environmental hazards due to industrial activities.

1.5 Green Value Stream Mapping (GVSM)

The sequence of activities to develop novel green products and services plays an important role in attaining sustainability. Therefore, a process mapping method named as the "value stream mapping (VSM)" is compiled to understand the sequence of activities, to identify and eliminate the non-value-added activities, and to streamline the process. Constrained environmental rules have extended the conventional VSM to include both eco-friendly green elements in its mapping technique. Green value stream mapping (GVSM) is an extension to achieve the same goal by ruling out the non-value-added materials, reducing costs, identifying ways to minimize waste generation, and ascertaining opportunities to improve performance. It focuses on excess usage of energy, water, and raw material; release of pollutants, and hazardous and solid waste discharge into the ecosystem; use of hazardous substances in the production process, etc.

1.6 "Greening the Waste"

"Greening the waste" concept refers to an approach in which industry shifts from less preferred methods of waste treatment including treatment with toxic chemicals and incineration toward the "three R: Reduce, Reuse, and Recycle" principle of sustainability. Further, researchers are aware of the fourth dimension of sustainability, Recovery. By integration of the construct of "recovery" in solid waste management, we can move in the desired upstream direction in solid waste management hierarchy.

This calls for the integrated solid waste management (ISWM) approach, based on appropriate waste treatment and recovery technology, to bridge the gap between community and waste management authorities. Considerations regarding the environmental impacts, social impact, and financial constraints have to factored in while designing the relevant ISWM strategy (Figure 1.4).

In order to apply the concept of "Greening the waste," the "three 3 R" principles of sustainability should be applied: reuse the resource wherever possible to abstain from the waste of virgin resources. Recycle the waste by converting waste into suitable products and recover the energy from waste using pertinent techniques. Use of resource optimization techniques assists in minimizing resource wastage. Proper collection and segregation of waste at source ensures efficient waste treatment. Further, resources should be reserved and be used prudently to avert their overconsumption. Landfilling of the waste should be avoided as leachates from it are hazardous to nearby environment. Resource optimization techniques need to be applied to improve the consumption of resources.

The greening of the waste sector should be facilitated by significant breakthroughs in technologies required for collection, reprocessing, and recycling of waste, extracting energy from organic waste, and efficient gas capture from landfills. Recovering energy and other useful products from waste is being enabled by considerable technological breakthroughs. Many countries have moved from incineration to energy generation as a part of waste treatment. Mechanical and biological treatment (MBT) and biomethanation are also being recognized as suitable for processing organic wet waste in developing countries. However, incomplete segregation of dry and wet organic waste has been a major barrier to the widespread successful adoption of these technologies in these countries. Techniques such as vermi-composting and rapid composting have led to conversion of organic waste into useful agricultural manure at a pace faster

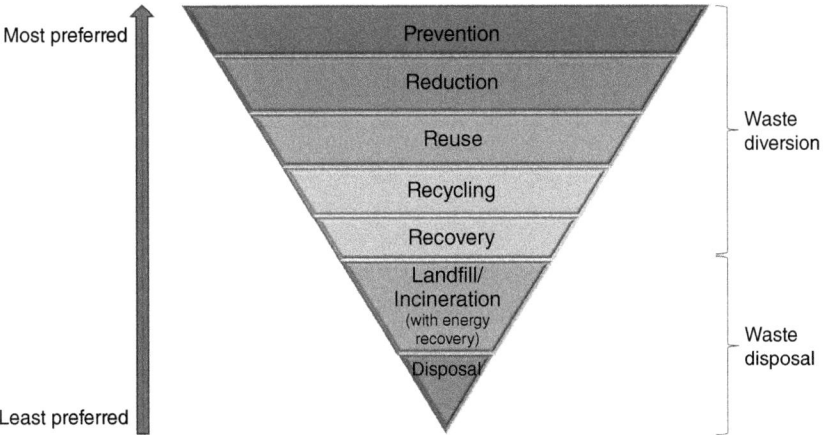

Figure 1.4 Waste management hierarchy.

than natural decomposition. Advanced technologies are employed to convert waste into value-added products. For instance, waste with high calorific value is used to generate refuse derived fuel (RDF).

In spite of the efforts made by researchers in the field of valorization of waste, no specific targets for greening the waste sectors have been established by international conventions. Appreciable efforts have been made by northern Europe, the Republic of Korea, and Singapore, by subjecting over 50% of waste materials to the material recovery process (Gaillochet and Chalmin 2009). Many Asian countries, such as Japan, have initiated efforts to evaluate the yearly developments in recycling rates by setting material flow indicators under three categories, i.e. "input," "cycle," and "output." Based on indicators such as resource productivity (increased from 210 000 yen/tonne in 1990 to 390 000 yen/tonne in 2010), recycle use rate (increased from 8% in 1990 to 14% in 2010), and final disposal amount (decreased from 110 million tonnes in 1990 to 28 million tonnes in 2010), recycling of material is encouraged (Ministry of Environment, Government of Japan 2008). Similarly, the concept of circular economy (CE) has been included in the 11th five-yearly plans of China. Efforts on the practice of three Rs policy – Reduce, Reuse, and Recycle are made in European countries to foster sustainable materials management. Earlier in 2011, cities of the United Kingdom, especially London, drafted a policy to manage waste by setting up 70% commercial/industrial waste's recycling/composting by 2020 and 95% reuse and recycling of waste by 2020 (Mayor of London 2010). However, The Netherland's waste management policy (also known as the "Lansink's Ladder") is based on the concept of the wise use of existing resources, along with the reduction of waste generation and recovery of resources and energy from waste (Zimring and Rathje 2012). The country uses stony waste that meets the criteria of Dutch Building Materials Decree for construction applications (Liu et al. 2015). The United States too, through the EPA, has laws to deal with release of waste that are toxic for public health, welfare, and environment. Quantitative risk assessment policy is followed in the country to evaluate the effect of waste on public health and environment.

Thus, global awareness about increasing waste generation and potential of resource recovery from waste has given rise to the term "Greening the waste." The various aspects of valorization of product are economic benefits (energy generation, new business avenues, creation of jobs, secondary sources of metals), environmental protection (resource conservation, reduction in greenhouse gases (GHG)), and social benefits such as compost production supporting organic agriculture, contributions to equity, and poverty eradication. Better management of waste improves public health and the condition of water bodies. This consequently leads to reduced health costs, which is an important stream of benefits of "Greening the Waste."

1.7 Green Chemistry Terminology

In green chemistry, environment factor (E-factor) is an important term to illustrate the process efficiency. Figure 1.5 illustrates a general process flow diagram demonstrating input of raw material, which in turn results in product and waste generation.

To define the environmental efficiency, E-factor is defined as the mass ratio of waste produced (excluding water) to the desired product. E-factor is related to the yield of the process as defined in Eq. (1.1):

$$\text{E-factor} = \frac{\text{mass of waste}}{\text{mass of desired product}} \quad (1.1)$$

It is recommended that a process should be designed to minimize the E-factor. There are various factors that affect the value of the E-factor, including the following:

i. Yield of the desired product
ii. Reagents used
iii. Number of steps and reaction sequences
iv. Extraction and purification solvents.

The average E-factor varies for the oil refining (<0.1) to bulk chemical (1–5), to fine chemical (5–50), to pharmaceutical industries (25–>100).

In addition, there is another term known as "atom efficiency." It is defined as the ratio of molecular weight of the desired product to the sum of molecular weights of all substances produced in a stoichiometric equation.

$$\text{Atom efficiency} = \frac{\text{molecular weight of desired product}}{\text{sum of molecular weight of all products}} \times 100\% \quad (1.2)$$

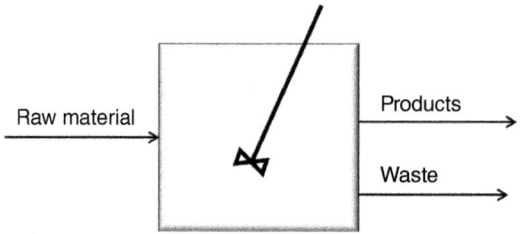

Figure 1.5 General chemical process showing input and output.

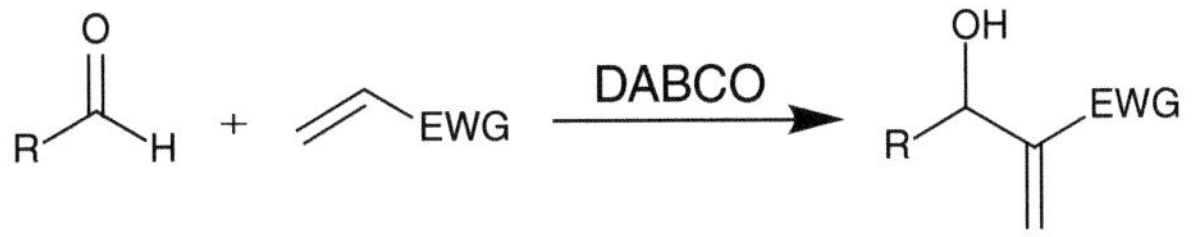

Figure 1.6 Example of an atom-efficient reaction.

Atom efficiency is a tool to show the efficiency of a process to utilize all atoms involved. It is useful to compare different pathways to obtain the desired product, without considering the yield.

Baylis–Hillman reaction, shown in Figure 1.6, is a 100% atom-efficient process. In this reaction, an activated alkene derivative reacts with an aldehyde, catalyzed by a tertiary amine (for example, DABCO = 1,4-diazabicyclo[2,2,2]octane). No co-products are known to be formed. All atoms are efficiently converted into the desired product, resulting in complete efficiency of the process.

Effective mass yield is also used to define the "greenness" of a reaction. This term can be explained as

$$\text{Effective mass yield} = \frac{\text{mass of desired products}}{\text{mass of non-benign reagents}} \times 100\% \quad (1.3)$$

1.8 Green Ways of Metal Extraction: Core of the Book

It is clear from the previous discussion that sustainable management is of utmost importance for the accomplished growth of any industry. Environment is one of the pillars of the sustainability triangle that is generally ignored by various industries in order to add financial benefits. However, these days, the constrained environmental regulations are forcing industries to pay sincere attention toward waste minimization and resource optimization. Waste minimization is a process of increasing resource utilization and extracting maximum from the waste before final disposal. With the new technological innovations, the concept of waste minimization can be directly referred to as Green Technology Development, which includes energy generation, resource extraction, pollution prevention, and recycling of resources and disposal with energy recovery.

The core of this book is based on the belief that transdisciplinary research can be considered as the executable direction to achieve the sustainable governance of natural resources and solid waste management. The dominant waste streams, i.e. electronic waste (waste electrical and electronic equipment [WEEE]), agro-residue waste, and industrial waste originated from petrochemical/fertilizer/textile industries, have been covered in detail in the Chapters 3–7 to illustrate various conventional, emerging, and future technologies for the recovery of metals from these waste streams.

The growing proportion of heavy metals in soil and water is a threat to human health. Some of the negative impacts of heavy metals on plants include decrease of seed germination and lipid content by cadmium, decreased enzyme activity and plant growth by chromium, the inhibition of photosynthesis by copper and mercury, the reduction of seed germination by nickel, and the reduction of chlorophyll production and plant growth by lead. The impacts on animals include reduced growth and development, cancer, organ damage, nervous system damage, and in extreme cases, death. In addition, metals are also an ineluctable contributor in the industrial development. Globally, scarcity of heavy metal resources is an important matter of concern, keeping in view the increasing rate of industrialization Therefore, recovery of metals from waste (industrial/agricultural/electronic) is of significant importance from the environmental as well as industrial perspectives. Numerous approaches have been widely reported in the literature for recovery of metals from contaminated sites. Adsorption has been considered a good alternative for efficient extraction of metals from industrial waste water. Membrane separation, electrotreatment, photocatalytic processes, and absorption by aquatic plants have also been employed to remove metals and other pollutants from water. Various techniques such as vitrification, excavation, removal of contaminated soil layer, phytoremediation, and soil washing using surfactants, acids, alkalis, and chelating agents have been investigated for soil remediation. Various hydrometallurgical (leaching, solvent extraction, and biological methods) and pyrometallurgical processes such as calcination followed by leaching or roasting at high temperatures have been employed for extraction of heavy metals from spent catalysts, soil, and other industrial wastes. All these processes have shown potential in removing metals from industrial wastes; however, use of hazardous chemicals, secondary pollution possibility by these chemicals, less acceptance of biological processes in the environment, and high process cost restrict their use on a large scale.

This excogitation is now assisting researchers to move toward greener approaches such as chelation technology, ionic liquids (ILs), and other green extractants. A new term "Green Adsorption" has also been introduced recently for metal extraction from waste water, which includes low-cost ecofriendly materials originated from agricultural sources and by-products (fruits, vegetables, food), agricultural residues, and wastes. Several authors employed natural bentonite and zeolites, rice/peanut husk and fly ash, banana and orange peel, and anaerobic granular sludges for metal extraction. Some authors also employed modified biopolymer adsorbents derived from chitin, chitosan, and starch for metal removal from contaminated sites. Chelation technology, evolved over the century for being used in metal intoxication, has also emerged as a green chemical engineering approach for metal extraction from contaminated sites (soil, waste water, spent catalyst) in recent years. It is believed that technical applicability, operational simplicity, and economic feasibility are the key factors in selecting the most suitable technology for metal extraction.

In the following chapters of this book, the technical applicability of various new greener approaches for metal removal from solid waste and contaminated sites will be identified. Recent developments related to these new technologies in order to accomplish the concept of "Greening the Waste" will also be discussed.

Questions

1. How do you define the terminology "Sustainability" and "Green chemistry"?

2. Explain the dimensions of sustainability on the basis of TBL?

3. Does the five dimensional framework define additional information to "Sustainability"? If yes, please explain how.

4. Explain how the four force model illustrates connecting dots between current USD and the future SD.

5. Discuss the concept of "Greening the Waste."

6. Following two different routes illustrate the formation of maleic anhydride:
    a. *Oxidation of benzene:*

    $C_6H_6 + \frac{9}{2}O_2 \longrightarrow$ (maleic anhydride) $+ 2CO_2 + 2H_2O$ (Yield: 65%)

    b. *Oxidation of but-1-ene:*

    $C_4H_8 + 3O_2 \longrightarrow$ (maleic anhydride) $+ 3H_2O$ (Yield: 55%)

    i. Assuming that each reaction is performed in the gas phase only and that no additional chemicals are required, calculate (i) the atom economy and (ii) the effective mass yield of both reactions. You should assume that O_2, CO_2, and H_2O are not toxic.
    ii. Which route would you recommend to industry? Outline the factors that might influence your decision.

7. Calculate the E-factor for the following reaction:

    $$K_2PtCl_4 + C_2H_4 + HCl \text{ (in water)} \xrightarrow{SnCl_2 \text{ (40 mg)}}$$
    (4.5 g)   (~2 g)   (8.21 g)

    $$K[PtCl_3(C_2H_4)] \cdot H_2O + KCl + C_2H_4$$
    (3.86 g)   (unreacted)

2

Waste Handling and Pre-treatment

"Pollution is nothing but the resources we are not harvesting. We allow them to be dispersed because we've been ignorant of their value."
R. Buckminster Fuller

2.1 Introduction

The rapid growth of industrialization has resulted in the human population shifting gradually from rural to urban lifestyle. There is a significant increase in world urban population in the past few decades, from 28.3% in 1950 to 55% in 2018 (United Nations Department of Economic and Social Affairs 2018). While the world hurtles toward its urban future, it is impossible to ignore the fact that the amount of municipal waste (municipal solid waste [MSW]) and electrical and electronics waste, which are a few of the most significant by-products of an urban lifestyle, has gained momentum at a considerable growth rate, even higher than the rate of urbanization itself. It is predicted (Table 2.1) that approximately 6 069 703 tonnes of urban waste will be generated everyday by 2025 across the world (What a waste 2012). These values, however, depend on a number of factors and vary from region to region. East Asia and Pacific region are currently generating most of the world's waste at 23%, although they only account for 16% of the world's population. On the other hand, the combined waste generation from high-income countries is over one-third (34%) of the world's waste (World Bank 2018).

Higher rates of resource consumption and generation of industrial effluent are other concerns to achieve sustainable development. The growing global environmental impact of unscientific and unorganized waste management is a matter of concern for researchers, process engineers, scientists, non-governmental organizations (NGOs), and government organizations. In 2016, 5% of global emissions were generated from solid waste management, excluding transportation. If no actions is taken soon, solid waste related emissions will likely increase to 2.6 billion tonnes of CO_2-equivalent by 2050 (World Bank 2018). Disposal of industrial effluent into the water and soil without any treatment may also affect the environment significantly. In addition, these industrial effluents may contain high amount of metals (and thus known as "secondary sources of metals")

Table 2.1 Projections of waste generation for the year 2025 by regions.

Region	Data for the year 2010			Projections for the year 2025			
	Total urban population (millions)	Urban waste generation		Projected populations		Projected urban waste	
		Per capita (kg/capita/d)	Total (tonnes/d)	Total population (millions)	Urban population (millions)	Per capita (kg/capita/d)	Total (tonnes/d)
Africa	260	0.65	169 119	1 152	518	0.85	441 840
East Asia and Pacific	777	0.95	738 958	2 124	1 229	1.5	1 865 379
Europe and Central Asia	227	1.1	254 389	339	239	1.5	354 810
Latin America and Carribean	399	1.1	437 545	681	466	1.6	728 392
Middle East and North America	162	1.1	173 545	379	257	1.43	369 320
OECD	729	2.2	1 566 286	1 031	842	2.3	1 742 417
South Asia	426	0.45	192 410	1 938	734	0.77	567 545
Total	2 980	1.2	3 532 252	7 644	4 285	1.4	6 069 703

Source: Hoornweg and Bhada-Tata (2012). https://openknowledge.worldbank.org/handle/10986/17388. Licensed under CC BY 3.0.

and other resources that can be recovered and reused. Thus, achieving resource sustainability by improving the resource intensity with amelioration in existing technology as well as minimizing the negative impact of industrial development on the environment should be compelling for sustainable development (Chauhan et al. 2013a).

Waste management is a vast field that encompasses different types of waste including organic waste, polymer, plastic, municipal solid waste, industrial waste, and electronic waste. The Resource Conservation and Recovery Act (RCRA §1004-27) defines waste as "Any garbage, refuse, sludge from a wastewater treatment plant, water supply treatment plant, or air pollution control facility, and other discarded material, including solid, liquid, semisolid, or contained gaseous material, resulting from industrial, commercial, mining, and agricultural operations and from community activities" (USEPA 2017).

The Environmental Management (EM) Act refers to waste as "Any material discarded or intended to be discarded which constitutes garbage, refuse, sludge, or other solid, liquid, semisolid or gaseous material resulting from residential community, commercial, industrial, manufacturing, mining, petroleum or natural gas exploration, extraction or processing, agricultural, health care, or scientific research activities."

Waste generation and management has been the subject of many books and reviews; however, resource (metal) recovery from a variety of wastes has not been widely laid out intricately in textbooks. With the view that the reader should not lose interest, this book focuses specifically on the recovery of metals from the dominant waste streams, i.e. electronic waste (waste electrical and electronic equipment [WEEE]), agro-residue waste, and industrial waste originated from petrochemical/fertilizer/textile industries. This chapter provides an overview of waste handling ranging from an analysis of relevance, various categories of wastes discussed in this book, consequences of untreated waste disposal into the environment, government initiatives, management strategies, and pretreatment of waste. A summary of the chapter and a brief overview of the emerging approaches for waste management are given at the end of the chapter.

2.2 Waste Categorization

This section covers the variety of wastes that will be addressed further for resource recovery in the coming chapters. Motivations to address these wastes include fast-paced generation of waste stream, presence of significant amount of metals, environmental risk associated with these waste materials, and economic benefits associated with the successful recovery of metals from the waste stream.

2.2.1 Waste Electrical and Electronic Equipment (WEEE)

Electronic waste (e-waste) is one of the fastest growing streams in the world. The relatively short life span and high elimination rate of electrical and electronic equipment have resulted in increasing amounts of discarded equipment, which could be a threat to the environment and human health due to its hazardous components. WEEE and e-waste are the two more frequently used terms for discarded electrical and electronic equipment (EEE) appliances (Chauhan et al. 2018). E-waste refers to discarded electronic goods (e.g. computers, mobile telephones), whereas WEEE additionally incudes non-electronic appliances (e.g. refrigerators, air conditioning units, washing machines). In this book, solid waste generated from any electrical and electronic equipment will be referred by the term "WEEE" inclusively. There are several ways to define WEEE, but there is no strict definition available. Some of the popular definitions of WEEE are as follows:

> "WEEE means electrical or electronic equipment which is waste within the meaning of Article 1(a) of Directive 75/442/EEC, including all components, subassemblies and consumables which are part of the product at the time of discarding."
> Directive 75/442/EEC defines "waste" as "any substance or object which the holder disposes of or is required to dispose of pursuant to the provisions of national law in force."
> EU WEEE Directive (EU, 2002)

> "E-waste encompasses a broad and growing range of electronic devices ranging from large household devices such as refrigerators, air conditioners, cell phones, personal stereos, and consumer electronics to computers which have been discarded by their users." (Puckett and Smith 2002)
> Basel Action Network

> "Any appliance using an electric power supply that has reached its end-of-life."
> OECD (2001)

The various parts/materials/composition of WEEE may be divided broadly into six categories:

- *Casting and framing materials*: Iron and steel
- *Non-ferrous metals*: Especially copper used in cables, and aluminum
- *Screens and windows*: Glass
- *Casing*: Plastic
- Electronic components
- *Others*: rubber, wood, ceramic, etc.

WEEE is a complex mixture of more than 1000 toxic substances that can be categorized into organic materials, metals, and ceramics (Kaya 2016). The major fraction of WEEE includes base metals and precious metals and contributes to nearly 60% of WEEE. Table 2.2 shows the average metal composition of WEEE samples. The total value is calculated in USD($) based on year 2015 prices. Provided that a suitable treatment and recovery process is applied, WEEE can serve as a secondary metal resource. Polycyclic aromatic hydrocarbons (PAHs), polychlorinated biphenyls (PCBPs), polybrominated dibenzo-*p*-dioxins and

Table 2.2 Average composition of metals present in WEEE generated from different sources.

Element	WEEE generated from mobile phone (wt%)	WEEE generated from computer (wt%)	WEEE generated from television (wt%)
Copper	12.8	20	10
Aluminum	—	5	10
Iron	6.5	7	28
Lead	0.6	1.5	1
Nickel	1.5	1	0.3
Tin	1	2.9	1.4
Silver	0.36	0.1	0.03
Gold	0.0347	0.025	0.0017
Palladium	0.0151	0.011	0.001
Total value ($)	23 000	16 900	2 300

Source: Tickner et al. (2016). Reproduced with permission of Springer Nature.

dibenzofurans (PBDD/Fs); persistent organic pollutants (POPs), dioxins, and glass fibers are included in organic materials. Polyethylene, polypropylene, polyesters, and polycarbonates are typical plastic components of e-waste. Silica, alumina, alkaline earth oxides, barium titanate, etc. are listed under the ceramic group.

The global volume of WEEE generated is anticipated to reach 130 million tonnes in year 2018 from 93.5 million tonnes in year 2016 at a compound annual growth rate (CAGR) of 17.6% (Assocham 2016). Globally, around 20–50 million tonnes (MT) of WEEE is disposed of every year, which accounts for 5% of all municipal solid waste. Figure 2.1 shows that total WEEE generation in 2014 was 41.8 million tonnes, which is comprised of 1.0 MT lamps, 3.0 MT of small IT equipments, 6.3 MT of screens and monitors, 7.0 MT of temperature exchange equipment (cooling and freezing equipment), 11.8 MT large equipment, and 12.8 MT of small equipment.

Ample generation of WEEE, non-rigorous management strategies, and lack of consumer awareness have led to a cavalier manner of WEEE disposal into the environment. Melting of computer chips can produce acids and sludge. Disposal of these chips on ground can cause acidification of soil, which can contaminate the water sources. Chauhan et al. (2018) suggested that processing of WEEE plastic using hot shredder/granulation equipment can produce dioxins and furans.

The lesser the particle size of plastic, the higher the amount of toxic components released into the environment. Landfilling and incineration, despite causing high risks to health and environment, are still being carried out in an unprofessional manner for WEEE disposal. The US Environmental Protection Agency (EPA) and the United Nations estimate that only 15–20% of WEEE is recycled; the rest of these consumer electronics go directly to landfills and incineration (United Nations University 2009). WEEE contains brominated

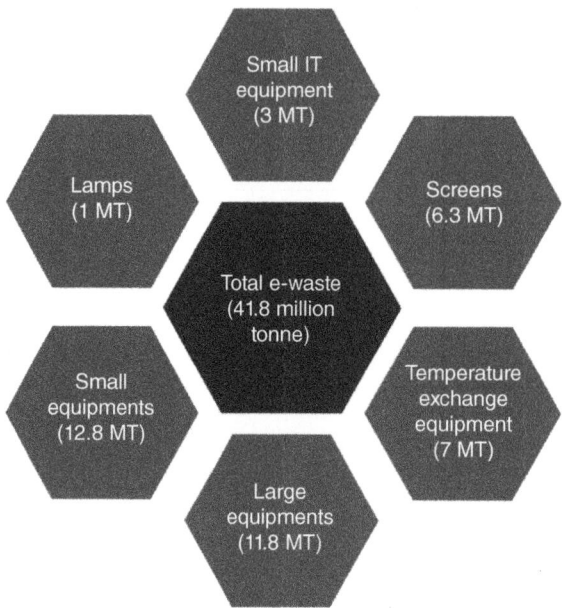

Figure 2.1 Contribution of various sectors to total e-waste of 41.8 Mt in 2014 globally. Source: Adapted from Baldé et al. (2015).

flame retardants (BFRs) that may lead to the formation of PBDD/Fs during the incineration process.

Considering the Reduce, Reuse, and Recycle (3R) policy, the highest emphasis is necessary on safe recycling of WEEE to minimize waste generation and optimize resource utilization. Various mechanical/thermal/hydrometallurgical processes are being employed for recovery of metals from electronic waste. Some of these techniques will be discussed in detail in the Chapters 3–7.

2.2.2 Agro-Residue Waste

Agricultural waste, by-products, and co-products are usually defined as plant or animal residues that are not (or not further processed into) feed (Gontard et al. 2018). It is comprised of animal waste (manure, animal carcasses), food processing waste, crop waste (corn stalks, sugarcane bagasse, pruning, etc.), and toxic agricultural waste (pesticides, insecticides, herbicides, etc.). These residues can be further categorized into primary residues and secondary residues. Agricultural residues, which are generated in the field at the time of harvest, are defined as primary or field-based residues (paddy straw, maize stalks, coconut empty bunches, etc.) whereas those co-produced during processing are called secondary or processing based residues (paddy husk, maize cob, coconut shell, saw dust, palm oil shell, etc.). Primary residues can be turned into resources using intensified conversion processes whereas secondary residues are usually available in relatively large quantities at the processing site and may cause additional environmental (soil erosion, prevent soil from drying) and economic (delay in farming operations, transportation cost) burdens in the primary processing sectors (Zafar 2019).

Weight (wt), volume (vol), moisture content (MC), total solids (TS), volatile solids (VS), fixed solids (FS), dissolved solids (DS), and suspended solids (SS) are the major physical properties of agricultural waste that describe the amount and consistency of the material to be dealt with in treatment and storage facilities. Of the chemical constituents, nitrogen (N), phosphorus (P), and potassium (K) are the principal chemical constituents for the planning of an agricultural waste management strategy. VS and biochemical oxygen demand (BOD_5) are used in the designing of biological treatment procedures. The definition and description of each physical/chemical term related to agro-residue waste characterization is given in Table 2.3.

Although agro-residue waste may contain a little fraction of metals due to the use of chemicals in modern agricultural economy, these wastes are still not rich in metal content and thus cannot be considered a secondary resource of metals. This book represents agro-residue wastes as potential resources rather than undesirable and unwanted, to recover metals from other contaminated sites and to avoid transmission of hazardous materials in the life cycle. Recent scientific advances now allow agricultural wastes to be a low-cost alternative for the treatment of effluents containing heavy metals in a sustainable way.

2.2.3 Industrial Waste

Industrial activities produce a significant amount of waste, and without a comprehensive waste disposal program, the health and safety of employees and residents in nearby areas may be at risk. A variety of industrial waste is generated during product manufacturing/processing that can be categorized on the basis of their source of origin, reusability, major contaminant, disposal methods, and harmful impacts on the ecosystem. Broadly, these industrial wastes can be categorized into nonhazardous industrial waste and hazardous industrial waste.

Nonhazardous industrial waste is generated by industrial or commercial activities but is similar to household waste by its nature and composition. It is not toxic, presents no hazard, and thus requires no special treatment. In particular, it includes ordinary waste produced by companies, shopkeepers, and trades people (paper, cardboard, wood, textiles, packaging). Owing to its nonhazardous nature, this waste is often sorted and treated in the same facilities as household waste.

The RCRA defines hazardous wastes, which may be in any state of matter, i.e. solid, liquid, or gaseous form, may cause danger to health or environment, either alone or when in contact with other wastes. Flammability, toxicity, reactivity, and corrosivity are the major factors responsible to make a waste hazardous. Hazardous wastes can be further classified as follows:

(a) *Nonspecific source wastes:* Industrial process wastes as a product of degreasing, solvent usage, electroplating, heat treatment, etc. are known as nonspecific waste.
(b) *Specific source wastes:* The waste obtained from wood preservation, petroleum refining, organic chemicals, etc. are specific source waste.
(c) *Discarded commercial chemical products, off-specification species, container residues, and oil spill residues:* Commercial chemical product hazardous

Table 2.3 Definitions and description of physical/chemical terms related to agro-residue waste characterization (Agricultural Waste Management Systems 2008).

Term	Unit	Definition	Method of measurements	Remarks
Physical characteristics				
Weight (wt)	Pounds (lbs); grams (g)	Quantity or mass	Scale or balance	
Volume (vol)	Liters (l); gallons (gal); ft^3	Space occupied in cubic units	Place in or compare to container of known volume; calculate from dimensions of containment facility	
Moisture content	%	That part of a waste material removed by evaporation and oven drying at 217 °F	Evaporate free water on steam table and dry in oven at 217 °F for 24 h or until constant weight	Moisture content (%) + total solids (%) = 100%
Total solids	%; % wet basis; % dry weight basis	Residue remaining after water is removed from waste material by evaporation; dry matter	Evaporate free water on steam table and dry in oven at 217 °F for 24 h or until constant weight	Total of volatile and fixed solids; total of suspended and dissolved solids
Volatile solids	%; % wet basis; % dry weight basis	That part of total solids driven off as volatile (combustible) gases when heated to 1112 °F (600 °C); organic matter	Place total solids residue in furnace at 1112 °F for at least 1 h	Volatile solids determined from difference of total and fixed solids
Fixed solids	%; % wet basis; % dry weight basis	That part of total solids remaining after volatile gases driven off at 1112 °F (600 °C); ash	Determine weight (mass) of residue after volatile solids have been removed as combustible gases when heated at 1112 °F for at least 1 h	Fixed solids equal total solids minus volatile solids

Dissolved solids	%;% wet basis; % dry weight basis	That part of total solids passing through the filter in a filtration procedure	Total dissolved solids (TDS) may be further analyzed for volatile solids and fixed dissolved solids parts %
Suspended solids	%; % wet basis; % dry weight basis	That part of total solids removed by a filtration procedure	Total suspended solids may be further analyzed for volatile and fixed suspended solids parts
Chemical properties			
Ammoniacal nitrogen (total ammonia)	mg/l; µg/l	Both NH_3 and NH_4 nitrogen compounds	Volatile and mobile nutrients; may be a limiting nutrient in land spreading of wastes and in eutrophication
Ammonia nitrogen (NH_3-N)	mg/l; µg/l	A gaseous form of ammoniacal nitrogen	
Ammonium nitrogen (NH_4-N)	mg/l; µg/l	The positively ionized (cation) form of ammoniacal nitrogen	Can become attached to the soil or used by plants or microbes
Total Kjeldahl nitrogen (TKN)	mg/l; µg/l	The sum of organic nitrogen and ammoniacal nitrogen	Digestion process which converts all organic nitrogen to ammonia
Nitrate nitrogen (NO_3-N)	mg/l; µg/l	The negatively ionized (anion) form of nitrogen that is highly mobile	Nitrogen in this form can be lost by denitrification, percolation, runoff, and plant microbial utilization

(Continued)

Pass a measured quantity of waste material through 0.45 µm filter using appropriate procedure; evaporate filtrate and dry residue to constant weight at 217 °F

May be determined by difference between total solids and dissolved solids

Common laboratory procedure uses digestion, oxidation, and reduction to convert all or selected nitrogen forms to ammonium that is released and measured as ammonia

Table 2.3 (Continued)

Term	Unit	Definition	Method of measurements	Remarks
Total nitrogen (TN)	%; pounds (lb)	The summation of nitrogen from all the various nitrogen compounds		Macronutrient for plants
Phosphorus (total phosphorus [TP], soluble reactive phosphorus [SRP])	mg (TP);	TP is a measure of all the forms of phosphorus, dissolved, or particulate present in a sample	Laboratory procedure uses digestion and/or reduction to convert phosphorus to a colored complex; result measured by spectrophotometer or inductive coupled plasma	Critical in water pollution control; may be a limiting nutrient in eutrophication and in spreading of wastes
SRP	mg/l (SRP)	SRP is a measure of orthophosphate, the filterable (soluble, inorganic) fraction of phosphorus		
5-d biochemical oxygen demand (BOD_5)	lb of O_2		Extensive laboratory procedure of incubating waste sample in oxygenated water for 5 d and measuring amount of dissolved oxygen consumed	Standard test for measuring pollution potential of waste
Chemical oxygen demand (COD)	lb of O_2	Measure of oxygen consuming capacity of organic and some inorganic components of waste materials	Relatively rapid laboratory procedure using chemical oxidants and heat to fully oxidize organic components of waste	Estimate of total oxygen that could be consumed in oxidation of waste material

Source: U.S. Department of Agriculture.

wastes include products with the generic names listed on the Code of Federal Regulation 40 CFR 261 (1990) from hospitals, research laboratories, photography laboratories, and analytical laboratories.

(d) *Characteristics wastes:* Characteristics wastes are the waste that are wastes but not specifically identified elsewhere. They too exhibit properties of ignitability, corrosivity, reactivity, or toxicity.

Spent catalysts are one of the primary industrial wastes from the processing units of petrochemical and fertilizer industries. Catalysts are responsible for facilitating hydrotreatment and hydrodesulfurization processes with high selectivity for converting heavy oils and residues to more valuable distillates (Kaufmann et al. 2000; Breysse et al. 2003; Rana et al. 2007). Activity of the catalysts deteriorates eventually after certain reaction cycles due to high temperature reactions and deposition of impurities on the catalyst surface. Therefore, they need to be replaced with fresh catalysts and the discarded catalysts are referred to as "spent catalysts." These spent catalysts contain significant amounts of heavy metals including vanadium (V), molybdenum (Mo), nickel (Ni), cobalt (Co), and aluminum (Al) and thus can be considered a potential "secondary source of metals." Table 2.4 lists various causes of catalyst deactivation that have been mentioned in the literature.

In addition, they are a threat to the ecosystem if not disposed of in a proper manner due to the high toxicity of these metals. Spent catalysts from the petroleum industry generally contain 4–12% Mo, 15–30% Al, 1–5% Ni, 0–4% Co, 5–10% S, 1–5% Si, and 0–0.5% V (Chauhan et al. 2015a). Spent catalysts of the fertilizer industry contain significantly high amount of nickel metal in the range of 18–28%, which needs recovery operations. It is reported in the literature that annual consumption of fertilizers in India has increased from 0.7 lakh million tonnes (MT) in 1951–1952 to more than 277.39 lakh MT in 2011–2012 (Department of Fertilizers 2013). This steady increase in resource consumption also accelerates the need for catalysts in the processing of heavy feedstock and consequently, the generation of spent catalysts increases significantly. It is estimated that 150 000–170 000 tonne/yr of spent catalyst is generated worldwide by the fertilizer industries because of the harsh reaction conditions in the primary and secondary reformers. Automobiles, transformers, or capacitor oil are also another important source for the generation of used/spent oil, containing high levels of various heavy metals such as Pb, Cd, As, and Cr, in addition to contaminants such as chlorinated solvents, polychlorinated bi-phenyls, and other carcinogens.

Keeping in mind the complex composition and associated toxicity of industrial waste, food chain contamination, and occupational exposure to toxic substances are the major associated risks to address, which can cause a threat to human health and severely contaminate the soil–air–water continuum. It can be remarked that significant content of toxic substances is present in the blood, serum, hair, scalp hair, human milk, and urine from the residents of waste recycling areas.

Table 2.4 Causes of catalyst deactivation.

References	Catalyst	Cause of deactivation
Myerson et al. (1987)	CoO 5–8%, 19–20% MoO_3	Deactivation because of exogenous form of nickel and vanadium on the surface of and in the pores of hydrodesulfurization (HDS) catalyst. Regeneration can be done if Brunauer-Emmett-Teller (BET) surface area of the regenerated catalyst is significantly enhanced over the value from that displayed by the spent catalyst
Neduein et al. (1998)	$NiO/MoO_3/$ γ-Al_2O_3	1. Rapid sintering in oxidizing conditions due to formation of free MoO_3 phase 2. Formation of reactive molybdate in the interaction of active phase and catalyst support 3. Intensive loss of active phase
Seki and Yoshimoto (2001)	HDS catalyst MoO_3 (10.5 wt%), NiO (0.7 wt%), and CoO (1.2 wt%) on alumina support	Coke deposition: a. If pressure in HDS reaction is low b. Asphaltene aromaticity c. Increase of operating temperature
Vogelaar et al. (2007)	Thiophene HDS, 10 wt% MoO_3 and 2.5 wt% NiO	Activity of Ni-Mo HDS catalyst was nearly 20 times faster than the unpromoted Mo catalyst. Coke did not deposit on the active sites of Ni-Mo, but deposited on the alumina support without any significant effect on the activity 1. The unpromoted Mo catalyst exhibited a partial deactivation, which was due to coke deposition on the active Mo phase 2. Main cause of deactivation for the Ni-Mo/Al_2O_3 catalyst was the loss of sulfur during the HDS reaction. The sulfur interaction with the Ni-Mo active phase was found fully reversible by the H_2S/H_2 regeneration 3. Coke deposition on the active sites was found to be a major cause for deactivation of the unpromoted Mo/Al_2O_3 catalyst, which could not be reactivated by the H_2S/H_2 treatment
Dufresne (2007)	Hydrotreating catalyst	1. Coke formation 2. Damaged active phase due to presence of permanent poisons or an unacceptably low surface area or HDS activity
Baghalha and Ebrahim-pour (2007)	CoMo/γ-Al_2O_3 HDS catalyst: MoO_3 23.8%, 4% CoO, rest alumina	1. Sulfur loss and coke deposition 2. Co-Mo-S permanent loss of activity a. Arsenic poisoning b. Co reaction with the alumina support c. Decrease of the catalytic active surface area

(Continued)

Table 2.4 (Continued)

References	Catalyst	Cause of deactivation

Kubička and Horáček (2011)	CoMo/γ-Al$_2$O$_3$ 3.0 wt% CoO and 13.5 wt% MoO$_3$	1. Presence of alkalis promoted catalyst deactivation due to their deposition on catalyst surface leading to blockage/poisoning of active sites
		2. The effect of phosphorus was twofold
		    a. When charge-compensating alkalis were present, corresponding phosphates were deposited above and at the beginning of catalyst bed leading to gradual buildup of deposits
		    b. On the other hand, in the absence of alkalis decomposition of phospholipids yielded phosphoric acid that catalyzed oligomerization reactions leading to rapid catalyst deactivation by carbonaceous deposits
		3. The changes in active sites structure caused by their desulfurization caused the irreversible loss of active sites

2.3 Legislations and Regulations for Hazardous Wastes

Developed nations have strict rules and regulations regarding waste handling and disposal system whereas developing countries, in general, face challenges on two fronts. While their rate of domestic waste generation is high, the illegal import of waste (for recycling or reprocessing) has further aggravated their problem. Transboundary shipment of industrial/electronic waste is considered as reallocation of resources and opportunities for developing countries to intensify the economic activity (Chauhan et al. 2018); nevertheless another side of this debate states that these wastes may contain toxic components and may cause hazardous health risk to the people who are dealing with them for recycling. Less stringent environmental regulations and lower labor cost in developing countries increase the propensity for transboundary movement, informal dismantling, and recycling of hazardous wastes. The European Environment Agency (2012) highlighted the fact that about 1.3 million tonnes of discarded EEE is exported every year from Europe to Africa and Asia. The Basel Action Network (BAN) and the Silicon Valley Toxics Coalition (SVTC) estimated that up to 80% of the WEEE produced in the United States and initially collected for recycling purposes is being exported to developing countries for informal recycling procedures (Ketai et al. 2008).

Several conventions and rules have been proposed to deal with the transboundary movement and disposal of hazardous wastes as well as other chemical wastes by regulating and controlling the movement of scheduled hazardous wastes from OECD countries to non-OECD countries. African countries have strong rules to restrict the transboundary movements of hazardous waste into their nations. The Bamako Convention bans the movement of goods into African countries. South Korea acted in the early stages of WEEE import and banned it in 1994. Pakistan had passed the National Environment policy in 2005 to strictly ban the inflow of solid waste. Singapore has a policy in place since 2008 to be strict not only with inflow but also outflow of e-waste and electronic goods. Japan has passed the law for no export, or import of specified hazardous wastes and other wastes. China has banned import of 24 kinds of waste items including mining slag and household waste plastics into its territory from January 2018. In India, The Ministry of Environment and Forests (MoEF) has promulgated Hazardous Wastes (Management & Handling) Rules, 1989, under the provisions of the Environment (Protection) Act, 1986, for management and handling, and import of hazardous wastes into the country. As per Rule 11 of the Hazardous Wastes Rules, 1989, import of hazardous wastes from any country to India shall not be permitted for dumping. More specific new regulations, codified as the "2010 E-waste Management and Handling Rules," became effective in 2012. These set hazardous substance limits on six Restriction of Hazardous Substances (RoHS) Directive materials including lead, mercury, cadmium, hexavalent chromium, polybromintaed biphenyls (PBBs), and polybrominated diphenyl ethers (PBDEs). Wastes containing Hg, As, waste asbestos (dust or fibers), waste oil, etc. are in the list of banned wastes for import and export. Apart from the MoEF, Central Pollution Control Board (CPCB), and State Pollution Control Boards (SPCBs)/Pollution Control Committees (PCCs) have been delegated certain powers for the control and regulation of hazardous wastes (Implementation of e-waste rules 2011). The Rotterdam and Stockholm Conventions also aim at reducing or eliminating various types of hazardous emissions. This comprehensive agreement, with 178 nations party to it (signed, though not yet ratified, by the United States), charges countries wherein the hazardous waste is produced with responsibility for the safe disposal of such products, while banning hazardous material exports to developing countries – except when written "prior informed consent" from the accepting nation has been obtained. The United States follows legislation of HR 2284: Responsible Electronics Recycling Act (2011) that has banned the export of certain electronic wastes from cathode ray tubes, batteries, switches from equipment such as computers, televisions, printers, copier, videogame systems, telephones, and similar used electronic products, which are hazardous in nature. To further protect children from the ill-effects of e-waste, the World Health Organization's Children's Environmental Health team is framing rules for the production of safe electronic goods. Table 2.5 lists some of the world initiatives for WEEE management.

However, the effects of the ruling seem to be easily evaded through such loopholes as mislabeling of wastes. The effect of the regulations on the informal recycling sector, or on the hazardous waste illegally imported each year, is likely to be modest. Lack and/or lax enforcement of existing laws has caused the leaching of toxins from waste to soil, water, and air. Disposal of hazardous wastes in an environmentally sound manner using a safe landfill site is generally ignored in most of these countries.

2.4 Handling/Management of Hazardous Waste

The toxins that solid waste can contain cannot be disposed of without any treatment. Owing to the complex and integrated nature of solid waste, there are

Table 2.5 World initiatives for WEEE management.

Initiative	Details
Restriction of Hazardous Substances Directive (RoHS)	Enacted along with EU WEEE, restricts amounts of lead, mercury, cadmium, hexavalent chromium, PBB, and PBDE used in manufacture. Versions adapted by many other countries, including China and India
EU WEEE Directive	Adapted by all EU members by 2007. Establishes systems of collection and recycling based on producer takeback, for 10 categories of electrical goods
Solving the E-waste Problem (StEP)	Instituted formally in 2007 by UN agencies, StEP partners with prominent academic and government organizations (e.g. Massachusetts Institute of Technology [MIT], United States Environmental Protection Agency [USEPA]) on promoting reuse of recycled materials and control of e-waste contaminants
Reduce, Reuse, Recycle (3Rs)	Promoted by Japan. Seeks to prevent creation of waste, and to further cooperation on recycling with developing countries. Allows waste export for remanufacture
US State laws and the Responsible Electronic Recycling Act (HR2284)	25 US states have laws for e-waste collection, some stipulating consumer payment. HR2284 is a proposed national law to control e-waste export and certify used electronic goods for export
US NGOs – Basel Action Network (BAN), Silicon Valley Toxics Coalition (SVTC), Electronics TakeBack Coalition (ETBC)	These three act together for workable national e-waste collection and recycling programs. They internationally promote the "Basel Ban," a more restrictive waste export amendment to the Basel Convention. BAN has produced documentaries, and much research
Bamako Convention	In force since 1998 in African Union countries. Sets more stringent waste import limits than the Basel Convention, and sets penalties. Seldom evoked

Source: Jadhao et al. (2016). Reproduced with permission of Elsevier.

many inherent complexities involved. Depending on the physical and chemical characteristics of hazardous wastes, these wastes are managed using any one or a combination of methods such as secured landfilling, incineration, or recycling of waste.

2.4.1 Secured Landfilling

Hazardous waste landfill refers to a waste disposal unit that is designed and constructed with the aim of least environmental impact. In principle, landfills must operate in a manner that protects the environment, particularly surface and ground waters, from leachate contamination. The hazardous wastes landfill site is designed scientifically to have an impervious stratum at the bottom to inhibit the percolation of leachates, and thus to avoid soil and water pollution/contamination in the vicinity of the landfill site. High-density polyethylene (HDPE) lining is used in making the landfill impervious. There are arrangements made for collection and treatment of leachates from the hazardous wastes. Once the landfill is filled up, it is covered with HDPE lining to make the landfill impervious from the top. The top is again covered with a thick layer of soil and finally the site is covered with vegetation cover. However, continuous monitoring of the reclaimed site is necessary for a longer period.

Landfilling of waste (though in a controlled manner) is a serious matter of concern as landfill leachate may potentially transport toxic metals and POPs into the food chain and cause acute and chronic health effects. There are more than 30 000 chemical and radioactive disposal sites in the United States and of these 1200–2000 sites are considered a threat to human health, as reported by the Environmental Protection Agency (EPA 2001). Industrial wastes can be disposed of into landfills only if it is ensured that the landfill meets non-hazardous criteria. Efforts were made to develop processes for the treatment of spent catalyst by suitable procedures that can reduce leachability. However, the cost of disposal is estimated to vary from $50/tonne to more than $2000/tonne. The disposal cost of one layer of catalyst used in a 500-MW unit can be $20 000 to $500 000 and thus economic constraints are another concern to adopt secured landfilling for disposal of hazardous waste.

2.4.2 Incineration

Incineration is the process of destruction of all high calorific and highly toxic wastes by burning the waste at high temperature. Incineration at 1200 °C breaks down organic matter present in the waste into basic nontoxic components. It serves the dual purpose of reduction of both the toxicity and the volume of the waste, which is an important consideration when the disposal of wastes is finally destined for landfills. Most of the process wastes from chemical unit operations can be treated in properly designed incinerators. However, the process of incineration releases toxic air pollutants such as dioxins, furans, etc. if the waste is not incinerated at very high temperature. Therefore, destruction efficiencies of toxic compounds during incineration (effectively 99.99%) with no generation of POPs should be the primary objective while designing an incinerator system. The combustion of the un-salvaged industrial/WEEE scraps is most often adopted to reduce the volume prior to the recovery of precious metals.

However, waste incineration is expensive and poses challenges of air pollution and ash disposal. High concentrations of trace metals, such as more than 20 g/kg Cu and 4 g/kg Pb and Zn, were reported in the WEEE open burning sites, and these locations have been identified as major pollution sources that may pose a significantly potential risk of metal leaching to the surrounding environment. Secondly, incineration requires waste placed outside for collection to be containerized to stay dry, and much of the waste stream is not combustible. Hazardous waste and electronic waste are often collected along with municipal waste in rural areas and burned without any sorting in order to reduce the volume before final disposal at unlined landfills (Labunsk et al. 2013; Tsydenova and Bengtsson 2011; Kahhat and Williams 2009). It increases mobility of heavy metals along with plastics, which are released to the atmosphere during burning while not being bioavailable during washing or landfilling. Incineration in a high temperature combustion chamber or burning of waste into open environment results in release of toxic fumes and therefore, should be completely prohibited.

2.4.3 Recycling of Hazardous Waste

Recycling of waste is another illustrious way to minimize hazardous waste generation and optimize the resource's useful life. The hazardous wastes may be categorized as recyclable when resource recovery is possible by reprocessing the waste. In recent years, the global market for recyclables has increased significantly. The world market for post-consumer scrap metal is estimated at 400 million tonnes annually and around 175 million tonnes annually for paper and cardboard. This represents a global value of at least $30 billion/yr. Producing new products with secondary materials can save significant energy. Production of aluminum from recycled aluminum requires 95% less energy than producing it from virgin materials. As the cost of virgin materials and their environmental impact increases, the relative value of secondary materials is expected to increase. Hazardous waste can also be used as alternate fuels. Fuel oil-based plants generate carbon waste since they are based on 80% carbon recycle whereas plants based on Texaco process recycle 100% carbon. The carbon waste generation can be completely eliminated if these plants switch over to gas feedstock. Used oil is also a precious and nonrenewable resource and can be recycled back to pure lube oil again. Fly ash, another hazardous waste liberated from chemical industries, can be utilized in the manufacturing of cement, bricks, reclamation of farmland, and also for landfilling.

Recycling of wastes can be designated as part of the "formal" and "informal" economic sector. Formal recycling employs specially designed facilities for the safe extraction of materials and protective working conditions; however, it is still not widely employed due to higher economics associated with specially constructed facilities and equipment. "Informal" recycling is characterized as unregulated, unregistered, illegal, and beyond the reach of official governance. Reckless practices such as child labor, cavalier incineration, uncontrolled emission of pollutants to air, and discharge of leachate in surface water negate all the positive environmental benefits through recycling and push the environmental benefit–impact balance in the negative direction (Chauhan et al. 2018).

2.5 A Call for Metal Recovery from Waste

The inconceivability of having economic and industrial development without concomitant increase in resource consumption has now stepped up the necessity of metals, which are considered as "key of industrialization." With increasing demand for ever more complex metallic composite and alloy materials in modern manufacturing processes, it becomes imperative to develop appropriate methods for the recovery of these valuable metals.

2.5.1 Threat to Human Health and Environment

Metals are bioaccumulative in the ecosystem and therefore it is difficult to get rid of metal contamination for a long time once it enters the soil–air–water continuum. Heavy metals are extensively used in the preparation of various catalysts in petrochemical and fertilizer industries in the form of metal oxides and metal sulfides. The global catalyst market is driven by factors such as growing consumption of fertilizer and refinery products, environmental regulations, technology developments, and emerging trends. Disposal of industrial waste containing heavy metals requires compliance with stringent environmental regulations. Spent hydroprocessing catalysts have been classified as hazardous wastes by the EPA in the United States. The EPA added spent hydrotreating catalyst (K171) and spent hydrorefining catalyst (K172) to its hazardous waste list in August 1998 because of their self-heating behavior and toxic chemicals content (Rapaport 2000) while spent hydrocracking catalyst was added to the list in 1999 (USEPA 2003; WHO 2006). Metals such as Cd, Cr, Pb, Cu, Ni, and V that are present in spent hydroprocessing catalysts are included in the list of potentially hazardous wastes published by the Environment Canada. Cadmium is associated with renal dysfunction and leads to lung disease; chromium can damage the kidney, liver, as well as the circulatory and nerve tissue; copper can cause anemia, liver and kidney damage, and stomach and intestinal irritation; lead is harmful to the nervous system; nickel causes cancer of lungs, nose, and bone (Parker 1980). These metals can be leached by water after disposal and pollute the environment. Besides the formation of leachates, the spent hydroprocessing catalysts, when in contact with water, can liberate toxic gases.

As mentioned in Section 2.2.1, WEEE is a complex mixture of metals, organic materials, and ceramics. Keeping in mind the complex composition and associated toxicity of WEEE, food chain contamination and occupational exposure to toxic substances in primitive recycling areas are the major issues to address, which cause hazardous health problems to the residents of WEEE recycling yards and nearby locations. Several unregulated e-scrap yards have been reported in Bengaluru (India), Lagos (Nigeria), Guiyu and Taizhaou (China), Accra and Agbogbloshie (Ghana), Thailand, the Philippines, and Vietnam (Chen et al. 2010; Grant et al. 2013; Sepulveda et al. 2010). Significant content of toxic substances were found in the blood, serum, hair, scalp hair, human milk, and urine from the residents of WEEEs recycling areas and thus confirm significant levels of human exposure to heavy metals and POPs released from WEEE treatment processes, which may pose significant health risk to workers and local inhabitants.

2.5.2 Waste: An Artificial Ore

Mobilization of heavy metals from waste is important not only from the viewpoint of preventing environmental pollution but also from the viewpoint of conserving and reusing the resources.

The market demand for hydrotreatment catalysts is estimated to increase with an annual growth rate of 4.4% (Dufresne 2007). Currently, the market for fresh hydrotreatment catalyst is around 120 000 tonne/yr, 50% of which is used for hydrotreatment of distillates to produce clean fuels and the other 50% is used for residue upgrading and purification. It is estimated that because of the high volume of fertilizer required, more than 3000 tonne/yr of spent catalyst is generated by China and India and 150 000–170 000 tonne/yr of spent catalyst is generated worldwide. Global consumption of petroleum and other liquid fuels grew by 1.4 million barrels/d in 2017, reaching an average of 98.4 million barrels/d for the year (U.S. Energy Information Administration 2018). To meet this demand, production has to be expanded and eventually, catalyst consumption will increase. Needless to say, demand for metals and other natural resources will also increase at an exponential rate.

The spent catalyst removed from residue desulfurization process at the end of each cycle run contains several metals including vanadium (V), molybdenum (Mo), nickel (Ni), cobalt (Co), and aluminum (Al). Monolithic automotive catalysts contain a large quantity of precious metals such as platinum, palladium, and rhodium. The average loading of platinum group metals (PGMs) per catalytic converter has been 1.555 g of platinum, 0.622 g of palladium, and 0.1555 g of rhodium. About 34% of total platinum, 55% of total palladium, and 95% of total rhodium demand is now used to produce automotive catalysts (Kim et al. 2000). Similarly, the market for WEEE is growing exponentially these days. Cadena et al. (2015) reported 10 times (500 million to 5000 million) increase in the number of mobile phone users across the globe from year 2000 to year 2011. A report suggests that WEEE from old computers in South Africa and China will jump by 200–400% and by 500% in India by the end of year 2020 compared to that of the 2007 levels (Shamim et al. 2015). India has emerged as the second largest mobile market with 1.03 billion subscribers, and the fifth largest producer of WEEE in the world, discarding 1.8 million metric tonnes of WEEE each year. It is also predicted that India is likely to generate 5.2 million metric tonnes of WEEE per annum by year 2020 at a CAGR of about 30% (Assocham 2016). These waste materials containing high metal concentrations should be considered as "artificial ores" instead of discarding them recklessly, since they can serve as secondary raw materials with a consequent reduction in the demand for primary mineral resources.

2.5.3 "Waste" to Wealth

Reuse of spent catalysts after regeneration is another option to minimize environmental pollution. Recycling of metal scraps is an eco-efficient way to manage 33% less energy consumption and 60% less pollutants generation than the production of fresh fuel (Marafi and Stanislaus 2008). Increasing demand for metals in industries also causes a significant rise in the prices of metals, which affects the economics of chemical processes. It would be worth transforming the waste into wealth by retrieving its metal content and conserving the resources. WEEE contains 13–26 times and 35–50 times higher Cu and Au content respectively compared to ores (Cui and Zhang 2008). Yu et al. (2010) suggested that 1250 tonnes of Cu, 13 tonnes of Ag, 3 tonnes of Au, and 2 tonnes of Pd, representing a market value of $105 million, would be recovered if all mobile phones discarded in year 2008 in China were recycled. Also, primary production of precious metals may pose significant environmental footprint due to high levels of greenhouse gas emissions, especially when the concentration of valuable components is quite low and thus requires the removal of large volumes of material involving several steps of treatment (Cayumil et al. 2016). Literature suggests that 1 tonne of gold produced through mining emits nearly 17 000 tonnes of CO_2, while palladium and platinum respectively produce ~10 000 and ~14 000 tonnes of CO_2 (Chancerel et al. 2009). Therefore, metal recovery from waste is advantageous from both environmental and economic perspectives.

2.6 Pretreatment of Waste

Recovery of metals from waste material requires pretreatment including disassembly, size reduction, shredding, sieving, and segregation of waste material before performing any metallurgical process. This section covers various unit operations for the pretreatment of industrial/electronic waste that necessitate employing any hydrometallurgical or pyrometallurgical approaches for mineral processing.

2.6.1 Disassembling the Waste

Disassembly can be performed in either a selective ("look and pick" principle) or a simultaneous ("evacuate and sort" principle) manner. Selective dismantling is used in case of reuse-oriented operations in which components (e.g. plastics, cables, and capacitors) are removed manually from the waste based on their reusability. Simultaneous disassembly refers to the "wiping off" of all components in one go by heating in a furnace. High efficiency is achieved in the simultaneous disassembly method; however, higher risk of desired component damage, additional sorting process, longer processing time, and high cost are the major limitations associated with this method. "Look and pick" disassembly operation is preferably performed for electronic waste dismantling. Various capacitors, wires, cables, resistors, and batteries are removed beforehand and then the electronic waste is shredded into pieces. The industrial/municipal waste is referred to simultaneous disassembly in which all the waste is directly sent to a high temperature furnace.

2.6.2 Size Reduction (Comminution)

Comminution is the process of reducing particle size by crushing/grinding in order to maximize the liberation of minerals present in the waste. In general, crushers and grinders are used commonly to perform size reduction. Crushers are used for larger particles (up to 1 m size) where reduction in size is achieved by compression and impact. Particles of size less than 50 mm are considered appropriate for grinding operations that work based on the attrition forces. A variety of crushers and grinders are available to reduce the particle size depending upon the waste material's strength and the desired particle size. The choice of crushers/grinders and the energy requirement to conduct this unit operation are discussed in more detail in Chapter 8.

The single stage open/closed circuit, multi-stage open/closed circuit designs, or combination of one or more designs is used in the process of size reduction. In a single stage open circuit comminution operation, the product consists of a range of particle sizes and therefore, it is tedious to achieve the desired degree of

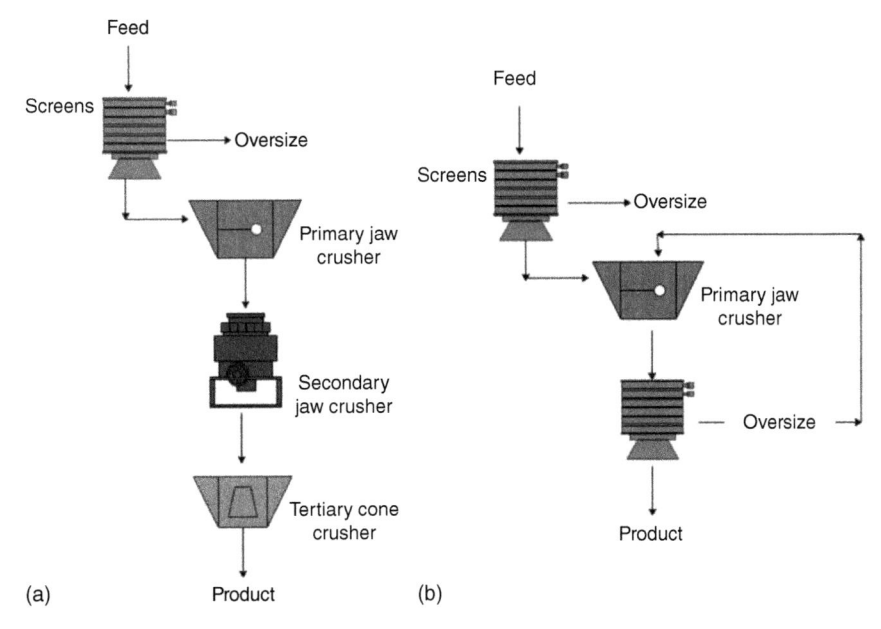

Figure 2.2 Design for comminution: (a) open circuit size reduction, (b) closed circuit size reduction.

liberation. Multiple stage size reduction operations are required to progressively reduce the remaining particle size to an acceptable range. In the multi-stage open circuit design, various types of crushers are used in series to obtain the desired size of the product (Figure 2.2a). In closed circuit, size reduction is performed in the same unit in multiple cycles. The products obtained from the first stage of operation are separated into relatively fine and coarse fractions through sieve operation. The coarser fraction is then collected and re-crushed in the same unit to attain the desired particle size (Figure 2.2b). In case of closed circuits, circulating load is established on the equipment during repeated cycles; nevertheless, it reduces the installation cost of multiple units to achieve the same degree of size reduction or sometimes even less than that from open circuit designs.

2.6.3 Screening/Sieving

Screens/sieves are used to separate particles of different sizes and shapes produced during the size reduction operation. Particles that are smaller than the aperture of the screen are allowed to pass through while retaining the others. This separation of mineral constituents can be considered an efficient and non-expensive method to reject the gangue constituents and to concentrate the desired minerals in mineral ores.

In the metallurgical industry a distinction is made between screening and sieving. The mechanism of size separation by both is the same, but screening generally applies to industrial scale size separations while sieving refers to laboratory scale operations. In this book, "screens" will be referred to illustrate size separation process. Various designs of screens are employed in metallurgical pretreatment operations depending upon the desired particle size. Coarse screen surfaces (grizzly screens) are generally fabricated by welding steel rails, rods, or bars forming grids of a desired pattern. The spacing in between is of the order of 5–200 mm. Medium screen surfaces are fabricated from woven wire/plates to produce a perforated pattern and are used to separate particles of 2–100 mm size. Choosing the right screen aperture to pass a specific size of particle depends on the angle of inclination of the screen, and the amplitude and frequency of the vibration. Therefore, the screens can be designed straight, curved, or cylindrical with vibrations or without vibrations depending upon the required particle size.

Separation of dry materials by screens and sieves is generally attempted down to about 75 mm. Finer materials tend to blind the sieve openings; therefore, wet screening is performed with the aid of water.

2.6.4 Classification

After initial liberation of a mineral constituent from its ore by crushing, grinding, and screening, separation of even finer sizes is attempted by "classification" pretreatment operation. This is, in general, carried out in a liquid environment (usually water) where particles are allowed to settle under gravity. The irregular shaped, coarser, heavier particles reach the bottom of the vessel at a faster rate due to their higher terminal velocity compared to smaller and lighter particles. Removing the settled particles while the others are still settling offers a simple means of separation. For very small particles (clay or silt), long times are required to settle, and the small difference in the settling rates of these fine particles leads to low separation efficiency. Therefore, centrifugal forces are employed such as in cyclones or hydrocyclones to accelerate the settling rate of fine particles.

2.6.5 Segregation

The process of grinding and classification may require the use of large quantities of water for the separation of fine particles as mentioned in the Section 2.6.2 to 2.6.4. It is essential to remove the bulk quantity of water for further metallurgical operations to extract metals/minerals from waste/ores. Thickeners are employed to separate fine solids from liquids by gravity sedimentation, which is a slow process. In general, 75–80% of the water can be separated and removed by thickeners. The thickener operation can be a batch or continuous process, with either co-current or counter-current flow of underflow and overflow slurries. For further water removal, filters are used either in batch or continuous mode and approximately 90% of the water can be removed during this filtration operation. Gravity separation methods are being employed in WEEE recycling due to the presence of particle fractions with different density range, e.g. e-scrap consists of plastics (density < 2.0 g/cm^3); light metals and glass (density ∼ 2.7 g/cm^3); and heavy metals (density > 7 g/cm^3) (Kaya 2016).

Various other segregation methods based on magnetic properties, electrostatic properties, and flotation principle are used for separating metallic and nonmetallic fraction. Magnetic separators are widely used for the recovery of ferromagnetic metals from nonferrous metals and other nonmagnetic wastes. These methods can also be used in combination to facilitate better separation efficiency. Yoo et al. (2009) performed gravity separation followed by two-stage magnetic separation to milled printed circuit boards (PCBs). Particle sizes > 5.0 mm were separated from the <5.0 mm PCBs particles by gravity separation. A low magnetic field of 700 G in first stage magnetic separation led to the separation of 83% of Ni and Fe in the magnetic fraction and 92% of Cu in the nonmagnetic fraction. Again, the second magnetic separation stage, conducted at 3000 G, resulted in a reduction in the grade of the Ni–Fe concentrate and an increase in the Cu concentrate grade.

Electrical conductivity based separation techniques (corona electrostatic separation, triboelectric separation, and eddy current separation) can also be used in those waste streams where difference in polarity and electrical conductivity of components is dominant. Eddy current separators are used widely for waste reclamation where they are particularly suited for handling the coarse feeds.

Very few studies have also been reported on the segregation based on the hydrophobicity of waste components. Mechanical separation from computer PCBs with jigging (for +0.59 to 1.68 mm size) and flotation (for <0.59 mm size) was illustrated by Sarvar et al. (2015). Collectorless flotation with gasoline was demonstrated as a promising method for separating nonmetals (i.e. polymers) of PCBs. 85% metal recovery from flotation and more than 95.6% metal recovery from jigging were obtained.

2.6.6 Calcination and Chemical Pretreatment

In addition to mechanical unit operations, chemical pretreatment and energy-based pretreatment methods can also be used to remove the undesired impurities present over the surface of waste particles. Calcination of waste streams is an energy-intensive process and is usually performed at high temperature to remove organic impurities. These processes may emit toxic gases to the atmosphere, mainly SO_2 and CO_2. In order to avoid the release of carbonaceous components and toxic gases to the atmosphere, washing with proper organic solvent over calcinations is another preferable way of pretreating the waste. The organic solvent can be recycled and reused easily after the washing process. The industrial solid wastes (spent catalysts) are in general subjected to calcination to remove carbon/organic coating or chemical washing in a Soxhlet extractor to remove the oil coating over them.

2.7 Summary and Outlook

This chapter provides a basic framework to illustrate the current scenario of waste generation, and its handling and management. Significant increase in the generation of waste and their poor handling is causing environmental concerns across the globe. Release of various kinds of toxins into the soil–water–air continuum has made it essential to treat waste streams in a responsible way using suitable techniques. In addition, the waste streams are a rich source of metals and therefore, it is important to recover these metals in order to optimize resources. An urgent necessity of substantial research is illustrated to pave the way forward for successful, environment friendly, and economic processes for resource extraction from waste. Greening the waste using eco-friendly technique is an upcoming field to treat the waste in a "win–win" situation from the environmental and economic perspectives. Various conventional methods to recover metals are discussed in the next chapter (Chapter 3) to make the readers aware of the major advantages/limitations associated with these processes. Some of the alternative processes are further illustrated in the following chapters as emerging and future technologies.

Improving waste management will help cities become more resilient to the extreme climate occurrences that cause flooding, damage infrastructure, and displace communities and their livelihoods. In an era of rapid urbanization and population growth, solid waste management is critical for sustainable, healthy, and inclusive cities and communities. If no action is taken, the world will be on a dangerous path to more waste and overwhelming pollution. Lives, livelihoods, and the environment would pay an even higher price than they do today.

Questions

1 What are the categories of solid waste on the basis of sources?

2 What is the difference in the management of solid waste in developing countries as compared to developed nations?

3 Comment on the statement that direct disposal of e-waste is hazardous for the eco-system.

4 Explain the steps essential for the treatment of solid waste before landfilling.

5 Which law was approved by the United Nations to restrict the transboundary movement and disposal of hazardous wastes?

6 Name the industries that are potential sources of metals in solid waste.

7 Discuss various legislations to restrict the import of hazardous waste into developing countries.

8 Discuss various unit operations for the pretreatment of spent catalyst before performing an extractive metallurgical process.

9 Write a short note on any two waste treatment methods.

3

Conventional Technologies for Metal Extraction from Waste

> *Important thing in science is not so much to obtain new facts as to discover new ways of thinking about them.*
>
> Sir William Bragg

3.1 Introduction

It is almost impossible to imagine what the world would be like if the effects of the industrial revolution were swept away. However, the uncontrolled growth rate of industrialization comes with its possible consequences. It has become difficult, and eventually will be even more difficult than ever, to make a strategic balance between opportunity and risk. One of the leading consequences of the current stage of urbanization/industrialization has been the increased rate of environmental pollution and consequently, high exposure to hazardous substances (heavy metals, chemicals and toxic gases) releasing from the industrial and urban wastes. In addition, these wastes are storehouses of numerous metals and thus can be considered secondary resources. The chemical constituents and metal composition of various kinds of industrial and electronic wastes have been discussed in Chapter 2. With the increasing demand for ever more complex metallic composites and alloy materials in modern manufacturing processes, it has become imperative to develop appropriate methods for the recovery of these metals from waste. Mobilization of heavy metals from waste material is not only important for conserving and reusing the resources but also essential in order to minimize pollution.

A variety of processing approaches have been reported heretofore to recover heavy metals from contaminated sites. Adsorption has been considered a good alternative for the efficient extraction of metals from industrial wastewater. Membrane separation, electrotreatment, photocatalytic processes, and absorption by aquatic plants have also been employed to remove metals and other pollutants from water. Toxic metals may also be present in soil, and may exhibit chemical interactions, mobility, and potential toxicity. Therefore, various techniques such as vitrification, excavation, removal of contaminated soil layers, phytoremediation, and soil washing using surfactants, acids, alkalis, and chelating agents have been investigated for soil remediation. Various

Sustainable Metal Extraction from Waste Streams, First Edition.
Garima Chauhan, Perminder Jit Kaur, K.K. Pant, and K.D.P. Nigam.
© 2020 Wiley-VCH Verlag GmbH & Co. KGaA. Published 2020 by Wiley-VCH Verlag GmbH & Co. KGaA.

hydrometallurgical (leaching, solvent extraction, and biological methods) and pyrometallurgical (calcination, roasting, smelting) processes have been investigated for the extraction of heavy metals from solid wastes such as spent catalysts, soil, electronic waste, and many others. Hydrometallurgical approaches involve aqueous chemistry for the process, easy implementation at laboratory scale, and less energy requirement, and thus these are preferred approaches at small scales. On the other hand, the pyrometallurgy approach involves thermal treatment and is, in general, performed at very high temperatures. It includes roasting, smelting, and calcinations, which all need thermal treatment and high energy consumption. These approaches are widely employed to process the waste at industrial scale. Depending upon the complex matrix of solid wastes, in some cases, the extraction processes may involve a combination of pyro- and hydrometallurgical processes. Figure 3.1 provides an overview of the conventional processes that have been widely employed for the extraction of metals from waste.

Considering that the focus area of this book is recovery of metals from industrial and waste electrical and electronic equipment (WEEE), hydrometallurgical and pyrometallurgical approaches along with their merits and demerits, in particular, will be discussed in the Sections 3.2 and 3.3 of this chapter.

The hydrometallurgical processes based on biological media (bioleaching and biosorption) have not been discussed in this chapter. These processes are at their emerging stage and therefore are considered in emerging metal extraction processes instead of conventional processes. Chapters 4 and 5 discuss these two processes in detail.

3.2 Pyrometallurgical Operations

The term "pyrometallurgy" typically means application of heat to bring about physical and chemical changes in the waste in order to extract metals. Conventionally, this process has been in practice for the extraction of metals from their ores. In the pyrometallurgical separation and removal process, metals present are subjected to either volatilization at temperatures of 200–900 °C as in the case of mercury, or melting, accompanied by slag formation, as for lead, arsenic, cadmium, and chromium. The metals in the spent catalyst are usually recovered either by smelting the spent catalyst to metal alloys and slag in the gaseous environment or by recovering the metals in the liquid environment, preceded by roasting or wet oxidation of the spent catalyst.

3.2.1 Pyrometallurgical Treatment of Industrial Waste

Several studies have been reported on the treatment of waste using pyrometallurgical approaches. Rabah (1998) performed pyrometallurgical treatment to recover high lead bronze alloy by melting of bronze turnings however the process failed to recover standard bronze alloys due to thermal volatilization of tin and escape of lead into the slag. Veldbuizen and Sippel (1994) reported the production

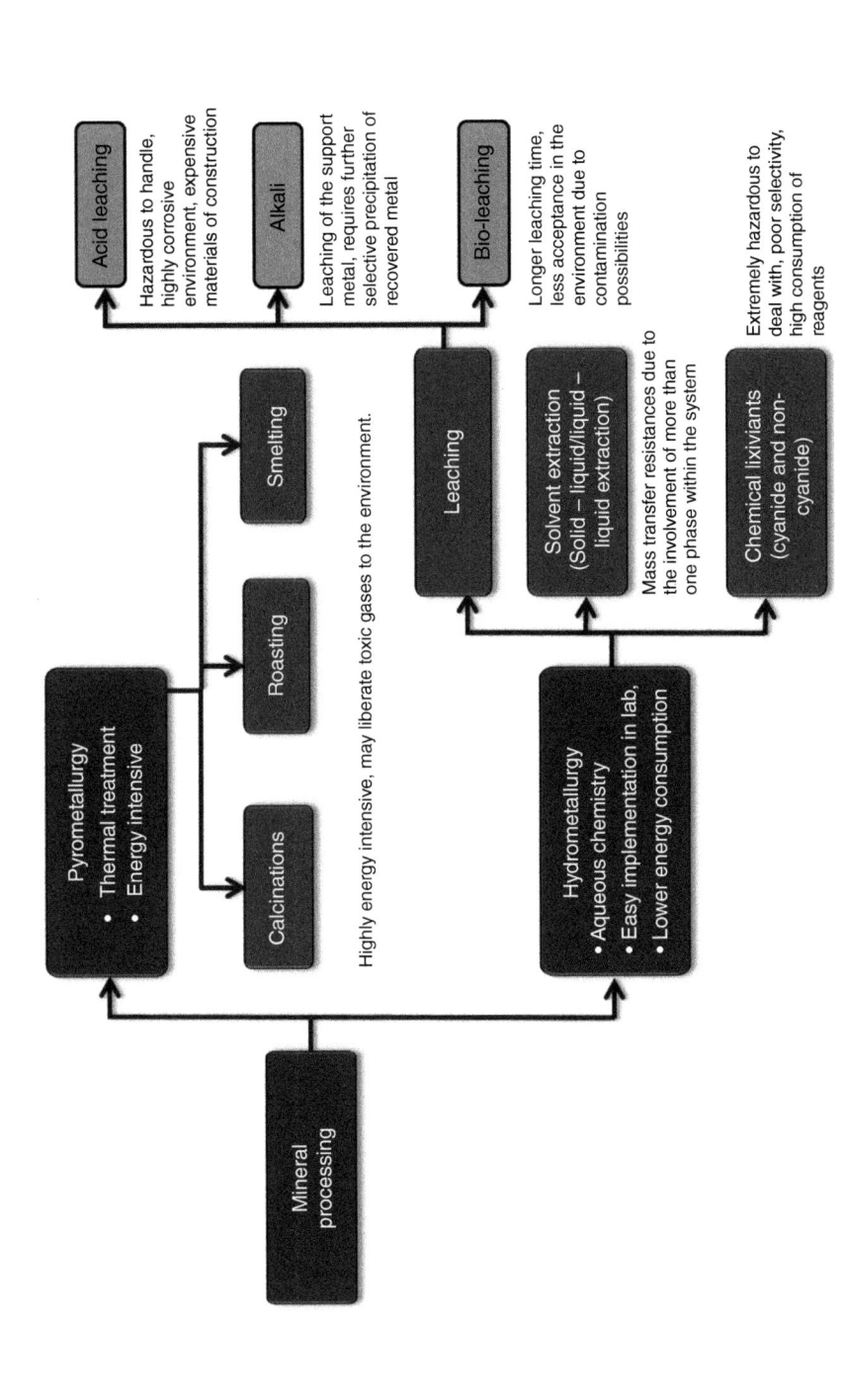

Figure 3.1 Conventional methods to extract metals from solid waste streams.

of 34 000 tonnes of Cu, 123 tonnes of Ag, 7.1 tonnes of Au, and 5 tonnes of Pt and Pd using copper smelter in Canada. The process uses a temperature of 1250 °C to form a molten bath of metal, churned by supercharged air containing about 40% oxygen. Metals such as Fe, Pb, and Zn get oxidized and fixed in a silica based slag. The slag is then cooled and processed to recover these metals from it. A mixture of copper sulfide and iron sulfide, known as copper matte, was used to recover 99.1% pure copper metal from it. Minor fractions of other metals such as gold, silver, platinum, palladium, selenium, tellurium, and nickel are also obtained along with copper.

In another experimental investigation, spent hydrorefining catalyst was roasted at 900 °C with sodium chloride (NaCl) to extract molybdenum in the form of sodium molybdate (Kar et al. 2005). NaCl and spent catalysts were kept in a mullite crucible to heat in a furnace to a predetermined temperature. The roasted material was then allowed to cool down to 100 °C, inside the furnace. Deionized water (pulp density: 20 wt%) was used at 70–90 °C for an hour to recover water-soluble sodium molybdate, shown by reaction (3.1). The proposed reaction between molybdenum and NaCl was as follows:

$$MoO_3 + 2NaCl + \tfrac{1}{2}O_2 \rightarrow Na_2MoO_4 + 2HCl(g) \tag{3.1}$$

The exemplary process flow sheet is shown in Figure 3.2. Nearly 90% recovery of Mo was reported at optimum reaction conditions, i.e. roasting temperature 900 °C, addition of 20 wt% NaCl to the feed, and a roasting period of one hour; nevertheless, high temperature requirement was one of the major limitations associated with the proposed process. To maintain a high temperature, this process needs more energy consumption in the industrial process; therefore, research is needed to reduce the energy consumption in order to perform pyro-treatment of waste material.

Some authors investigated the multistage processes in which pyrometallurgical approach is followed by hydrometallurgical treatment and thus the required time duration and temperature for the pyrometallurgical treatment can be possibly reduced to a certain extent. Spent desulfurization catalyst obtained from the vacuum residue desulfurization unit was roasted under atmospheric oxygen flow in a muffle furnace and dissolved in variety of leachants. Roasting of spent catalysts resulted in some major alterations such as molybdenum loss from the surface, migration of vanadium to the surface, and formation of some alloy phases due to sintering when roasting was performed at more than 650 °C. Na_2CO_3 was used as a leachant for the preferential dissolution of V and Mo in the liquid phase while H_2SO_4 solution was used to dissolve the solid phases obtained sequentially. Selective recovery of V and Mo was achieved using organic extractants; however, recycle or disposal of spent sulfuric acid solution as well as spent organic extractants would result in economic and/or environmental problems to be dealt with further (Kim and Cho 1997).

Chen et al. (2006) used roasting followed by precipitation for the recovery of valuable metals from alumina based spent catalyst. Roasting at 750 °C converted V, Mo, and Ni into oxide form and then these oxides reacted with alkali solution for recovery of V and Mo by precipitation while Ni–Co was sent to acid leaching for Ni and Co recovery. This process was useful for the recovery of

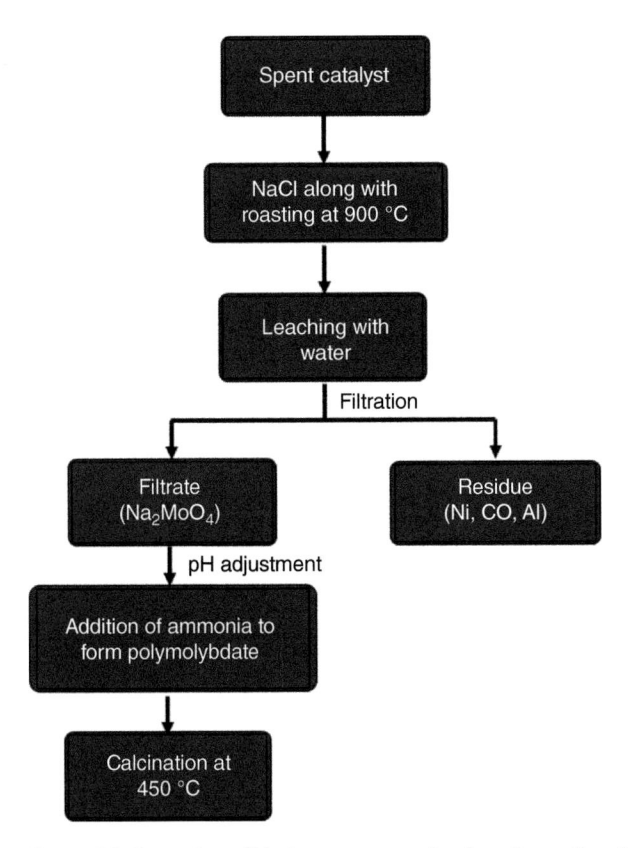

Figure 3.2 Extraction of Mo from spent catalyst by salt-roasting. Source: Kar et al. (2005). Reproduced with permission of Elsevier.

alumina to a certain extent with high purity. Allain and Gaballah (1994) designed an extraction procedure for strategic metals such as Nb and Ta by combining hydro- and pyrometallurgical processes. The planning of successive alkaline and acid leaching of industrial slag based on the "pseudo-structure" of these amorphous materials permitted selective dissolution of the economic metals, lower consumption of chemical reagents, especially HF, less toxic residues, and high recovery rate (>85%).

Plasma arc furnace is also being used in recent investigations for the extraction of metals from waste. Plasma is an electrified gas with a chemically reactive species such as electrons, ions, and neutrals. It can be characterized based on the thermal range into high temperature and low temperature plasma. Figure 3.3 demonstrates the various types of plasma and corresponding examples. Thermal plasmas are characterized by quasi-equilibrium between electrons, ions, and neutrals, which gasify organic waste to valuable fuel gas (synthetic gas), whereas inorganic waste is vitrified into ingot and slag.

Wong et al. (2006) employed the plasma arc furnace for extraction of metals. The reactions were performed at a high voltage AC power supply, with a maximum voltage of 10 kV, frequency of 10–20 kHz, and power of 10–15 kW

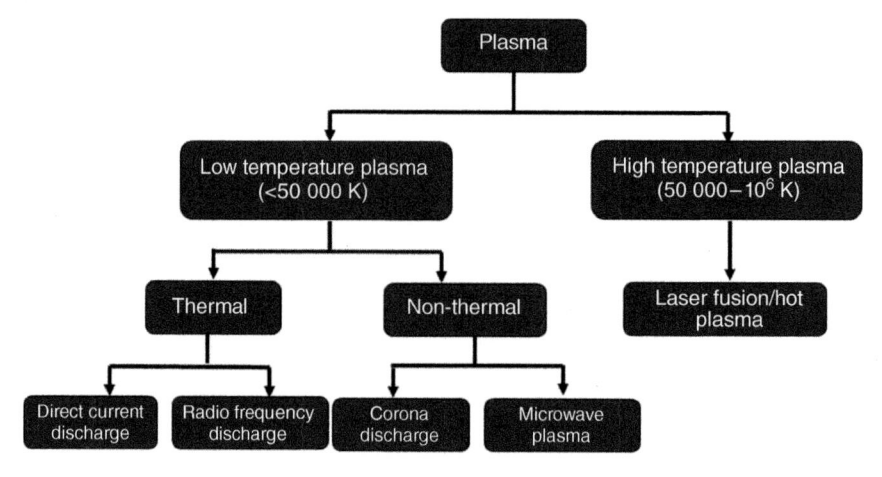

Figure 3.3 Types of plasma and corresponding examples.

under the flow of nitrogen as plasma gas. Waste spent catalysts obtained from petrorefinery industry were maintained at a temperature of about 1500 °C for two to three hours and the NiO/SiO_2 catalyst was reduced along with the liberation of syngas (Co, CO_2, H_2). The possible mechanism of the proposed scheme during the process is listed in Scheme 3.1:

(i) Stage I -
   (a) Active species in plasma at the initial step -

   $N_2 + e^* \rightarrow N_2^+ + 2e$ (3.2)

   $N_2 + e^* \rightarrow 2N + e$ (3.3)

   (b) Active species in the presence of water -

   $H_2O + e^* \rightarrow OH + H + e$ (3.4)

   $N + H \rightarrow NH$ (3.5)

   $OH + OH \rightarrow H_2O + O$ (3.6)

(ii) Stage II -
   (c) Active species formed with the organic waste in the spent catalysts -

   Organic waste material $\rightarrow$ Char + Volatile (3.7)

   Char & Volatile + $e^* \rightarrow C_mH_n + H + C + C_2 + e$ (3.8)

(iii) Stage III -
   (d) Reduction of NiO and gases formed in the system

   $C + OH \rightarrow CO + H$ (3.9)

   $H + H + M \rightarrow + H_2 + M$ (3.10)

   $CO + OH \rightarrow CO_2 + H$ (3.11)

   $CO + O + M \rightarrow CO_2 + M$ (3.12)

   $H + CH \rightarrow H_2 + C$ (3.13)

   $H + C + H_2 + e + NiO \rightarrow OH + CO + Ni$ (3.14)

Scheme 3.1 Possible mechanism for the reduction of NiO in nitrogen medium under plasma conditions.

The proposed technique was reported to be useful for industrial reduction and recovery of spent metallic oxide catalyst. Spent nickel oxide catalyst (NiO/SiO_2) was successfully converted into nickel metal. The process however requires further detailed investigations on recovery efficacy and bench scale studies of the process.

Kim and Park (2004) used the thermal plasma process in order to vitrify a mixture of wastewater sludge and fly ash and to confine heavy metals in vitreous stable slag for leaching. Leaching of heavy metals from vitrified slag was found to be below the regulatory limits and thus considered an effective means for detoxification of industrial sludge. Treatment of electroplating sludge in thermal plasma arc was also investigated by Ramachandran and Kikukawa (2002). A batch reactor was designed to investigate the effect of arc plasma on removal efficiency of heavy metals from sludge and the conversion of the residual to slag. A power input of 7–16 kW and different plasma gas regime (Ar, H_2, N_2 and mixture of two at a time) were used during the reaction. It is reported that heavy metals (Cr, Ni, Cu, and Zn) present in the sludge were separated into a metal ingot while the resulting slag was found to be inert to leaching test. Several other studies have also been performed on a variety of industrial waste using plasma arc furnace as a pyrometallurgical method (Cubas et al. 2014; Fabry et al. 2013).

3.2.2 Pyrometallurgical Treatment of WEEE

Pyrometallurgical methods have also been widely employed for the recovery of metals from WEEE. In general, pyrometallurgical treatment is followed by hydrometallurgical or electrometallurgical process to recover precious metals in pure form. Figure 3.4 shows the general approach for the extraction of metals from WEEE using metallurgical operations. It depicts the route for WEEE treatment including pretreatment, thermal treatment, hydrometallurgical processing, and post processing for the extraction and separation of base/precious metals from plastic and other components.

Table 3.1 lists some of the typical pyrometallurgical processes that are being employed for the recovery of metals from WEEE and other wastes.

Noranda process at QC, Canada, performs smelting of about 100 000 tonnes of WEEE per year, representing 14% of the total throughput while the remaining 86% comprises mined copper concentrates. Materials are treated at 1250 °C in a molten bath followed by churning using a mixture of supercharged air (Veldbuizen and Sippel 1994). Significant reduction in energy consumption is reported since the same is compensated by the energy produced through combustion of plastics present in WEEE.

In Boliden Ltd. Rönnskar Smelter, Sweden, personal computers (PCs) with only selective removal of components are treated using pyrometallurgical approach. The PC scrap is sent to a fuming furnace after crushing it to 30 mm size where fuming of Zn, Pb, As, other related metals and halogens is facilitated by coal fines. The cleaned slag is tapped from the fuming furnace to a settler furnace where copper alloy and copper sulfide are separated. The process flowsheet is adapted from Cui and (2008) and redrawn here in Figure 3.5 for the reader's ready reference. As shown in Figure 3.5, the waste material can be fed into the process in different steps depending on their purities. High copper containing scrap is fed into

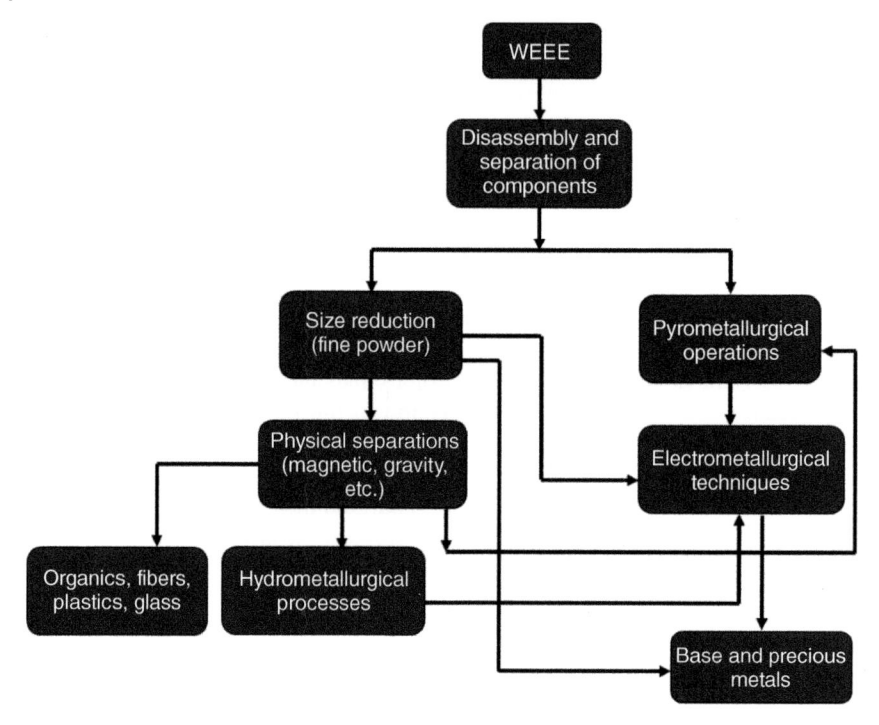

Figure 3.4 Metallurgical process routes for e-waste recycling. Source: Akcil et al. (2015). Reproduced with permission of Elsevier.

converting process directly, whereas low grade WEEE and lead concentrates are fed into the Kaldo furnace.

The energy consumption in the pyro-treatment of WEEE is relatively less than the industrial waste due to its plastic content. The thermal value of the contained plastics adds 8.8 million BTU heating value to each tonne of WEEE; therefore, these polycarbonates supply the heat needed for pyrometallurgical operations. Recovery of precious metals from WEEE has been investigated widely at Umicore Plant in Hoboken, Belgium (Heukelem et al. 2004). Various industrial by-products from nonferrous industries (e.g. drosses, mattes, speiss, anode slimes), precious metals sweeps, spent catalysts, as well as consumer recyclables such as printed circuit boards (PCBs) are acceptable for the integrated metals smelter and refinery process. The plant treats 2 50 000 tonnes of wastes every year, out of which nearly a 10% fraction of the feed is covered by WEEE. Recovery of metals at Umicore Pland follows a complex flowsheet with the inclusion of various pyrometallurgical, hydrometallurgical, and electrochemical processes. The waste is sent to smelting furnace for the separation of precious metals in a copper bullion while other metals are concentrated in a lead slag. Leaching and electrowinning are performed further on the copper bullion in order to recover precious metals and copper. The lead slag is further treated in a lead blast furnace at the base metals operations (BMO) where the oxidized lead

Table 3.1 Typical pyrometallurgical methods for recovery of metals from electronic waste.

Techniques	Metals recovered	Process features	Results obtained	References
Noranda process at QC, Canada	Cu, Au, Ag, Pt, Pd, Se, Te, Ni	Feeding to copper smelter with copper concentration (14% of the total throughput); upgrading in converter and anode furnaces; electrorefining for metal recovery	High recovery of copper and precious metals	Veldbuizen and Sippel (1994)
Boliden Rönnskar Smelter, Sweden	Cu, Ag, Au, Pd, Ni, Se, Zn, Pb	Feeding to Kaldo reactor with lead concentrates; upgrading in copper converter and refining; precious metals refining	High recovery of copper and precious metals	Cui and Zhang (2008)
Umicore's precious metal refining process at Hoboken, Belgium	Base metals, precious metals, platinum group metals and selenium, tellurium, indium	IsaSmelt, copper leaching and electrowinning and precious metal refinery for precious metals operation (PMO); offgas emission control system is installed at the IsaSmelt furnace	Recovering base metals, precious metals, and special metals such as Sb, Bi, Sn, Se, Te, In	Heukelem et al. (2004)
Full scale trial at Umicore's smelter	Metals in electronic scrap	Plastics-rich materials from WEEE were tested to replace coke as a reducing agent and energy source for the IsaSmelt	The smelter operation and metal recovery are not negatively affected by using 6% WEEE plastics and 1% of coke to replace 4.5% coke	Brusselaers et al. (2005)
Dunn's patent for gold refining	Gold	Gold scrap reacted with chlorine at 300–700 °C; acid washing to dissolve the impurity metal chlorides; ammonium hydroxide and nitric acid washing respectively to dissolve the silver chloride	Gold with 99.9% purity was recovered from gold scraps	Dunn et al. (1991)
Day's patent for refractory ceramic precious metals scraps	Precious metals such as platinum and palladium	Plasma heat treatment at 1400 °C; a molten metallic phase containing precious metals and collector metal was produced; ceramic residues went into a slag phase; silver and copper are suitable collector metals in the process	platinum and palladium were recovered with recovery of 80.3% and 94.2%, respectively	Day (1984)

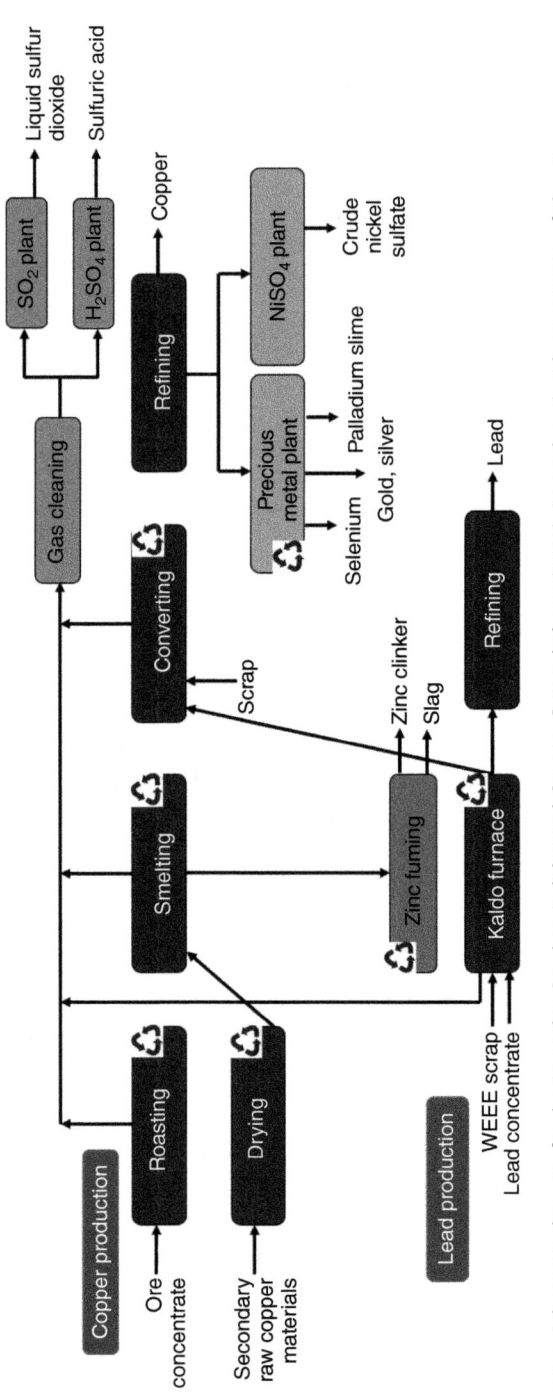

Figure 3.5 Schematic diagram for the Rönnskar Smelter, Boliden Ltd. Source: Cui and Zhang (2008). Reproduced with permission of Elsevier.

slag is reduced in the presence of high lead containing third party raw materials in order to transform the slag into impure Pb bullion, Ni speiss, Cu matte, and depleted slag. The impure lead bullion along with some other metals is further treated in the lead refinery and consequently, pure metals are recovered in a metals refinery process.

The applicability of plasma arc furnace has also been investigated for the recovery of precious metals from refractory ceramic materials (38.1 kg WEEE). The scrap was treated at 1400 °C in a plasma arc furnace for 15 minutes to produce a molten metallic phase with a significant proportion of the precious metals and collector metal (Day 1984). The slag phase mainly contained the ceramic residues. This process resulted in the recovery of 80.3% platinum and 94.2% palladium from the scrap.

3.2.3   Major Challenges Associated with Pyrometallurgical Operations

Although pyrometallurgical treatment offers a significant recovery of metals in industrial scale processes, still there are certain challenges associated with the use of these techniques that inhibit their wide applicability in waste treatment. One of the major limitations is the high temperature requirement to carry out the process, which makes it an energy intensive and expensive choice to adopt. Loss of metals and release of toxic gases during the process are other associated concerns. Ceramic components and glass in the waste stream can lead to increased concentration of slag, and poor metal extraction efficiency (Khaliq et al. 2014). In addition to this, metals such as iron and aluminum enter the slag phase as oxides and cannot be removed using this technique. Therefore, only partial separation of metals is achieved using pyrometallurgy and incorporation of hydrometallurgical techniques and/or electrochemical processing are subsequently necessary to achieve recovery of metals. Combustion of waste streams also causes loss of metals and releases CO_2/SO_2 into the environment. If the waste contains plastics and halogenated flame retardant as in the case of WEEE, the smelting of these waste streams may produce dioxins and other toxic fumes. This is certainly an environmental threat. In addition, huge investment cost is required to make special installations and measures to handle the release of these toxins in a permissible limit and to effectively recover metal, along with being environment friendly. Smelting of nonmetallic parts of e-waste (fine particles) can pose serious health risks to the worker handling the waste.

It is evident that although pyrometallurgical process could be used for metal extraction, it cannot be considered a beneficial decision for industrial growth because of highly energy intensive operations. Economic as well as environmental concerns associated with these operations do not suggest pyrometallurgical methods as an appropriate choice for metal recovery. Various laboratory level, small scale investigations are going on to further improve the process and allow the process to comply with environmental regulations prescribed by nations.

3.3 Hydrometallurgical Treatment of Waste

Hydrometallurgy is a traditional metal recovery process using chemical processes combining water, oxygen, and various chemicals to dissolve metals from their source. The main steps in hydrometallurgy consist of a series of acid or caustic leaching of solid materials followed by organic/inorganic lixiviants to extract precious metals. The solutions are then subjected to separation and purification procedures such as precipitation of impurities, solvent extraction, adsorption, and ion exchange to isolate and concentrate the metals of interest. Consequently, the solutions are treated by electrorefining process, chemical reduction, or crystallization for metal recovery.

Many researchers investigated the best suited method for the recovery of metals. Despite their great variety, the proposed methods in the literature can be grouped into the following categories based on their applicability in a variety of reaction mediums: (i) Acid/alkali leaching; (ii) leaching with lixiviants (cyanide, thiourea, thiosulfate, etc.); (iii) halide leaching (chlorination); (iv) bioleaching. These methods involve aqueous chemistry for the process, can be easily implemented in the laboratory, and energy requirements are less for the process. Each of the aforementioned hydrometallurgical processes has certain pros and cons, which will be discussed along with various research works related to the corresponding processes in the following sections.

3.3.1 Leaching of Metals in Acidic Medium

There are numerous studies reporting on the extraction of metal from spent catalyst resulting from industrial processes using mineral acids and bases. Organic and inorganic acids with varying concentrations have been the major sources for leaching process. Compared to pyrometallurgical operations, leaching is easier to perform and less hazardous since no/less toxic pollutants are released into the ecosystem and the reaction can be carried out at a lower temperature. Figure 3.6 illustrates the general process flow sheet to carry out leaching experiments for removal of metals from waste using hydrometallurgical processes. It is evident from Figure 3.6 that multiple leaching cycles can be carried out using the same waste in order to improve the extraction efficiency.

Inorganic and organic acids such as sulfuric acid, nitric acid, hydrochloric acid, citric acid, and oxalic acids have been used as leaching agents for the recovery of metals from solid wastes. Ivascanu and Roman (1975) reported 99% recovery of nickel in the form of $NiSO_4$ from spent catalyst (obtained from ammonia plant) by performing acid leaching in sulfuric acid medium under optimum process conditions. Al-Mansi and Abdel Monem (2002) also recovered nickel sulfate from a spent catalyst generated in the steam reforming industry using a similar H_2SO_4 leaching process. Lee et al. (2010) used sulfuric acid leaching process for nickel recovery from spent Raneynickel catalyst, which is generated during the hydrogenation reaction of sulfolene to sulfolane in the pharmaceutical industry. This process was useful to reduce the contamination of Al in the leach liquor, which was reduced drastically from 83% to 39%.

3.3 Hydrometallurgical Treatment of Waste | 51

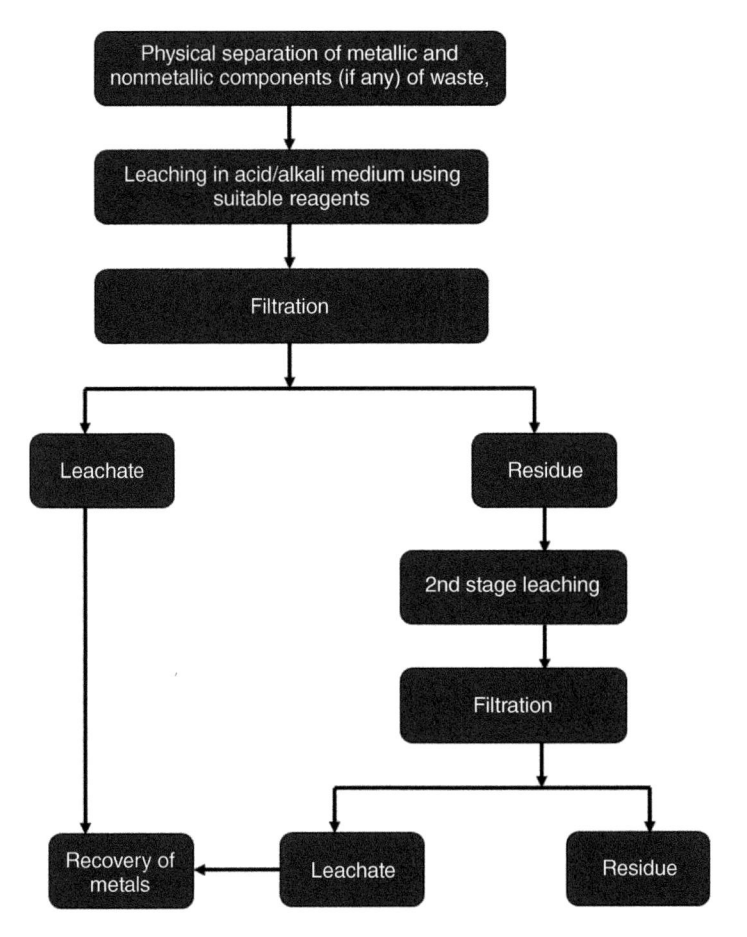

Figure 3.6 Generalized process flow for leaching of metals from waste.

Idris et al. (2010) achieved 85% nickel recovery from spent catalyst (produced during the palm oil hydrogenation process) at 67% sulfuric acid concentration, with an optimum reaction digestion time of 140 minutes, solid-to-liquid ratio of 1 : 14, and reaction temperature of 80 °C. Speciation of nickel as a function of free sulfate (from H_2SO_4) and chloride (from HCl) concentration suggested that Ni can form stable $NiSO_4$, $Ni(SO_4)_2^{2-}$, $NiCl^+$, and $NiCl_2$ complexes at concentrations higher than 1 M. It was observed that at low concentrations of sulfate (0.01–1 M), relatively high percentages (30–95%) of $NiSO_4$ complex were formed while for corresponding concentrations of chloride (0.01–1 M), complexes can only be formed at very low percentages (<10%). Nickel extraction increases from 35% at 1 M to an optimum extraction of 63% at 11 M using H_2SO_4 while in the case of hydrochloric acid, the optimum extraction is approximately 49% at 6 M. Higher sulfuric acid concentration resulted in increased nickel extraction whereas for hydrochloric acid, minor fluctuations in nickel extraction due to changes in acid concentration were observed and extraction was found comparatively consistent (about 40%) even with increases in acid concentration.

Table 3.2 Optimum reaction condition for recovery of metals by acid leaching from spent catalyst.

Spent catalyst	Acid concentration	Solid: liquid	Temperature (°C)	Time (min)	Recovery (%)	References
Low-grade spent catalyst (Ni)	28.8% HCl	—	80 °C	—	73%	Chaudhary et al. (1993)
Spent catalyst (NiO/Al$_2$O$_3$) from steam reforming process	50% H$_2$SO$_4$	1 : 12	100 °C	>300	99%	Al-Mansi and Abdel Monem (2002)
Spent catalyst (NiO/Al$_2$O$_3$) from fertilizer process	50% H$_2$SO$_4$	1 : 20	85 °C	150	94%	Abdel-Aal and Rashad (2004)
Spent catalyst (NiO/Al$_2$O$_3$) from fertilizer process	18% H$_2$SO$_4$	1 : 800	70 °C	60	77%	Mulak et al. (2005)
Ni–Co concentrate of a spent alumina-based catalyst	30% concentrate H$_2$SO$_4$	1 : 10	80 °C	180	93.5% Ni 94.2% Co	Qi-ming et al. (2009)
Spent catalyst (NiO/TiO) from hydrogenation process (Ni)	67% H$_2$SO$_4$	1 : 14	80 °C	140	85%	Idris et al. (2010)
Raneynickel catalyst from pharmaceutical industry	12% H$_2$SO$_4$	1 : 20	30 °C	120	98.6%	Lee et al. (2010)

Thus, sulfuric acid provides better nickel recovery as compared to hydrochloric acid due to the presence of twice the amount of H$^+$ ions, which enables greater ion exchange capabilities with nickel ions.

Chaudhary et al. (1993) also reported that the effect of hydrochloric acid strength on nickel extraction was insignificant and maximum 73% recovery of nickel could be achieved by carrying out the leaching process with 28.8% HCl at 80 °C from a spent catalyst containing 17.7% Ni. Mulak et al. (2006) compared the leaching efficiency of solutions of sulfuric and oxalic acids with addition of H$_2$O$_2$, NaNO$_3$, NH$_4$NO$_3$, and (NH$_4$)$_2$S$_2$O$_8$ to choose the best leachant among all these and concluded that oxalic acid as a leachant forms soluble metal complexes with molybdenum and vanadium, resulting in the extraction of Mo and V being higher than that in the leaching with H$_2$SO$_4$ in the presence of H$_2$O$_2$. Table 3.2 lists the optimum reaction conditions for the effective leaching of metals from spent catalyst in acidic medium. All these studies highlighted the importance of acid concentration, metal to leachant ratio, reaction time, and reaction temperature in order to improve the extraction efficiency of metals.

Mishra et al. (2010) used leaching followed by solvent extraction to recover the metal values from spent petroleum catalyst containing mainly Ni, V, and Mo.

Leaching tests were carried out using 1 M sulfuric acid while LIX-84I was used as an organic extractant for solvent extraction studies. Metal recovery was higher for sulfur free catalyst (by CS_2 + acetone washing) than the catalyst washed only with acetone as the sulfur layer works as a diffusion barrier for the reaction. Leaching studies were carried out in both acid and alkali media and acid leaching showed relatively high metal recovery. More than 95% of Ni, V, and Mo were recovered using 1 M H_2SO_4 followed by $(NH_4)_2CO_3$ washing within one hour of reaction time. Mo extraction efficiency was higher than that of V at low pH while at the equilibrium pH of more than 2, the solvent extracted both V and Mo. This process was useful in enhancing the Mo kinetics and recovery of sulfur values by distillation; however, the process was not eco-friendly as well as quite expensive.

Marafi and Furimsky (2005) compared the efficiency of four organic acids, water, and an aldehyde for extracting metals from a spent hydrodemetallization catalyst. Prior to the leaching experiment, the coke deposited on the catalyst was removed by combustion using 4% O_2 in N_2 at temperatures in the range of 350–450 °C. The decoked catalyst contained 4.4 wt% Mo, 11.6 wt% V, 5.9 wt% Ni, and 2.5 wt% Fe, and the balance Al_2O_3. Leaching experiments were conducted at 50 °C by ultrasonic agitation in an ultrasonic bath. The following order in leaching efficiency was found for six reagents.

$$(\text{Tartaric acid} = \text{Citric acid}) > (\text{Glyoxylic acid} \sim \text{Lactic acid})$$
$$> \text{glycolic acid} > \text{water} > \text{glyoxal}$$

In this process nearly 95% of Mo and V as well as more than 85% of Ni were dissolved during the four hours agitation of decoked spent catalyst in an ultrasonic bath with a solution containing 5 wt% of either tartaric acid or citric acid.

Qi-Ming et al. (2009) investigated the shrinking core model in order to determine the kinetic parameters and rate controlling steps for leaching of nickel from Ni–Co concentrate. Particle size, stirring speed, concentration of acid, and leaching temperature are the major process parameters that may affect the reaction kinetics. It was suggested that the proposed leaching process was controlled by diffusion through product layer. The activation energy of the overall reaction was 15.8 kJ/mol (Qi-ming et al. 2009). Abdel-Aal (2000) also calculated 13.4 kJ/mol activation energy for sulfuric acid leaching of low-grade zinc silicon ore whereas Mulak et al. (2005) demonstrated nearly 15.7 kJ/mol activation energy for nickel leaching from spent catalyst in sulfuric acid solution. It was concluded that the reaction was controlled by diffusion through the catalyst network. In addition, leaching rate of nickel from a spent catalyst in sulfuric acid was found to be independent of the stirring, which indicates that the reaction is not controlled by the diffusion in the liquid phase. On the contrary, Abdel-Aal and Rashad (2004) suggested that the sulfuric acid leaching of a spent nickel oxide catalyst was controlled by chemical reaction on the surface of particles, and the activation energy was calculated to be 41.1 kJ/mol.

Acid leaching has been used for the recovery of base metals from WEEE. A few reports also suggested the recovery of precious metals using inorganic acids; however, the percentage extraction efficiency is not significant in these cases. Leaching in the presence of other lixiviants such as cyanide, thiosulfate, thiourea, etc. are widely employed for the recovery of metals from WEEE and will be discussed in detail in the Section 3.3.3. Nitric acid is used successfully to extract lead, tin, and copper from electronic waste. Copper can be oxidized to copper nitrate using nitric acid solution, which can be further converted into copper sulfate solution for the extraction of copper metals from it. Park and Fray (2009) illustrated successful extraction of 98% Ag and 93% Pd from WEEE in the presence of aqua regia with a metal to leachant ratio of 1 : 20. Jadhao et al. (2016) performed sulfuric acid leaching to extract metals from PCBs of computer desktop. Temperature, reaction time, solid to liquid ratio, concentration of leachant, and stirring speed were varied over a wide range in order to optimize the leaching efficiency; however, copper could not be extracted more than 30% at optimum reaction conditions in the absence of any oxidants. Table 3.3 lists some of the studies based on acid leaching for the extraction of metals from WEEE.

Jadhav and Hocheng (2015) investigated the effect of various inorganic (HCl, HNO_3, H_2SO_4) and organic acids ($C_2H_4O_2$ and $C_6H_8O_7$) on the extraction of metals from large size PCBs. The idea of using large pieces of PCB instead of pulverized samples was probably to illustrate the effect of mass transfer during the acid leaching experiments. In addition, it was claimed that directly using large pieces of WEEE for processing facilitates the recycling of the remaining boards and avoids the problem of precipitate contamination when recovering the metals from leach liquors. Among the leaching reagents examined, HCl showed great potential for the recovery of metals. Stirring speed showed a pronounced positive effect on metal recovery; however, temperature did not show any significant effect on metal recovery. It was observed that with the increase in stirring speed, metal recovery could be improved and at 150 rpm, nearly 100% copper recovery was attained from PCB pieces. Havlik et al. (2010) explained this observation by the fact that copper is significantly leached only in an oxidative environment. Since HCl is a non-oxidizing acid, ideally no significant reaction should occur between Cu and HCl; nevertheless, the oxygen from the surrounding atmosphere enters the solution and acts as an oxidizing agent at higher stirring speed. The probable reaction between Cu and HCl can be given as follows:

$$4Cu + 4HCl(aqueous) + O_2(aqueous) \rightarrow 4CuCl(aqueous) + 2H_2O \quad (3.15)$$

Similarly, lead forms an insoluble species, lead chloride ($PbCl_2$), while reacting with HCl (Xiu et al. 2013). However, Pb was leached out efficiently by applying HCl (Jadhav and Hocheng 2015). It could be due to the formation of differently coordinated Pb^{2+} complexes with Cl (Zhang et al. 2015), as shown in Eq. (3.16):

$$Pb(solid) + 4HCl(aqueous) \rightarrow PbCl_4 + 2H_2(gas) \quad (3.16)$$

Dehchenari et al. (2017) performed extraction of gold from composite CPUs of computer circuit board in the presence of nitric acid and aqua regia medium. Nearly 82% gold was recovered with 83.9% purity. Maguyon et al. (2012) have also performed the simultaneous extraction and deposition of copper from waste PCBs in nitric acid and aqua regia media. The results suggested that the nitric acid treatment yielded higher Cu recovery than the aqua regia treatment. However, the leaching solution could not be used directly for electrodeposition of Cu. Sheng and Etsell (2007) proposed a five-step process for recovery of gold

Table 3.3 Extraction of metals from electronic wastes using acid leaching.

Electronic waste	Reagent	Metals of interest	Reaction conditions	% Metal recovery	References
Waste PCB	H_2SO_4 as an acid leachant and H_2O_2 as an oxidant	Cu	15 wt% sulfuric acid solution; 30% hydrogen peroxide; S/L: 1/10 for 3 h; 23°C, PCBs smaller than 1 mm	Shredding pieces of waste PCBs smaller than 1 mm was efficient and suitable for copper leaching while effect of reaction temperature and initial Cu concentration were found to be insignificant	Yang et al. (2011)
Waste PCB	2 M H_2SO_4, 98% w/v	Cu	5% H_2O_2 at 25°C, 1:10 S:L ratio and 200 rpm	100% Cu Recovery of precious metals was not discussed	Birloaga et al. (2014)
Waste PCB from computers	2.18 N H_2SO_4, 2.18 N H_2SO_4 + 3.0 N HCl, 3.0 N HCl, and 3.0 N HCl + 1.0 N HNO_3	Sn and Cu	Leaching time 120 min	H_2SO_4: 2.7% for Sn and lower than 0.01% for Cu; 2.18 N H_2SO_4 + 3.0 N HCl: 59.3 ± 2.3% Sn and 8.9 ± 0.4% Cu; 2.18 N H_2SO_4 + 3.0 N HCl: 90.5 ± 3.0% Sn and 12.3 ± 0.4 Cu; 3.0 N HCl: 89.1 ± 3.5 Sn and 33.2 ± 1.1 Cu; 3.0 N HCl + 1.0 N HNO_3: 98.1% Sn and 93.2% Cu	Castro and Martins (2009)
Spent lithium ion batteries	H_2SO_4, HNO_3	Li, Co, Ni, Mn	Acid baking at 300°C using H_2SO_4, 60 min leaching at 75°C and 25% pulp density; reductive leaching using mixture of H_2SO_4 and HNO_3 in 45 min	93.2% Li, 90.52% Co, 82.8% Ni, 77.7% Mn	Meshram et al. (2016)

(Continued)

Table 3.3 (Continued)

Electronic waste	Reagent	Metals of interest	Reaction conditions	% Metal recovery	References
PCB	NaOH for pretreatment and HCl, HNO_3, H_2SO_4, $C_2H_4O_2$, and $C_6H_8O_7$ for metal recovery	Base metals and precious metals	150 rpm, room temperature	HCl and HNO_3 solubilized 100% of all metals present in PCB in 22 and 96 h respectively. Nearly 8.8%, 9.8%, and 19.57% Cu was recovered using H_2SO_4, $C_2H_4O_2$ and $C_6H_8O_7$ respectively	Jadhav and Hocheng (2015)
PCB from desktop computer	H_2SO_4	Cu	H_2SO_4 concentration 0.7 M, L/S ratio = 15:1 (v/w), stirring speed = 700 rpm, temperature = 100 °C, time = 4 h, particle size = 150 μm	<30%	Jadhao et al. (2016)
PCB	Aqua regia	Au, Ag, Pd	Leachant to waste ratio is 1:20, multistage process, dodecanethiol and sodium borohydride were used in the form of nanoparticles	Au (98 wt%), Ag (97%), Pd (93%)	Park and Fray (2009)
Computer circuit board scrap	HNO_3, epoxy resin and aqua-regia	Au	Multistage process	Au (500 mg/l)	Sheng and Etsell (2007)
PCB	Aqua regia and H_2SO_4	Cu	Dissolution in solvents	Cu (97%), minor fractions of Sn, Pb, Fe	Veit et al. (2006)

from scrap computer circuit boards and reported 500 mg/l of gold extraction at a temperature of 23 °C. The steps are as follows:

(i) Leaching in a nitric acid–water system
(ii) Mechanical crushing of computer chips and coagulated resin
(iii) Leaching in a nitric acid–water system
(iv) Leaching of solid residue in aqua regia
(v) Precipitation of gold with ferrous sulfate

Sarioglan (2013) used a solution containing hydrochloric acid and hydrogen peroxide along with sodium borohydride as a reducing agent for palladium to recover palladium. It was found that the mixture of hydrochloric acid and hydrogen peroxide was able to completely extract palladium from the activated carbon-PTFE organic matrix.

It is evident that acid leaching has been successfully employed for the treatment of waste and for the recovery of valuable metal resources from the waste streams.

However, acid is hazardous to handle at higher temperatures and therefore, care must be taken while performing acid leaching experiments. Secondly, the leaching process needs to be carried out in specially designed reactors due to the corrosive nature of acids, which increases the installation and operating cost of the setup. The leaching reagent cannot be recovered/reused in the process; therefore, it may cause economic constraint to the operations. In addition, single stage leaching is, in general, not very effective and therefore, multistage recovery process is designed, which certainly adds to the operating cost of the process.

3.3.2 Leaching of Metals in Alkali Medium

Alkali leaching has also been widely explored to extract metals from a variety of waste streams. NaOH, Na_2CO_3, and other alkali agents have been used very frequently for leaching, while aqueous solutions of ammonia and ammonium salts have also proved their importance as a leaching agent for metal recovery from industrial wastes. There are very few studies reported on WEEE treatment using alkali/ammonical leaching methods, although this is an open area of research and has not been explored yet.

Abas (2008) investigated the caustic leaching of vanadium from spent catalyst using 20% NaOH solution and recovered 85% of vanadium pentoxide (V_2O_5) under optimum process conditions at 85 °C reaction temperature and eight hours reaction time. Villarreal et al. (1999) compared two solutions of NaOH, i.e. one containing 10% and the other 40% of NaOH in order to study the effect of alkali concentration. It was observed that the less concentrated solution resulted in higher recovery of metals (92% Mo and 89% V). However, Ni and Al could not be leached out using NaOH solutions. Park et al. (2006) investigated the selective extraction of high purity MoO_3 from spent HDS catalyst using a mixture of Na_2CO_3 and H_2O_2. The results indicated that the recovery of molybdenum was largely dependent on the concentrations of Na_2CO_3 and H_2O_2 in the reaction medium, which controls the acidity (pH) of the leach liquor and carryover of impurities such as Al, Ni, P/Si, and V. The leaching process was exothermic and leaching efficiency of molybdenum decreased with increasing solid to liquid ratio. 84% Mo was recovered from the spent catalyst under optimum leaching conditions of 20% pulp density, 85 g/l Na_2CO_3, 10 vol% H_2O_2 and 60 minutes. reaction time. Recovery of molybdenum from the leach liquor as MoO_3 with a purity of 97.3% was achieved by employing ammonium molybdate precipitation method. Conversion of molybdenum content of the spent catalyst into soluble form with H_2O_2 and Na_2CO_3 mixtures was described by the following equation(s):

$$MoS_2 + 2H_2O_2 + 3.5O_2 \rightarrow MoO_3 + 2H_2SO_4 \qquad (3.17)$$

$$MoO_3 + Na_2CO_3 \rightarrow Na_2MoO_4 + CO_2 \qquad (3.18)$$

$$2H_2SO_4 + 2Na_2CO_3 \rightarrow 2Na_2SO_4 + 2CO_2 + 2H_2O \qquad (3.19)$$

Overall: $MoS_2 + 3Na_2CO_3 + 2H_2O_2 + 3.5O_2$
$$\rightarrow Na_2MoO_4 + 2Na_2SO_4 + 2H_2O + 2CO_2 \qquad (3.20)$$

Literature suggests that alkali leaching process, in general, is employed in combination with pyrometallurgical operations such as calcination to remove the hydrocarbons followed by leaching with alkali reagent (Sun et al. 2001; Shariat et al. 2001) or roasting of spent catalyst with suitable reagents at high temperatures followed by water leaching (Kar et al. 2004, 2005). A schematic representation of the two-stage leaching process is given in Figure 3.7 in which roasting of spent catalyst was combined with alkali leaching to complete the extraction process.

Spent catalyst obtained from petroleum refining industries was roasted with soda ash to convert molybdenum into a water-soluble compound. Results indicated that 92% of molybdenum could be extracted from the spent catalyst at 600 °C with 30 minutes of retention time and using 12 wt% soda ash, in the form of sodium molybdate. Roasting with soda ash was found very efficient for low concentrations of plant catalyst and allowed recovery of molybdenum as trioxide (Kar et al. 2004).

Alkali fusion–leaching process is another method for the recovery of nonferrous metals. Guo et al. (2015) investigated the applicability of the process for extraction of metals in crushed metal enrichment (CME) originated from waste PCBs. The fusion experiments were performed in a non-corrodible nickel crucible, using NaOH and $NaNO_3$ for 60 minutes leaching duration. Sodium salts of amphoteric metals such as Sn, Pb, Zn, and Al were formed in the process, which were thus separated from the residue containing copper and other precious metals. Recoveries of 98.66% for Cu, 91.08% for Sn, 91.25% for Zn, and 78.78% for Pb were reported at optimal reaction conditions. Precious metals were enriched in the final residue with grades of Au 613 g/t and Ag 2157 g/t.

Alkali leaching can be used to selectively dissolve metals from spent catalysts, but it is essentially used for the removal of aluminum from the support, leaving other metals in solution for further precipitation. The leaching of the support metal (e.g. aluminium) is yet another irritant that requires further selective precipitation of the recovered metal, all of which can add to the cost and batch processing time of the treatment.

Ammonia and ammonium salt solutions are known to be effective extractants and have been used by many researchers to extract metals such as Mo, V, Co, and

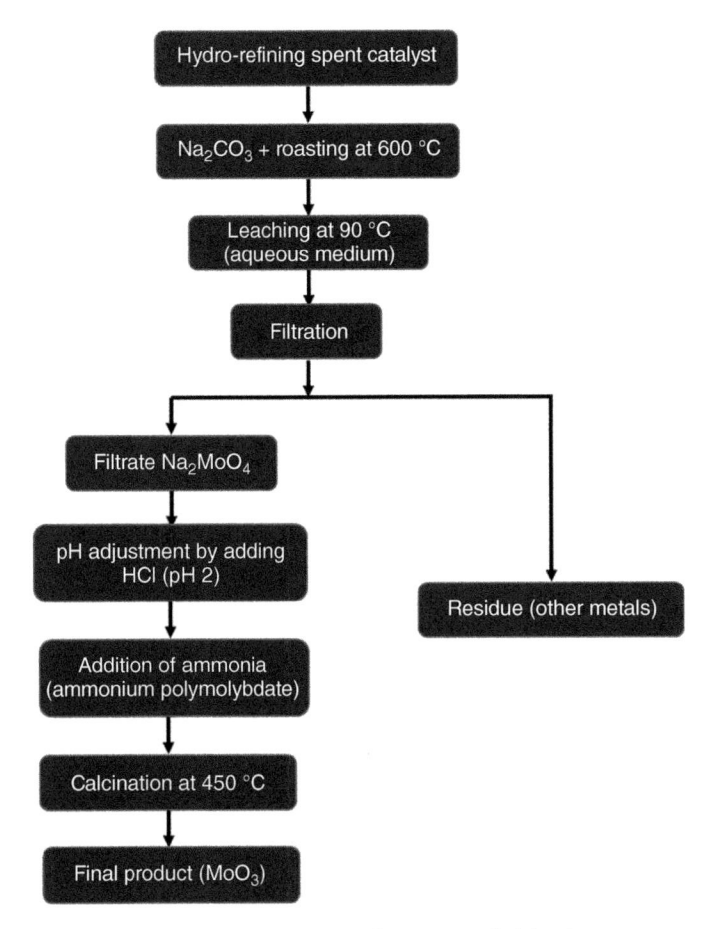

```
┌─────────────────────────────────┐
│  Hydro-refining spent catalyst  │
└────────────────┬────────────────┘
                 ▼
┌─────────────────────────────────┐
│  Na₂CO₃ + roasting at 600 °C    │
└────────────────┬────────────────┘
                 ▼
┌─────────────────────────────────┐
│        Leaching at 90 °C        │
│        (aqueous medium)         │
└────────────────┬────────────────┘
                 ▼
┌─────────────────────────────────┐
│           Filtration            │
└────────────────┬────────────────┘
```

Figure 3.7 Schematic presentation of two-stage alkali leaching process.

Flow continues: Filtrate Na_2MoO_4 → pH adjustment by adding HCl (pH 2) → Addition of ammonia (ammonium polymolybdate) → Calcination at 450 °C → Final product (MoO_3); side branch from Filtration → Residue (other metals).

Ni from waste streams. Leaching by ammonical solutions is generally used for the extraction of nonferrous metals. The dissolution of a divalent metal oxide (MO) in ammonia solution can be expressed as follows:

$$MO + nNH_4^+ \rightarrow M(NH_3)_n^{2+} + H_2O + (n-2)H^+ \qquad (3.21)$$

There are certain advantages associated with this process such as high selectivity, high leaching rates, low toxicity, less cost, easy recyclability, and minimum concerns related to neutralization and acid runoff. The use of ammonia for leaching-out of nickel relies on the fact that nickel hydroxides readily react with ammonia to give stable complexes of various compositions. The US Patent 3,567,433 illustrated recovery of more than 90% of Mo and V, and 60–70% of Ni from spent catalysts using ammonium carbonate solution as a leachant at 150 °C reaction temperature for one hour duration (Gutnikov, 1971). Millsap and Reisler (1978) used ammonium carbonate solution for extracting metals from spent Ni–Mo/Al_2O_3 catalysts. The catalyst was first roasted and then leached with a solution of ammonium hydroxide and ammonium carbonate to extract Mo and Ni, leaving Al_2O_3 as a residue. Ni was precipitated as carbonate after stripping the excess ammonia from the leach solution. Molybdenum was recovered by solvent extraction and precipitated as calcium molybdate. In another study, Yoo et al. (2004) used ammonium sulfate solution for leaching Ni from spent HDS catalyst. About 94% of Ni was leached from the catalyst with $2.6\,mol/dm^3$ $(NH_4)_2SO_4$ solution at the temperature 368 K.

Transition metals were recovered from spent petroleum catalyst by leaching with alkali solutions after preliminary treatment of catalyst with CS_2 or organic solvents (Villarreal et al. 1999). For the pretreatment of catalyst CS_2 was found to be the best over ethyl alcohol and benzene due to its low boiling point (easy for distillation) and low cost while ethanol and benzene gave poor extraction of sulfur. Leaching tests were performed using aqueous NH_3 and NaOH as leaching agent and NH_3 was found to be good for the selective extraction of vanadium while dilute NaOH gave about 92% recovery of both the transition metals Mo and V. Steemson (1999) has examined the ammoniacal leaching of nickel and cobalt hydroxides and observed a greater recovery of nickel and cobalt with faster leaching kinetics for ammonia/ammonium carbonate systems in comparison to ammonia/ammonium sulfate systems. *Wang et al.* (2010a,b) effectively concentrated vanadium along with the molybdenum from sodium molybdate solution by adsorbing with a weak base anion exchange resin in the pH range of 2.5–3.5 and eluting with ammonia liquor. Most of the vanadium in the eluted solution was precipitated by adjusting the pH value to 7.0–8.0 and standing for 24–48 hours. The vanadium that remained in the eluted solution was further separated by adsorption with strong base anion exchange resin at a pH value of about 7.0. By washing with hydrochloric acid and then ammonia solution, the treated vanadium precipitate could be used to produce V_2O_5 expediently.

Sun et al. (2015) designed a two-stage leaching process in order to selectively extract copper and enrich precious metals. More than 95% copper extraction and selectivity were obtained using ammonia-based leaching solutions. A size-screening step was introduced at this point to separate residue I into two fractions, a fine fraction (<1 mm) and a coarse fraction (>1 mm). The coarse fraction will be mixed with raw material for the next ammonia leaching cycle whereas the fine fraction of residue I is processed using sulfuric acid to remove the metals (e.g. Fe, Al, Ni) that are present in low concentrations and thus, to further concentrate precious metals in the residue. The schematic diagram of the proposed two-stage leaching process is given in Figure 3.8.

However, the volatile nature and high evaporation capacity of ammonia makes the process difficult. The low extraction efficiency for a few metals makes the process less attractive.

3.3.3 Leaching with Lixiviants (Cyanide, Thiourea, Thiosulfate)

Cyanide and non-cyanide lixiviants such as thiourea and thiosulfate leaching reagents are preferably employed for the recovery of precious metals from WEEE due to its high efficiency and lower cost. Cyanide anions (CN^-) are, in general, present in the form of complexes, free cyanide (CN^-), or simple compounds in

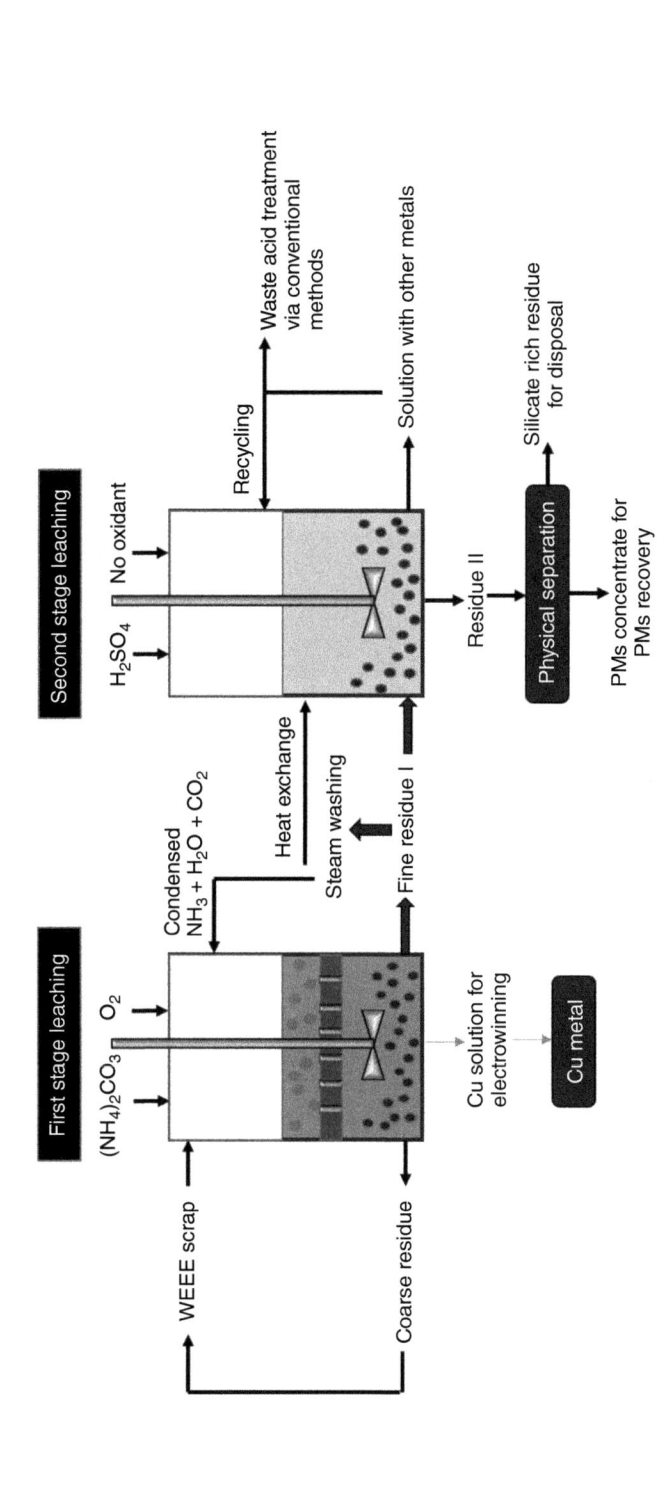

Figure 3.8 Development of a two-stage alkali leaching process for metals recovery from WEEE Source: Sun et al. (2015). Reproduced with permission of American Chemical Society.

solutions. Overall cyanidation process includes oxidation of cyanide to cyanate and passivation due to formation of metal hydroxides/oxides as shown below:

$$4Au + 8(K/Na)CN + O_2 + 2H_2O \rightarrow 4(K/Na)[Au(CN)_2] + 4(K/Na)OH \quad (3.22)$$

$$\text{Overall: } 4Au + 8CN^- + O_2 + 2H_2O \rightarrow 4Au(CN)_2^- + 4OH^- \quad (3.23)$$

Effective cyanide leaching of metals from waste streams depends upon the mass transfer of oxygen or cyanide from bulk solution to the particle's surface by intrinsic leaching reactions. Therefore, increasing the concentration of dissolved oxygen increases the rate of dissolution. Reaction temperature, the particle's surface area, solution pH, stirring rate, and other competitive anion/cations also play a dominant role in the cyanide leaching process.

Several studies have been reported in the literature to extract base metals and precious metals from WEEE using cyanide as a leaching reagent; however, this method has not been used widely for industrial waste treatment. Despite showing good recovery of metals, this method is not preferable for treatment of spent catalyst and other industrial wastes containing copper as a major fraction of their composition.

Column experiments were conducted using cyanide solution to extract precious metals from waste PCBs. Nearly 48% Au and 52% Ag were recovered in this study when a high cyanide concentration of 4 g/l was employed (Montero et al. 2012). Considering the toxicity of the cyanide reagent, high concentration is not desirable for metal extraction. In addition, simultaneous dissolution of Cu (77%) was reported during leaching, which affects the selectivity of the process. Marsden and House (2006) suggested that native copper readily dissolves in cyanide solutions even at low temperature reactions; therefore, direct cyanidation increases the consumption of reagent and decreases the process selectivity. Dissolution of copper can be avoided by performing pre-leaching with sulfuric acid in the presence of suitable oxidant (H_2O_2 or O_2) before cyanidation of WEEE (Kamberovic et al. 2011). Some other studies on the metal recovery from WEEEs using cyanide leaching are listed in Table 3.5.

The International Cyanide Management Code provides detailed information about the toxicity issue associated with cyanide and safe handling. Still, the safety of cyanide handling and its environmental consequences limit the use of this reagent on an industrial scale. The residual cyanide content in industrial effluents may leach into the soil and ground water if not treated properly, and eventually may cause chronic health problems. Considering the fact that the toxicity issues associated with cyanide cannot be avoided even if it is used in very low concentration, the process is now being phased out and active research has been shifted toward the development of less toxic non-cyanide lixiviants (thiourea, thiosulfate) for precious metals leaching from WEEE.

Thiosulfate is primarily used for the dissolution of gold in the presence of cupric ion (Cu^{2+}) and ammonia (NH_3). Cu^{2+} and NH_3 form a stable cupric tetra-amine complex, which stabilizes the gold–thiosulfate complex $[Au(S_2O_3)_2^{3-}]$ in thiosulfate solution and results in significant Au recovery (Aylmore and Muir 2001; Breuer and Jeffrey 2000). The chemical reaction for thiosulfate leaching (Eq. (3.24)–(3.25)) suggests the reduction of Cu^{2+} to Cu^+ and then oxidation of Cu^+ to Cu^{2+} to form the stable cupric tetra-amine complex in the presence of oxygen (Byerley et al. 1973a,b).

$$2Cu(NH_3)_4^{2+} + 8S_2O_3^{2-} \rightarrow 2Cu(S_2O_3)_3^{5-} + S_4O_6^{2-} + 8NH_3 \qquad (3.24)$$

$$2Cu(S_2O_3)_3^{5-} + 8NH_3 + 0.5O_2 + H_2O \rightarrow 2Cu(NH_3)_4^{2+} + 6S_2O_3^{2-} + 2OH^- \qquad (3.25)$$

The concentration ratio of ammonia to thiosulfate plays a dominant role in thiosulfate leaching process. An increase in gold leaching efficiency with the increase in thiosulfate concentration from 0.10–0.14 M was reported whereas a further increase in thiosulfate concentration resulted in the formation of unwanted by-products such as sulfate, trithionate, and tetrathionate, thus reducing the recovery efficiency (Tripathi et al. 2012; Ha et al. 2010). Petter et al. (2014) investigated the influence of addition of hydrogen peroxide and Cu^{2+} ion for recovery of precious metals (Au and Ag) using $Na_2S_2O_3$ and $(NH_4)_2S_2O_3$ on the crushed PCB and reported that the best leaching efficiency of gold was obtained by the addition of Cu^{2+} ion (between 0.015 and 0.03 M) to $Na_2S_2O_3$. However, addition of H_2O_2 did not affect the leaching efficiency significantly for both Au and Ag extraction. Some other studies on metal recovery from WEEEs with thiosulfate solution are listed in Table 3.4.

Thiourea has also been widely employed for metal leaching due to its low toxicity, promising recovery rates, faster kinetics, and high selectivity. Electron pairs between nitrogen and sulfur atoms are believed to have higher potential for a coordination bond between Au and Ag, compared to cyanide lixiviant (Chauhan et al. 2018). The associated chemical reactions for thiourea leaching are shown below (Murthy et al. 2003):

$$Au + 2CS(NH_2)_2 + Fe^{3+} \rightarrow Au(CS(NH_2))^+ + Fe^{2+} \qquad (3.26)$$

$$Ag + 3CS(NH_2)_2 + Fe^{3+} \rightarrow Ag(CS(NH_2)_2)^{3+} + Fe^{2+} \qquad (3.27)$$

The reaction is performed in acidic medium due to high stability and fast dissolution. Ferric ion oxidizes the thiourea reagent and forms formamidine disulfide, which quickly breaks down into sulfur and cyanimide and finally converts into a stable ferric sulfate complex as shown in Eq. (3.28)–(3.30) (Akcil et al. 2015):

$$2CS(NH_2)_2 + Fe^{3+} \rightarrow (SCN_2H_3)_2 + 2Fe^{2+} + 2H^+ \qquad (3.28)$$

$$(SCN_2H_3)_2 \rightarrow CS(NH_2)_2 + NH_2CN + S \qquad (3.29)$$

$$Fe^{3+} + SO_4^{2-} + CS(NH_2)_2 \rightarrow (FeSO_4 \cdot CS(NH_2)_2)^+ \qquad (3.30)$$

Gurung et al. (2013) investigated the effect of Fe^{3+} ions on the extraction of gold from WEEE and observed nearly four times higher extraction efficiency in the presence of Fe^{3+} ions than that in the absence of Fe^{3+} ions. However, concentration of ferric ion (Fe^{3+}) should be controlled cautiously as higher concentration may oxidize thiourea to formamidine disulfide (FDS), which may further oxidize to the undesirable products sulfur and cyanamide (NH_2CN) as shown in Eq. (3.28) and (3.29). The formation of sulfur affects the metal transfer by covering the surface with a thin film and consequently results in decrease in the extraction

Table 3.4 Recovery of metals from WEEEs using cyanide/non-cyanide lixiviant.

WEEE	Reagents	Experimental conditions	Recovery (%)	References
Cyanide leaching				
Waste PCB	Sodium cyanide solution	Column leaching, sodium cyanide concentration 4 g/l, flux 20 l/(d kg) PCBs day, pH between 10.5 and 11 and leaching time 15 d	Au (48%) and Ag (52%), simultaneous dissolution of Cu (77%)	Montero et al. (2012)
E-scrap of PCB and mobile phones	Cyanidation	pH > 10 and reaction temperature 25 °C	>90% Au, Ag, and Pd	Quinet et al. (2005)
PCB of mobile phones	Potassium cyanide (6–8%)	pH 12.5 and reaction temperature 25 °C	60–70% Au	Petter et al. (2014)
Thiosulfate leaching				
PCB	72.7 mM thiosulfate, 10 mM Cu^{2+} ion, and 0.266 M ammonia concentrations	At 20 °C and in pH ~ 10 (400 rpm)	>90% Au	Ha et al. (2014)
PCB of cell phones	0.1 M $(NH_4)_2S_2O_3$, 0.2 M NH_4OH, 0.015–0.03–0.05 M $CuSO_4$, 0.01–0.05–0.1 M H_2O_2	At 25–26 °C, in pH 9.0–10.1, 1/20 S/L ratio, in 4 h	~15% Au, 3% Ag	Ha et al. (2014)

PCB of mobile phones	$Na_2S_2O_3$ leaching: 0.1 M $Na_2S_2O_3$, 0.2 M NH_4OH; 0.1 M H_2O_2. $(NH_4)_2S_2O_3$ leaching: 0.1 M $(NH_4)_2S_2O_3$, 0.2 M NH_4OH, 0.03 M Cu^{2+} ion	At ambient temperature, pH 9–11; S/L = 1/20; time: 4 h	$Na_2S_2O_3$ leaching: 11% Au; $(NH_4)_2S_2O_3$ leaching: 21% Ag	Petter et al. (2015)
PCB	0.1 M $(NH_4)_2S_2O_3$, 40 mM $CuSO_4$	In 10 g/l solid–liquid ratio, in pH 10–10.5, at 25 °C, in 8 h (250 rpm)	79% Au	Tripathi et al. (2012)
Thiourea leaching				
PCB	20 g/l $CS(NH_2)_2$; 6 g/l Fe^{3+}; 10 g/l H_2SO_4	<2 mm particle size, stirring speed = 600 rpm	69% Au, 75% Cu	Birloaga et al. (2013)
PCB	0.5 M $CS(NH_2)_2$; 0.01 M Fe^{3+}; 0.01 M H_2SO_4	53–75 µm particle size, 5% S/L rate, in 2 h, at 50 °C (500 rpm)	68% Au	Gurung et al. (2013)
Mobile phone PCBs	24 g/l thiourea solution, 0.6% Fe^{3+}	2 h reaction time, 154 µm particle size, 40 ml solution at room temperatures	90% Au and 50% Ag	Jing-Ying et al. (2012)
PCB	14 g $CS(NH_2)_2$, 2.6 g $Fe_2(SO_4)_3$, 3.6 N H_2SO_4	0.84 mm particle size, in 200 ml leaching solution, in 24 h, at 20 °C (150 rpm)	100% Au, 100% Ag	Lee et al. (2011)

Table 3.5 Chlorination based hydrometallurgical processes.

Leaching agent	Types of waste and technology used	Metals extracted	References
$CuSO_4$–NaCl–HCl	E-waste	Cu	Herreros et al. (1999)
NaCl–HCl–$CuCl_2$	E-waste using electro-deposition process	Cu	Yu et al. (1999)
$CuSO_4$–NaCl–H_2SO_4	Electro-oxidation of e-waste		Zhu and Gu (2002)
Chlorine gas	E-waste treatment	Cu	Kim et al. (2010)
Chlorination	Alluvial material	Au	Ojeda et al. (2009)
Chlorination	Spent catalyst	Rh, Pd, and Pt	Shen et al. (2010)

of metals. The presence of copper also negatively affects precious metal recovery during the thiourea leaching process because of its simultaneous dissolution. It is recommended to remove base metals by employing an oxidative leaching pretreatment prior to precious metal extraction. Some other studies on metal recovery from WEEEs with thiourea solution are listed in Table 3.5.

3.3.4 Halide Leaching

Halide leaching method includes chloride, bromide, and iodide leaching, although chlorination has been widely used and proved to be more efficient and cheaper for the extraction and refinement of precious metals such as Ti, Zr, Nb, Ta, and Mo (Gaballah et al. 1994; Gaballah and Djona 1995; Jena and Brocchi 1997). In the chlorination method, precious metals present in the waste stream are chlorinated with Cl_2 at an elevated temperature. It requires temperatures higher than 1200 °C to vaporize these metals. Ojeda et al. (2009) extracted 98.23% gold present in alluvial material by chlorination at 873 K in one hour reaction time by using chlorine as a reactive agent, with very low attack of the matrix containing the metal. The influence of the reaction temperature and reaction time has been studied and concluded that extraction of gold was facilitated by increasing both the temperature and the reaction time.

Dry chlorination methods for extracting platinum group metals (PGMs) from auto-catalysts were also developed in recent years. In the methods, the powder of crushed spent auto-catalyst containing PGMs was usually first calcined and then chlorinated at 500–800 °C with chlorine in the presence of reducing agents such as carbon monoxide. After that, diluted hydrochloric acid was used to leach PGMs from the chlorinated powder of spent auto-catalyst. Finally, a chloride solution containing Rh, Pd, and Pt was obtained as filtrate by separating solids residue by filtration (Shen et al. 2010). The adsorption efficiencies of Rh, Pd, Pt, Al, Fe, Si, Zn, and Pb were 89.89%, 90.58%, 91.53%, 0.69%, 0.63%, 9.98%, 17.58%, and 55.14%, respectively, and the adsorptions of Mn, Ca, Mg, Ni, Cu, La, and Ba on the resin were negligible under adsorption conditions.

Figure 3.9 Process scheme of leaching of waste printed circuit board using chlorination process. Source: He and Xu (2015). Reproduced with permission of Royal Society of Chemistry.

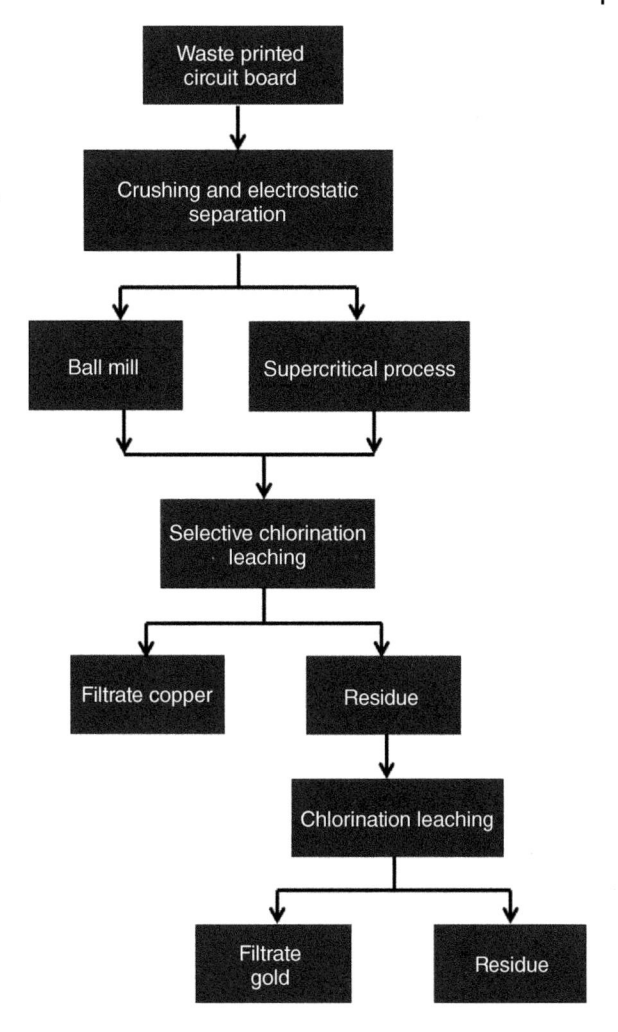

```
         Waste printed
         circuit board
              │
              ▼
      Crushing and electrostatic
            separation
          ┌────┴────┐
          ▼         ▼
       Ball mill  Supercritical process
          └────┬────┘
               ▼
       Selective chlorination
            leaching
          ┌────┴────┐
          ▼         ▼
    Filtrate copper  Residue
                      │
                      ▼
              Chlorination leaching
                ┌────┴────┐
                ▼         ▼
             Filtrate    Residue
              gold
```

Chlorination process was investigated to recover copper from waste PCBs using $CuSO_4$–$NaCl$–H_2SO_4 solution. The copper metals oxidized to Cu^{2+} during the process and these ions reacted with the solution to form copper chloride complex. However, longer reaction time (5.5 hours) was considered as the major drawback of the proposed reaction (Zhu and Gu 2002). It was suggested that carrying out the process in the presence of oxidant (O_2 and ClO^- as oxidizing agents) may reduce the reaction time with improved efficiency (Kim et al. 2010).

Chlorination is the most recommended method for gold extraction from waste in small scale operations. There are laboratory scale investigations that have attempted to improve the rate of recovery of metals using chlorination process in combination with different pretreatment methods. The process scheme shown in Figure 3.9 illustrates the studies of metal extraction (gold) using two types of pretreatment of waste PCBs.

The first set of samples were treated using supercritical water at 400 °C and 23 MPa, for 60 minutes, while other sets were ball milled for 10, 20, and 30 minutes before chlorination using a solution of sodium chlorate. Higher leaching efficiency (99% Cu and Au) was reported when supercritical pretreatment was coupled with the chlorination process (He and Xu 2015).

Several studies reported significant metal extraction from WEEE using various halide lixiviants (Zhang and Zhang 2013; Altansukh et al. 2016; Sahin et al. 2015). Iodination of electronic scrap was investigated for the recovery of Au at the ambient temperature of 35 °C and leaching time of four hours. It was observed that the presence of oxidant significantly enhances the metal recovery.

Approximately 90% Au recovery was obtained using 3% iodine concentration in the presence of 2% hydrogen peroxide as oxidant, whereas no significant gold recovery was observed in the absence of the oxidant (Sahin et al. 2015). In the iodide leaching process, concentration ratio of iodine to iodide (I^-) is an important parameter to recover precious metals from waste. Insufficient amount of I^- or excess concentration of I^- could lead to a decrease in metal extraction. Low concentration of I^- hinders the formation of I^{3-}, which results in weak complexation ability and thus leads to precipitation of AuI, AgI, and PdI_2. On the other hand, higher concentration of I^- may react with other base metals to form CuI and PbI_2 (Xiu et al. 2015; Chauhan et al. 2018). The formation of PbI_2 and CuI on the surface of Au, Ag, and Pd causes lower leaching of these metals. Table 3.5 lists a few reports on halide leaching for the extraction of metals from different kinds of waste.

The chlorination-based hydrometallurgy processes are known to have certain advantages. It is believed that the energy required to carry out the process is

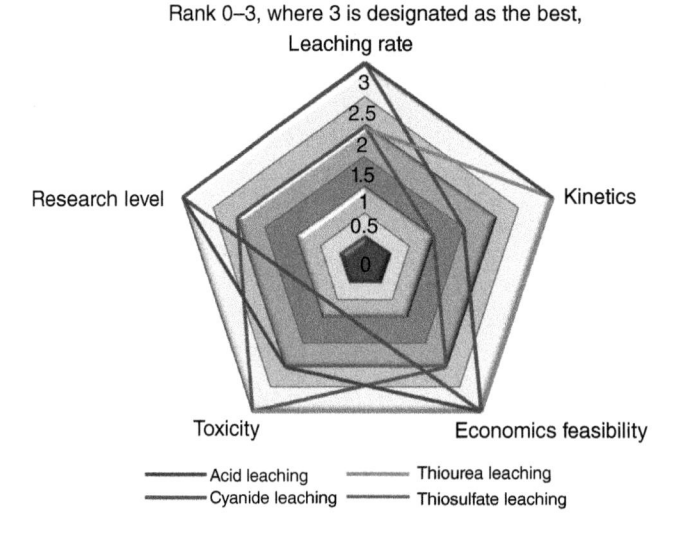

Figure 3.10 Ranking of leaching chemistries among five categories, leaching rate, research level, kinetics, toxicity, and economic feasibility. Source: Hsu et al. (2019). Reproduced with permission of Royal Society of Chemistry.

minimum, and the rate of metal recovery is considerably high; however, high consumption of reagent and high economics are the major drawbacks associated with the process (Chauhan et al. 2018). More research work is needed to resolve these issues. Chloride leaching needs special stainless steel and rubber-lined equipment to resist highly corrosive conditions (Awasthi et al. 2017). Similarly, specific instrumental equipments are required in bromide leaching process for safety and to avoid health risks.

Figure 3.10 illustrates the rankings of acid, cyanide, thiourea, and thiosulfate leaching among five categories, leaching rate, kinetics, economic feasibility, toxicity, and research level, where a ranking of 3 is the most optimal. Studying leaching more closely reveals several drawbacks associated with different types of acids and extraction chemistries; some of them are mentioned above.

3.4 Summary and Outlook

This chapter emphasizes the urgent necessity for substantial research to be done to pave the way forward for successful, environment friendly and economic processes for resource extraction from waste streams. All the abovementioned physicochemical processes have shown potential in achieving the target of recycling of metals from solid wastes; however, each one of them has its own merits and limitations. Process economics, large amount of chemicals needed, the disposal of the chemicals, secondary pollution possibility, and environmental constraints certainly restrict the use of these metallurgical processes on large scale. In addition, some of the hydrometallurgical processes (discussed in this chapter) have already been phased out (cyanide leaching) or are still being explicated on laboratory scale; therefore, more studies are needed to prove their effectiveness with respect to large scale applications.

There are certain environmental constraints on the disposal of waste containing acid and ammoniacal solutions. High energy requirement during pyrometallurgical operations inhibits the long-term use of these processes for resource recovery. On the other hand, the secondary waste generated during hydrometallurgical processes needs special arrangements to discard them in an eco-friendly manner and thus affects the process economics. The hazardous health effects associated with the prolonged use of these chemicals should also be addressed.

It is evident that none of the existing conventional metallurgical processes can be considered an efficient eco-technic process. More R&D efforts are still needed to provide a sustainable solution for waste management and resource optimization. The following chapters discuss some of the emerging technologies (adsorption, bioleaching) and future technologies (chelation technology; ionic liquids) that have the potential of providing a sustainable way of metal extraction. Although these processes have been investigated at small scale for other applications, they have not yet been explored widely for the metal extraction from industrial waste or electronic waste, and therefore considerable research efforts are needed in this direction.

Questions

1. Explain the conventional methods of metal extraction from waste. What are the major limitations associated with these processes?

2. "The conditions required for recovery of metals from WEEE are different than from spent catalyst." Justify the statement.

3. Discuss the applicability of pyrometallurgical processes for treatment of industrial waste.

4. Define the parameters that affect the recovery efficiency of hydrometallurgical process.

5. Explain the general flow sheet of recovery of metals from WEEE using metallurgical processes.

6. Name any two pretreatment methods that can be used to improve the efficiency of chlorination process for the extraction of gold from e-waste.

7. Why is cyanide leaching preferred for WEEE treatment than spent hydroprocessing catalyst?

8. Write short note on recovery of metals from solid waste using thiourea lixiviant.

9. Compare various hydrometallurgical leaching methods in terms of reaction efficiency and environmental toxicity.

4

Emerging Technology for Metal Extraction from Waste: I. Green Adsorption

As we continue to improve our understanding of the basic science on which applications increasingly depend, material benefits of this and other kinds are secured for the future.

Henry Taube

4.1 Introduction

Adsorption is a widely used chemical process in the laboratory as well as on a commercial scale for numerous applications and has drawn the attention of researchers for a long time. It is a well-defined process representing an ideal amalgamation of multiple branches of science such as chemistry, physics, and biology, in addition to engineering. New classes of solid adsorbents such as activated carbon fibers and carbon molecular sieves, fullerenes and hetero-fullerenes, microporous glasses, and nanoporous carbonaceous and inorganic materials have been explored for their adsorption capacities. Nanostructured solids due to high sorption, catalytic, magnetic, optic, and thermal properties are illustrated as economic and eco-friendly adsorbents in the literature. Inexpensive materials, such as chitosan, zeolites, and other natural adsorbents, have also been investigated widely in order to make the adsorption process more cost-effective and environment friendly.

This chapter presents, in brief, the basic concept and diverse applicability of various adsorbents for the recovery of metals from waste streams. The technical feasibility of low-cost green adsorbents has also been discussed for heavy metal removal from contaminated sites and other applications.

4.2 Adsorption

"Adsorption" is defined as the change in concentration of a given substance at the interface, instead of at the neighboring bulk phases (Dąbrowski 2001). This results in accumulation of that substance at the contacting surface, giving rise to higher concentration of that species on the interface relative to the bulk phase.

Sustainable Metal Extraction from Waste Streams, First Edition.
Garima Chauhan, Perminder Jit Kaur, K.K. Pant, and K.D.P. Nigam.
© 2020 Wiley-VCH Verlag GmbH & Co. KGaA. Published 2020 by Wiley-VCH Verlag GmbH & Co. KGaA.

In other words, when gas or liquid molecules come in contact with a solid surface, they may get accumulated or concentrated at the surface. This phenomenon of disposition of molecules of a gas or liquid at a solid surface is called adsorption. This is a well-established process, especially for effluent treatment and water treatment. Adsorption can be considered in liquid–gas, liquid–liquid, solid–liquid, and solid–gas systems depending on the type of phases in contact; however, major work on the industrial scale deals mainly with the solid–fluid interfaces. Being a surface phenomenon, the process is different from "absorption," which involves the penetration by the adsorbate molecules into the bulk solid phase.

Fluid in the adsorbed state is defined as the "adsorbate," while the bulk fluid prior to being adsorbed is called the "adsorptive." The surface-active material that adsorbs a solute from its solution is referred as the adsorbent. The specific area of an adsorbent is the surface area available for adsorption per gram of the adsorbent. Higher the surface area available, the more molecules are trapped on its surface. Generally, this means that a good adsorbent is very porous. In addition, a desirable adsorbent should have abrasion resistance and thermal stability. Adsorbents can be of any shape from spherical pellets, rods, to monoliths, on which a new substance accumulates. Most commonly used adsorbents are activated carbon, oxides of silicon, aluminum, titanium, and various types of adsorbent clays. In general, industrial adsorbents are classified into three categories.

4.2.1 Hydrophilic Compounds

These compounds are capable of making hydrogen bonds that enable them to dissolve more readily in water than in oil or other hydrophobic solvents. A hydrophilic molecule is charge-polarized, which corresponds to their affinity with polar substances such as water or alcohols; therefore, polar adsorbents come into the category of "hydrophilic" adsorbents. These are also known as "universal" adsorbents due to their negligible resistance toward any liquids and thus soak up water-based and petroleum-based liquids alike (Master 2010). Aluminosilicates such as zeolites, porous alumina, silica gel, or silica-alumina are examples of adsorbents of this type.

4.2.2 Hydrophobic Compounds

Hydrophobic adsorbents attract molecules via London dispersion or van der Waals forces. The induced dipole moment depends on the polarizability of adsorbate molecules and adsorbent atoms. Thereby, dispersion forces affect the adsorbent–adsorbent and adsorbent–adsorbate interactions equally. It is illustrated widely that adsorption on hydrophobic solids mainly depends on the polarizability of molecules, rather on their polarity. On the other hand, Halasz et al. (2002) presented polarity as the dominant factor than polarizability in the adsorption of vapors on hydrophobic microporous zeolites. This group of adsorbents are based on carbon and are nonpolar in nature. Carbonaceous adsorbents, polymer adsorbents, and silicalite are typical examples of hydrophobic adsorbents. These adsorbents have higher affinity with oil or hydrocarbons than with water.

4.2.3 Polymer Matrix

Polymer based compounds can be both polar and nonpolar in a porous polymer matrix. For example, nano-magnetic polymers (NMPs) and polysaccharides, due to low cost and ease of availability, are widely used for the removal of contaminants.

Adsorption is a mass-transfer operation, in which the adsorbate is adsorbed on the surface of the adsorbent. The strength by which adsorbate molecules are attached with the adsorbents determines the nature of adsorption. Figure 4.1 demonstrates the general mechanism for the adsorption process. The adsorbate can deposit on the surface in different ways. If it forms a single layer, the phenomenon is called monolayer adsorption, while deposition of various layers, one on the other, is termed as multilayer adsorption as shown in Figure 4.1.

In Figure 4.1, the adsorption process is illustrated to take place in definable steps:

(1) *Bulk solution transport*: The adsorbate moves across the fixed film boundary around the adsorbent media.
(2) *Diffusive transport*: This moves the adsorbate across the fixed film boundary.
(3) *Bounding process*: The adsorbate is attached to the absorbent media surface.

The rate controlling step of the overall process is the rate of diffusion of solute molecules within the pores of adsorbent particles, which is a function of adsorbate concentration, temperature, and molecular weight of the solute.

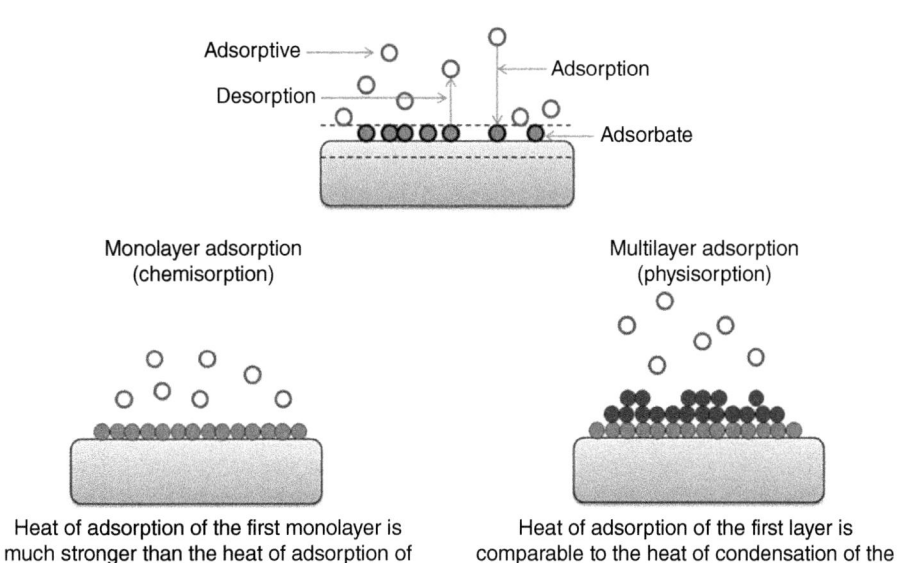

Figure 4.1 General representation of adsorption.

Depending upon the binding energy of the adsorbate to the substrate, the adsorption can be categorized into physisorption and chemisorption. Physisorption is a type of adsorption in which the adsorbate adheres to the surface only through van der Waals (weak intermolecular) interactions, which are also responsible for the nonideal behavior of real gases. The electronic structure of the atom or molecule is barely perturbed upon adsorption during the physisorption process. Chemisorption is a type of adsorption whereby a molecule adheres to a surface through the formation of a chemical bond, as opposed to the van der Waals forces that cause physisorption. In comparison with chemisorption, in which the electronic structure of bonding atoms or molecules is changed and covalent or ionic bonds form, physisorption can only be observed in the environment of low temperature (thermal energy at room temperature ~26 meV) and the absence of the relatively strong chemisorptions. Since the adsorptive and the adsorbent often undergo a chemical reaction, the chemical and physical properties of the adsorbate are not always just the sum of the individual properties of the adsorptive and the adsorbent, and often represent a phase with new properties (Christmann 2010).

Adsorption capacity is defined as the amount of adsorbate adsorbed per unit mass (volume) of adsorbent and depends on the type of adsorbent and adsorbate, concentration of adsorbate, and temperature of the process. The surface area and pore volume of the adsorbent limit the size of the molecules adsorbed.

4.3 Green Adsorption

A new term "Green Adsorption" has been introduced recently, which includes the low-cost materials originated from agricultural sources, agricultural by-products, agricultural residues, and wastes. Industrial waste/by-products are also drawing the attention of researchers as non-expensive adsorbents from which most complex adsorbents will be produced. These "green/low-cost adsorbents" have been recognized for their adsorption capacity that is lower yet comparable to that of the conventional super-adsorbents reported in the literature (activated carbon, structurally complex inorganic composite materials, etc.); nevertheless, their economic potential makes them competitive in the present context. Certain low-cost materials such as agricultural wastes (tea and coffee waste), hazelnut shells, peanut hull, sawdust, different tree barks, coconut husk, modified cellulosic materials, chemically modified corncob, and different agricultural by-products are investigated for their adsorption properties. These studies demonstrate the high efficiency of the green adsorbent to remove metals from wastewater.

Biosorption is also an emerging physicochemical adsorption process in which inactive or dead microbial biomass is used as a sorbent to concentrate heavy metals from even very dilute aqueous solutions. Biosorbents have also been referred as green adsorbents in some studies; however, this chapter does not illustrate biosorbents as green adsorbents. Only agricultural/industrial waste and by-products are referred to as green adsorbents in this chapter.

Searching for good green adsorbents among various agricultural wastes and by-products is a tedious task for researchers and therefore much research effort has been made in order to search for new low-cost adsorbents with good adsorption capacities. Park et al. (2010) suggested availability and low cost as the key factors in choosing green adsorbents for practical application. On the other hand, Chojnacka (2010) mentioned adsorption capacity as the decisive factor in the adsorbent selection process. Abbas et al. (2014) suggested that a reasonably good green adsorbent should meet several requirements, including low cost, abundant availability, high cost-effectiveness, high rate of reaction, easy desorption, high regeneration capability, and less formation/release of unexpected compounds into aqueous solutions.

4.4 Parameters Affecting the Adsorption Capacity of Green Adsorbents

As discussed in the Section 4.2, the extent of adsorption depends primarily on the nature of the adsorbent, especially its porosity and surface area. High porosity and large surface area are the desirable properties of adsorbents for high adsorption capacity.

The adsorption of heavy metals from contaminated sites may also be influenced by several physical and chemical factors, such as pH, solution temperature, initial concentration of heavy metals in solution, amount of the adsorbent, and presence of co-ions in solutions. Ionic strength of the solution, particle size of the adsorbent, and selectivity of adsorbents to metal ions may significantly affect the adsorption process and therefore, extensive research has been undertaken to investigate the effects of these operating parameters. Process parameters determine the overall adsorption by affecting the uptake rate, selectivity, and removal efficiency. This section briefly discusses the effects of various process factors on the adsorption process.

4.4.1 Influence of pH

Among process factors, pH seems to play a significant role in controlling the adsorption of heavy metals. The value of pH can affect the surface charge of the adsorbent, the degree of ionization, speciation of heavy metals, and the competition of target metal ions with coexisting ions in solution (Park et al. 2010). pH dependency could be explained by the involvement of functional groups in metal uptake and metal chemistry (Singh et al. 2017). At lower pH, the overall surface charge of the solid surface will be positive. The H^+ ions compete effectively with the metal cations causing a decrease in sorption capacity. With the increase in pH values, the adsorbent surface becomes negatively charged, which favors the metal ions uptake due to electrostatic interaction. At very high pH the adsorption stops, and hydroxide precipitation starts (Njoku et al. 2011; Taha et al. 2018). Since the pH dependence of metal uptake is associated with the surface functional groups, different types of green adsorbents attain maximum adsorption capacity at different pH values depending upon the presence of various functional groups on their cell walls.

4.4.2 Influence of Temperature

Adsorption process can be affected by temperature in different ways depending on the exothermic or endothermic nature of the process. If the metal removal efficiency increases with increasing temperature, then the process is considered endothermic. This could be due to increased availability of active sites, thereby increasing the mobility of metal ions for adsorption with increasing temperature. The endothermic nature of the adsorption process was reported in many research studies (Singh et al. 2017; Park et al. 2010). Removal of metal improves with the increase in temperature due to improved surface activity and kinetic energy. On the other hand, Ofomaja and Ho (2006) suggested a decrease in adsorptive forces between the target species and the active sites on the adsorbent surface due to increase in temperature, which may cause decrease in metal removal efficiency. Several other researchers corroborated the arguments that sorption capacity is inversely proportional to process temperature, and thus adsorption processes are exothermic (El-Sayed et al. 2015; Kumar et al. 2012). Damage of active adsorption sites at high temperature or escape of increasing number of metal ions from the adsorbent surface to the solution are reported to be the major cause of decrease in metal removal efficiency with increase in temperature. High temperatures might result in the physical damage of adsorbents (Park et al. 2010); therefore, adsorption experiments are recommended to be conducted at room temperature.

4.4.3 Effect of Initial Concentration

Transport of target species from the solution to the surface of adsorbents takes place due to the concentration gradient of the initial metal ion concentration. Several studies have been reported to investigate the effect of initial solute concentration on the adsorption capacity and metal removal efficiency. It is widely observed that adsorption capacity improves with increase in solute concentration; however, removal efficiency decreases significantly with increase in target solute concentration (Junior et al. 2013; Kumar et al. 2012). Increase in metal uptake capacity may be related to an increase in concentration gradient with increase in initial metal concentration, which provides a driving force to minimize mass transfer resistances of solute between solution and adsorbent phase(s) (Singh et al. 2017). Saturation of adsorption sites on the solid surface may result in decrease in adsorption efficiency. Sorption isotherms can be studied to elucidate the specific relation between equilibrium concentration of adsorbate in the bulk and the adsorbed amount at the surface.

4.4.4 Effect of Adsorbent Dosage

Adsorbent dose plays a significant role in determining the adsorption capacity at a given initial concentration of metal ions in aqueous solution. Several studies suggest that the removal percentages of metal ions increase with increase in the amount of adsorbent. It can be clearly attributed to a higher number of available adsorption sites on the adsorbent surface with increase in adsorbent dose. Singh et al. (2017) reported a decrease in adsorption capacity with the increase in adsorbent dose and argued that it may be attributed to the overlapping or aggregation of the adsorption sites (Nuengmatcha et al. 2016), which decreases the total surface area of the adsorbent, thus limiting the availability of active sites during the adsorption process.

4.4.5 Effect of Co-ions

Different researchers report diverse effects of co-existing ions on the adsorption of heavy metals. Goyal and Srivastava (2009) found that the removal percentages of heavy metals by *Zea mays* in single metal solutions (Pb 87.3%, Cd 79.3%, Ni 71.9%, and Cr 76.4%) were higher than those in a multimetal solution (Pb 81.2%, Cd 73.7%, Ni 64.0%, and Cr 68.9%). They attributed this effect to the competition between cations. On the contrary, García-Mendieta et al. (2012) reported that the removal percentages of Mn and Fe from a binary system were similar to the values found in single systems, which clearly indicated that Fe and Mn did not compete for the adsorption sites on the green tomato husk. It is recommended that adsorption studies in binary and ternary systems are essential to promote the real application of green adsorbents at large scales.

In order to minimize the mass transfer limitations, the effect of particle size and effect of agitation speed have also been investigated, although these parameters have not been discussed in this chapter.

4.5 Adsorption Kinetic Models

The term "modeling" in physicochemical processes study is quite general and ranges from simple fitting of the experimental data to detailed models based on basic principles. Especially in the field of adsorption, quite different modeling approaches can be found. The kinetics of the adsorption process is evaluated in order to determine the rate of adsorption (Singh et al. 2017). To review the kinetics models available in the literature, a typical experimental setup has been considered in this chapter that includes an initial concentration of adsorbate (C_0), a beaker volume (V), and a mass of adsorbent (M). The amount of adsorbate adsorbed on the adsorbent (per unit adsorbent mass) is expressed as "Q" and evolves from $Q = 0$ to an equilibrium value $Q = Q_e$. Thermodynamic equilibrium concentration C_e is given as $C_0 - M(Q/V)$. The modeling issue is focused on the quantitative description of the evolution curve $Q(t)$, which is deduced by measuring experimentally the evolution of adsorbate concentration $C(t)$ from C_0 to C_e.

The adsorption kinetic models that have been employed in green adsorption are listed in Table 4.1. Several researchers modified the kinetic models listed in Table 4.1 based on their experimental investigations. Some of the prominently used models are Langmuir kinetic model, shrinking core model, reversible and irreversible reactions model, pseudo-n order, etc.

Table 4.1 Kinetic models for adsorption process.

Kinetic model	Equation	References
Lagergren equation (pseudo-first order model)	$Q = Q_e(1 - e^{-k_1 t})$	Lagergren (1898)
Pseudo-second order model	$\left(\dfrac{t}{Q}\right) = \left(\dfrac{1}{k_2 Q_e^2}\right) + \left(\dfrac{t}{Q_e}\right)$	Blanchard et al. (1984)
Elovich model	$Q = \dfrac{\ln(\alpha\beta) + \ln(t)}{\beta}$, α and β are initial adsorption and desorption rate constants	Cheung et al. (2001)
Intraparticle diffusion model	$Q = Q_e\left(1 - \dfrac{6}{\pi^2}\sum\limits_{i=1}^{\infty}\dfrac{1}{i^2}\exp\left(-\dfrac{Di^2\pi^2}{R^2}t\right)\right)$ D = intra particle diffusion coefficient, R = particle radius	Ruthven (1984)
McKay model	$\ln\left(\dfrac{C}{C_0} - \dfrac{1}{1+KM}\right) =$ $\ln\left(\dfrac{KM}{1+KM}\right) - \left(\dfrac{1+KM}{KM}\right)hSt$ S is the total adsorbent particles' external area, h is the mass transfer coefficient, and K is the linear isotherm constant	McKay et al. (1981)

4.6 Mechanism of Metal Uptake

Various mechanisms can be responsible simultaneously for metal uptake at varying extents of their contribution. Ion exchange and electrostatic interaction are considered as the principal mechanisms for metal sorption, which includes formation of metal complexes and carboxyl groups followed by interaction of metal complexes with carboxyl groups to form metal–carboxylate complexes (Netzahuatl-Muñoz et al. 2015). Feng and Guo (2012) reported that ion exchange might be the dominant mechanism in the removal of Cu(II), Zn(II), and Pb(II) using orange peel based on the X-ray fluorescence analysis for modified orange peel. Several other reports suggested that ion exchange might be one of the dominant mechanisms during the adsorption process (Njoku et al. 2011; Taha et al. 2011; Panday 2008). In contrast, another possibility was demonstrated as a mass transfer limited process, in which rapid adsorption of metal ion to the external surface followed by intra-particle diffusion into the interior of the adsorbent could be the basis for the adsorption process (Sen and Bhattacharyya. 2011). Witek-Krowiak and Reddy (2013) explained that the dominant mechanisms for the adsorption of Cr(III) and Cu(II) onto soybean meal waste included (i) ion exchange, (ii) chelation by carboxyl and hydroxyl groups, and (iii) precipitation. The authors also highlighted the role of various functional groups in the adsorption process. It was reported widely that carboxyl and hydroxyl groups are the major contributors in order to conduct an effective sorption process. However, the existence of these functional groups on the surface of agriculture waste does not guarantee an effective removal of heavy metals as the biosorption process may be influenced by several other factors, such as (i) the number of active sites, (ii) accessibility of the sites, (iii) chemical state of the sites, and (iv) affinity between the sites and the target metal ions (Nguyen et al. 2013; Park et al. 2010).

4.7 Green Adsorbents: Relevant Literature

A variety of adsorbents are being used in the literature that are originated from agro-based sources or from industries either as a waste material or undesirable by-products. This section categorizes the green adsorbents based on their origin into the following: (i) agricultural resources, (ii) natural zeolites, (iii) clay, and (iv) industrial waste/by-products.

4.7.1 Agricultural Resources

Agriculture is an important activity across the globe. Although the type of crop grown varies from region to region, every region has developed its own methods to utilize the waste of the field left after primary activities such as harvesting or pruning of the field. By-products after processing of the crop in agricultural industries also act as a secondary source of agricultural residue. India and China have been reported to have 0.5×10^9 and 1.75×10^9 tonnes of agricultural residue respectively (Devi et al. 2017; Dai et al. 2018) from the crops of wheat, rice, maize, millets, sugarcane, jute cotton, and pulses. Traditionally this has been used as feedstock for animals, fuel, and manure. In the past two decades, agricultural waste and by-products have gained attention as adsorbent for the removal of heavy metals from industrial effluent and other contaminated sites (Figure 4.2). Biochar and activated carbon obtained from agricultural residue are proved to possess high surface area, with uniform pore size distribution, high porosity, and strong adsorption capacity.

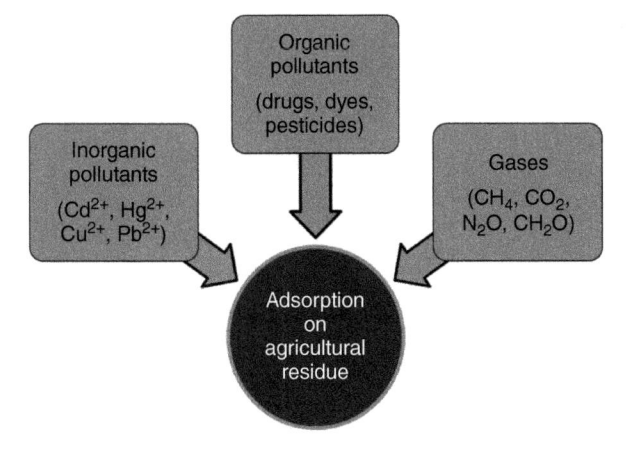

Figure 4.2 Agricultural residue as adsorbent for multiple components.

The major components of agricultural residue are lignin, cellulose, and hemicelluloses with minor fractions of starch and water soluble components. Cellulose is a polymer made up of monomer $C_6H_{10}O_5$ (Kaur et al. 2016). Carbonyl, hydroxyl, and ether groups of cellulose can bind with metal ions and can act as adsorbent for metal ions. Various traditional resources such as wheat straw and rice straw and new resources such as hazelnut shell, pecan shells, jackfruit, maize cob, or husk can be used as an adsorbent for heavy metal uptake after chemical modification or conversion by heating into activated carbon. Studies have been performed to evaluate the metal removal efficiencies of agricultural residue for the removal of metals such as nickel, lead, chromium, and cadmium from wastewater. Agricultural residue such as orange peel has been found to be effective for the removal of Ni(II) from simulated wastewater at pH 6. In general, untreated agricultural waste is used to eliminate heavy metals from wastewater. Table 4.2 lists some of the recent studies where agro-residue waste(s) have been used for the removal of metals from the aqueous stream.

The greatest hindrance to the wider application of agricultural waste is their relatively lower adsorption capacity. However, low yield and high operational temperature add to the overall cost of the process and turn it into a less profitable venture. The investigations performed by various researchers globally have highlighted that the modified agricultural residue possesses greater metal adsorption capacity. Feng et al. (2011) recommended pretreated agricultural waste to improve the adsorption efficiency of metals, remove soluble

Table 4.2 Latest developments in agricultural waste for metal extraction from waste.

Type of agro-residue	Type of metal and waste treated	Optimum reaction conditions	References
Orange peel	Ni from aqueous solution	Ni extraction: 97% at pH 6: extracted for 2 h at 50 °C	Ajmal et al. (2000)
Rice husk	Cu from water	Cu(II) was 17.0358 mg/g at pH 4: contact time: 180 min	Zhang et al. (2014)
Walnut shell	Cd from aqueous solution	Contact time: 80 min; pH 5 for Cu and Zn; pH 6 for Cd; adsorption capacity Cu(II): 14.5 Zn(II):7.4 Cd(II): 7.2 mg/g	Najam and Andrabi (2016)
Rice straw	Pb from aqueous solution	for 30 min contact time at a pH of 5.5, particle size 75–150 µm and a dose of 4 g/l, lead removal was 94%	Amer et al. (2017)
Banana peels and peanut shells	Pb and Mn extracted from aqueous solution	Adsorbent particle size: 600, 425, and 300 µm; contact time: 100 min	Maduabuchi (2018)
Orange peel	Cr from aqueous solution	1.1 g of adsorbent at 34 °C, yield 97%	Khalifa et al. (2019)

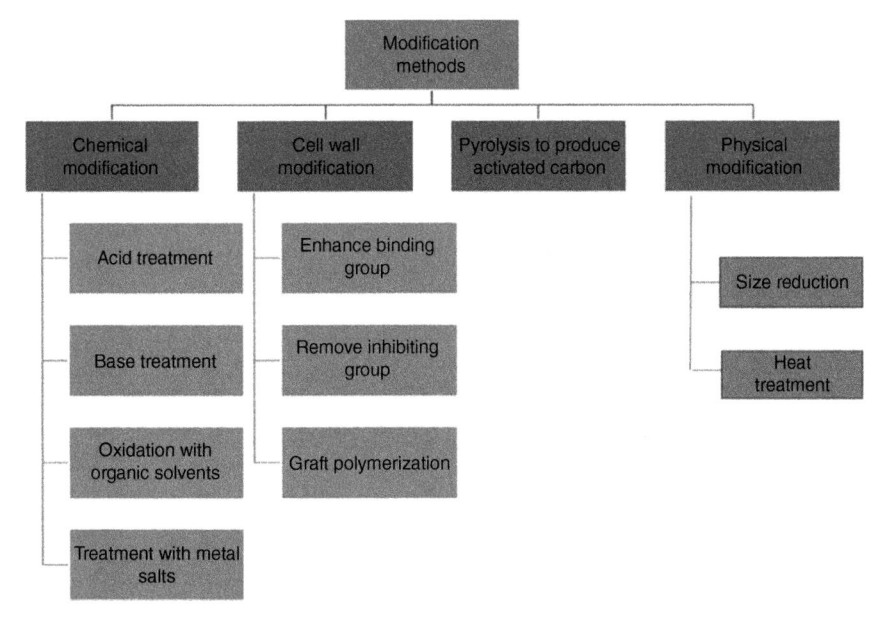

Figure 4.3 Modification methods for improving agricultural biosorbents.

[Flowchart showing Modification methods branching into: Chemical modification (Acid treatment, Base treatment, Oxidation with organic solvents, Treatment with metal salts); Cell wall modification (Enhance binding group, Remove inhibiting group, Graft polymerization); Pyrolysis to produce activated carbon; Physical modification (Size reduction, Heat treatment).]

organic compounds, and eliminate coloration of the solutions (Wan Ngah and Hanafiah 2008). Coconut shell charcoal on modification with an oxidizing agent and chitosan was used for the extraction of Cr(VI) (Babel and Kurniawan 2003). Similarly, rice hull on modification with ethylenediamine had adsorbed Cr(VI) ions from a simulated solution (Tang et al. 2003). Heavy metals such as Cu(II) and Zn(II) were removed from wastewater stream using activated charcoal obtained from pecan shells (Bansode et al. 2003) and potato peels charcoal (Amana et al. 2008). The modification methods include physical modifications, chemical modifications, and other methods as shown in Figure 4.3. Physical modifications are usually considered simple and inexpensive. However, they are not widely used because of their low effectiveness. Conversely, chemical modifications are preferred, due to their simplicity and efficiency (Park et al. 2010). Modifying agents can be classified as bases, mineral and organic acids, organic compounds, oxidizing agents, etc.

Table 4.3 summarizes metal adsorption capacities of various agricultural wastes modified using different agents.

The studies on agro-residue in natural form as well as modified form are limited to laboratory scale level for wastewater treatment only. Pilot scale studies of the same for other types of waste, especially spent catalyst and waste electrical and electronic equipment (WEEE), are required to further expand their field of application.

4.7.2 Zeolites

Zeolites are three-dimensional structures of amino-silicate minerals (SiO_4 and AlO_4 linked together). The large surface area of multidimensional interconnected

Table 4.3 Metal removal efficiency of various modified agricultural wastes.

Biosorbents	Modifying agents	Metal ions	Change in adsorption capacity (%)	References
Parsley	Pyrolysis + $FeCl_3$	As(V), As(III)	↑ 9463.16 (As(V)), ↑ 416.67 (As(III))	Jiménez-Cedillo et al. (2011)
Rosa bournobiaphyto biomass	Acetic acid, benzoic acid, citric acid 0.1 N	Pb(II), Cu(II)	↓ 39.64, 43.89, 23.01 (Pb(II)), ↓ 14.10, 15.19, 20.24 (Cu(II))	Manzoor et al. (2013)
Green tomato husk	Formaldehyde 0.2%	Fe(III), Mn(II)	↑ 5.09 Fe(III), ↑ 10.89 Mn(II)	García-Mendieta et al. (2012)
Orange peel	HNO_3 0.1 M	Cd(II)	↑ 61.38	Lasheen et al. (2012)
Orange peel	NaOH 0.8 M and $CaCl_2$ 0.8 M	Cu(II), Pb(II), Zn(II)	↑ 59.73 Cu(II), ↑ 84.84 Pb(II), ↑ 164.38 Zn(II)	Feng and Guo (2012)
Orange peel	The grafted polymerization	Pb(II), Cd(II), Ni(II)	↑ 420 Pb(II), ↑ 460 Cd(II), ↑ 1650 Ni(II)	Feng et al. (2011)
Pineapple peel fibber	Succinic anhydride	Cd(II), Cu(II), Pb(II)	↑ 374.05 Cd(II), ↑ 336.26 Cu(II), ↑ 242.7 Pb(II)	Hu et al. (2011)
Corncobs	Thermal treatment (180 °C); H_3PO_4	Zn(II)	↑ 93.96	Kumar et al. (2011)
Wheat straw	Urea	Cd(II)	↑ 822.82	Farooq et al. (2011)

cavities of zeolites serves for excellent adsorption properties. The internal structure of zeolite is a large network of cavities, which gives them high surface area. Zeolites are of around 40 types, out of which clinoptilolite ((K_2, Na_2, Ca)$_3Al_6Si_{30}O_{72} \cdot 21H_2O$) is the most widely used. Being cheap and abundant in nature, it is widely available for multiple purposes. Silicon or aluminum atoms are present, being surrounded by four oxygen atoms. Aluminum, with one less positive charge than silicon, gives the compound a net negative charge at the site of each aluminum atom, which is balanced by exchangeable cations.

It has strong ion exchange capabilities, which makes it a material of significant scientific research. For instance, Turkey clinoptilolite is reported to have a composition of SiO_2: 70.90%, Al_2O_3: 12.40%, Fe_2O_3: 1.21%, CaO: 2.54%, and has a cation exchange capacity (CEC) of 1.6 and 1.8 meq/g (Alpat et al. 2008). It has been used successfully to extract heavy metals such as Pb(II), Cd(II), Cr(III), Cu(II), Mn(II), Zn(II), and Ni(II) from wastewater.

Similarly, Bulgarian zeolites are useful for the removal of metals such as Ni^{2+}, Cu^{2+}, Cd^{2+}, Zn^{2+}, and Pb^{2+}. Adsorption capacities of some of the zeolites are mentioned in Table 4.4.

For the removal of silver metals, limited studies are available. It has been found that natural zeolite with a surface area of 16.76 m^2/g and pore volume of 39.71 Å has an adsorption capacity of 3.23 mg Ag^+/g zeolite. The factors that affect the

Table 4.4 Adsorption capacity of zeolites for heavy metals.

Metals	Zeolite	Adsorption capacity (meq/g)	References
Pb^{2+}	Sardinian clinoptilolite	0.27–1.2	Cincotti et al. (2006)
	Turkish clinoptilolite	0.299–0.730	Oter and Akcay (2007)
	Natural phillipsite	0.234–0.345	Ali and Hunaidi (2004)
	Mexican clinoptilolite	1.5	Monter et al. (2007)
Cd^{2+}	Sardinian clinoptilolite	0.5–0.19	Cincotti et al. (2006)
	Natural clinoptilolite	0.082	Ayuso et al. (2003)
	Clinoptilolite	0.12–0.18	Gedik and Imamoglu (2008)
Ni^{2+}	Turkish clinoptilolite	0.017–0.173	Oter and Akcay (2007)
	Natural clinoptilolite	0.068	Ayuso et al. (2003)
Zn^{2+}	Sardinian clinoptilolite	0.1	Cincotti et al. (2006)
	Turkish clinoptilolite	0.108–0.251	Oter and Akcay (2007)
	Natural clinoptilolite	0.106	Ayuso et al. (2003)
Cu^{2+}	Sardinian clinoptilolite	0.34	Cincotti et al. (2006)
	Turkish clinoptilolite	0.022–0.227	Oter and Akcay (2007)
	Natural clinoptilolite	0.186	Ayuso et al. (2003)
CO^{2+}	Scolecite	0.31	Bosso and Enzweiler (2002)
	Turkish clinoptilolite	0.448	Erdem et al. (2004)
Cr^{3+}	Natural clinoptilolite	0.237	Ayuso et al. (2003)
	Brazilian scolecite	5.81	Bosco et al. (2005)

adsorption capacity of zeolites are pH, adsorption time, initial metal concentration, and adsorption process.

The adsorption capacity of natural zeolites is limited and takes a longer time to completely remove the metals from waste. Efforts have been made by researchers to improve their adsorption capacity. Artificially prepared zeolites have also been investigated under laboratory conditions to enhance the metal removal capacity. Calcinated kalabsha Kaolin mineral on reaction with sodium hydroxide forms a gel and forms a synthetic zeolite labeled as 4A zeolite. This synthetic zeolite is able to completely remove copper and zinc ions along with 96% and 55% removal of manganese and chromium ions at different pH. While for Mn(II), basic solution at pH 11 was effective, acidic solution with pH 3 was optimum for Cr(VI) removal, with a contact time of one hour in each case (Barakat 2008).

Researchers have synthesized zeolites using different chemical solutions with enhanced adsorption capacity. A solution of sodium hydroxide and sodium silicate along with sodium aluminate in aqueous solution can be used to prepare synthetic zeolite, with the label of NaA zeolite. It is reported to have higher chromium removal efficiency than natural zeolites (Basaldella et al. 2007).

Modifications have been proposed in the literature to further exacerbate their adsorption capacity. The surface area and micropore volume of natural zeolite are $1\,m^2/g$ and $0.23\,cm^3/g$, respectively. It has been found that modification of synthetic zeolite with iron oxide (a magnetically modified zeolite [MMZ]) produces a much enhanced surface area and micropore structure. The surface area and micropore volume of synthetic zeolite modified with iron oxide is 2.2 m^2/g and 0.516 cm^3/g. The adsorption capacity of synthesized zeolite is reported to be 123 mg/g, much higher than that of natural zeolites, and possesses high adsorption capacity of lead ions in water (Nah et al. 2006).

4.7.3 Clay

Clay is defined as a colloidal fraction formed by combination of soils, sediments, and rocks with water. They are hydrous aluminosilicates, formed by quartz, carbonate, and metal oxides. The sorption capacity of clay is a function of its chemical composition, nature, and volume and structure of pores. For the sorption to take place in gaseous phase, pore structure is the dominating factor. In liquid phase, the adsorption capacity of clay depends mainly on the chemical composition of the clay.

The chemical structure of clay is made up of various layers on tetrahedral silica sheet, i.e. Si(O, OH)$_4$, and octahedral sheet of M(OH)$_6$, where M is a divalent or trivalent cation like alumina.

Based on the types of silicate layers, clay is classified into five categories. Kaolinite is the type of clay with one tetrahedral silica sheet followed by one octahedral alumina sheet (1 : 1). Its CEC is low (1–10 mol/kg) and has low surface area (20–50 m^2/g). Montmorillonite and vermiculite clay are 2 : 1 types where two tetrahedral silica sheets enclose one octahedral alumina sheet. The CEC of montmorillonite is 80–120 mol/kg and the surface area is 800 m^2/g, much higher than that of kaolinite clay. Chlorite clays are 2 : 1 : 1 type where the 2 : 1 layer contains a metal hydroxide sheet in between. Bentonite is a mixed type of clay, consisting of both the crystalline structure of montmorillonite and other additional crystalline structures (Table 4.5).

The properties of clay depend on the arrangement of these sheets. There are various arrangements possible. Tetrahedral Si can be replaced by Al or octahedral Mg can be replaced by Al in both sheets. A cation can replace the excess of permanent negative charge on the outer surface of the clay. Broken edges of the clay minerals also carry variable charges. The edges of the 2 : 1 clay mineral contains

Table 4.5 Surface area and CEC of various types of clay.

Type of clay	Surface area (m^2/g)	CEC	References
Kaolinite	20–50	1–10	Huang et al. (2011)
Montmorillonite	800	80–120	Huang et al. (2011)
Chlorite	20	1.5	Vdović, et al. (2010)
Bentonite with 80% montmorillonite	63	—	Mishra and Patel (2009)

hydroxyl Al or Si group. The octahedral Al–OH layer may exhibit a negative and positive charge in basic medium and acidic medium, respectively.

Thus, there are predominating types of ions, mainly Ca^{2+}, Mg^{2+}, H^+, K^+, NH_4^+, Na^+, SO_4^{2-}, Cl^-, PO_4^{3-}, and NO_3^{1-}, available on the surface for exchange.

The adsorption capacity is therefore because of the presence of these ions on the surface. Equilibrium, adsorption capacity, and kinetics of the process, however, are a factor of the pH of solution, adsorbent dose, particle size of the adsorbate, and contact time of the process. Depending on the type of the clay, metals such as As, Cd, Cr, Co, Cu, Fe, Pb, Mn, Ni, and Zn can be adsorbed in their ionic forms. Various studies have highlighted its usage in wastewater treatment processes.

A few laboratory level studies highlight the adsorption capacity of different types of clay for extraction of metals from industrial effluents as well. Table 4.6 lists some of the adsorption studies using various types of clay as an adsorbent.

The automobile effluent was collected in a pretreated plastic bottle from the discharge outlet of automobile industries. Adsorption capacity was found to be highly dependent on the pH of the solution. The concentration of metal in the effluent played a very important role in the adsorption capacity of the clay. With higher concentration of metals in the solution, higher rate of adsorption was observed. At pH 8, about 18% Zn, 10% Cu, 12% Mn(II), 11% Cd(II), 12% Pb, and 14% Ni were recovered on the clay. Ionic radii of metals in effluents was related to the rate of sorption and equilibrium constant of the process. The energy of the Dubinin–Radushkevich isotherm reveals the nature of the process as physical adsorption, with film diffusion being the rate controlling step.

Clay is one of the potential alternatives to activated carbon as well. Similar to zeolites, clay minerals are also important inorganic components in soil. Their sorption capabilities come from their high surface area and exchange capacities. The negative charge on the structure of clay minerals gives clay the capability to attract metal ions. The United States and the former Republics of Soviet Union such as Lithuania, Georgia, and Kazakhstan are well known for their large deposits of natural clay minerals. In order to further enhance their adsorption capacity, various modifications in natural clay have been proposed in the literature. Acidized as well as alkaline treatments are reported to be useful to improve their cation exchange capabilities. It has been reported in the literature that modification of the natural clay minerals with a polymeric material, known as clay–polymer composites, can significantly improve their capability to extract heavy metals from aqueous solutions.

4.7.4 Industrial Waste

There are various types of industrial waste with high adsorption capacity that can be used as a low-cost adsorbent for heavy metal removal. This effort will not only solve their disposal and handling problem but will also add value to them. Industrial wastes are generated as by-products along with the main product in an industry. Since these materials are locally available in large quantities, they are inexpensive. The technical feasibility of use of industrial wastes such as waste slurry, lignin, and iron(III) hydroxide has been explored by researchers under laboratory conditions. Industrial by-products such as fly ash, waste iron, iron slags,

Table 4.6 Metal adsorption capacity of various types of clay.

Type of clay	Type of solution	Metal adsorption capacity	Reaction conditions	References
Montmorillonite clay	Synthetic solution	5.45 mg tungsten/g of natural clay at pH 4	Optimum pH 4	Gecol et al. (2006)
Montmorillonite coated with chitosan	Synthetic solution	23.9 mg tungsten/g of biosorbent	Optimum pH 4	Gecol et al. (2006)
Kaolin and Bentonite clay	Synthetic solution	—	Optimum contact time: 3 h; pH: no effect	Mishra and Patel (2009)
Alkaline-modified montmorillonite clay (NaOH and KOH treated)	Synthetic solution	>70% removal of Ni(II) and Mn(II)	Concentration: 100 mg/l, adsorbent dose: 0.1 g, temperature: 300 K, particle size: 100 m, contact time: 180 min	Akpomie and Dawodu (2014)
Alkaline-modified montmorillonite	Synthetic solution	Zn > Cu > Mn > Cd > Pb > Ni	Contact time: 180 min; Ni (0.069 nm) < Zn (0.074 nm) < Mn (0.080 nm) < Pb (0.12 nm) related to adsorption capacity	Akpomie and Dawodu (2015)
Acid treated montmorillonite clay	Industrial effluent	Zn > Cu > Mn > Cd > Pb > Ni	0.1 g of the adsorbent to 50 ml of the effluent contact time: 180 min solution	Akpomie and Dawodu (2016)
Mixed clay	Synthetic solution	70% removal of Cd^{2+}, Cr^{3+}, Cu^{2+}, Pb^{2+}, and Zn^{2+} ions	pH of solution varied with type of metal to be removed; not effective for removal of arsenic ions	Sajidu et al. (2006)

Table 4.7 Significant research for metal extraction process from industrial waste.

Type of industrial waste	Metal removal efficiency	Reaction conditions	References
Green sand from iron foundry industry	Zn(II)	pH 2.6, 3.0, and 4.8 tested for initial zinc concentration: 100 mg/l	Lee et al. (2004)
Iron slag	Cu(II) and Pb(II)	pH 3.5–8.5 pH 5.2–8.5	Feng et al. (2004)
Fly ash from bagasse of sugar industry	Cd(II) and Ni(II)	pH 6–6.5	Gupta et al. (2003)
Fly ash from coal-burning	Cu(II) and Pb(II)	—	Alinnor (2007)
Saw dust treated with 1,5-disodium hydrogen phosphate	Cr(VI)	pH 2	Uysal and Ar (2007)
Iron(III) hydroxide from fertilizer industry	0.47 mg of Cr^{6+}/g at pH of 5.6	pH 5.6	Namasivayam and Ranganathan (1993)
Fly ash modified with chitosan	Zn	Freundlich isotherm model was followed	Adamczuk and Kołodyńska (2015)
Untreated coffee residues from cafeteria	Cu(II) adsorption: 70 mg/g; Cr(VI): 45 mg/g	Contact time: 120 min	Kyzas (2012)

and hydrous titanium oxide can be chemically modified to enhance their performance in metal removal from wastewater. For instance, fly ash has a negatively charged surface, and can adsorb metal from other types of waste using electrostatic adsorption and precipitation. Metal solutions are neutralized by fly ash and can be adsorbed on its surface.

Table 4.7 shows significant research in the area of development of metal extraction process from different types of industrial waste.

Fly ash, a waste from industries, is characterized by its unique metal adsorption capacity. The chemical composition of fly ash is about 36% SiO_2, 36% Al_2O_3, 6% CaO, 15% S, and 3% C (Liskowitz et al. 1980). The composition however varies with the source from which fly ash is obtained. The adsorption capacity of fly ash is reported to be related positively with its carbon content.

An adsorption capacity of 1.39 mg of Cu^{2+}/g at pH of 8.0 shows its potential to be explored for the adsorption of other metals as well (Pandey et al. 2009). A mixture containing fly ash and clay exhibited an adsorption capacity of 0.31 mg of Cr^{6+}/g at pH of 2.0. Further, fly ash mixed with wollastonite (1 : 1) showed an adsorption capacity of 2.92 mg of Cr^{6+}/g at pH 2.0. The adsorption process was reported to follow the Langmuir model of isotherm. In addition to this, some interesting studies have shown the potential use of coffee waste as an adsorbent for copper and chromium removal.

The waste materials investigated in these studies, which have economic potential to extract metals, are otherwise going to further pollute the environment.

Research work in this field is in the preliminary stages. With the help of scale-up studies, if these waste products are further utilized as the secondary source of precious metals, not only will this act as a source of metal but will also reduce the overall amount of industrial waste.

4.7.5 Modified Biopolymers

Biopolymers are a group of compounds derived from raw materials that are either bio-based or fossil fuel-based sources. Natural biopolymers include lignin, protein, and chitosans and are reported to be effective for adsorption of heavy metals from waste streams. They have a number of functional groups such as amines, hydroxyls, and acetamido, and have high selectivity to adsorbed aromatic compounds and multiple metals. Chemical stability, high reactivity, easy availability, and nontoxicity make them novel and green adsorbents. Various polysaccharides and their derivatives are known to be hydrophobic, biocompatible, and biodegradable, and have gained the attention of researchers globally. The second most abundant biopolymer chitin is obtained as a waste product from animals having crustacean shells, such as fish, shrimp, lobster, prawns, and crabs. Chitosan is a deacetylated form of chitin. The nitrogen present in chitosan and cellulose is 6.89% and 1.25%, respectively. This higher percentage of nitrogen and reactive amine group provides it with the capacity of adsorption. Chitosan is a cationic polymer and possesses electrostatic properties and is therefore capable of biosorption through anion exchange mechanisms. Metals such as copper, nickel, and zinc were found to be biosorbed effectively by chitosan from a wastewater stream (Juang and Shiau 2000). With excellent metal binding capacity, both chitin and chitosans are powerful chelating agents and are found to be effective in removing transition metal ions from waste streams.

Palladium is an expensive metal present in large proportion in spent catalyst. Ruiz et al. (2000) used chitosan with molecular weight 125 000 cross-linked with glutaraldehyde to extract it from aqueous solution. The sorption kinetics of adsorption of palladium from aqueous solution depends on the particle size of the adsorbent, cross-linking ratio, and concentration of palladium metal in the solution. The time required for the process varies from one to three days.

4.8 Innovative Applications of Adsorption

Increasing evidence suggests that adsorption process displays numerous advantages over conventional processes due to the ease of operation, low energy demand, and regeneration capacities of adsorbents. The process therefore has gained attention from various fields ranging from environment protection, water purification, medicine, and food to agro-chemical industries. Their capacity to remove pollutants even from low concentration wastewater further makes them a process of choice. Further, the process is considered to be green and eco-friendly due to the use of no toxic chemicals and gases, with no toxic by-products generation. The field of application depends entirely on the type of adsorbate and adsorbent. Materials such as activated carbon can be used to adsorb polluting gases, dyes and colors, and organic as well as inorganic contaminants. Adsorption of methylene blue dyes on the different surfaces is widely used for industrial wastewater treatment.

Adsorption can also be used for cleaning of cane, beet, and corn sugar solutions, fats, and oils from vegetable and animal oil. Its ability to adsorb poisonous gases makes it an important material to prepare gas masks for military and industrial purposes. In order to control odor in restaurants and airports, activated charcoal based adsorbents are used. Adsorption is used successfully for the adsorption of antioxidant phenolic components from food industries as a by-product. Enzymes are extensively used in chemical and biochemical industries and are preferred over chemical catalysts in food industry. To improve their catalytic ability adsorption of solid surface is used to immobilize them. Immobilization using adsorption uses van der Walls forces of attraction and thus allows the enzymes to retain their activity and enhances their efficiency. Urease is an important enzyme used mainly in food, urea production, and agricultural industry, as it is used for the conversion of urea to carbonic acid and ammonia. The immobilization of this enzyme using adsorption makes it useful in the field of medicine for the treatment of kidney as an alternative to dialysis.

4.9 Case Study

Endowed with high surface area and low cost, activated charcoal has been explored for the extraction of metals from various kinds of waste. In this chapter, the application of adsorption for removal of metals from industrial waste and WEEE has been explored. Commercial plants based on adsorption process alone for metal extraction have not been reported in the literature. However, there are instances of coupling of adsorption with hydrometallurgical process to extract metals from these waste materials. In one of such investigations, e-waste with around 27% copper, 0.52% silver, 0.06% gold, and 0.04% palladium was crushed to small size particles that constituted around 13% copper, 0.49% silver, 0.08% gold, and 0.04% palladium. These particles were leached using sulfuric acid (1.2, 1.5, and 2 M) at 80 °C. Chloride leaching was done using 2 M NaCl solution. Further, HCl (2 and 1.4 M) was used for another set of samples. The oxidizing agent, HNO_3/H_2O_2, was continuously fed to the process. The sulfuric acid leach residue, with pulp dilution 3 : 1, was used for treatment for three hours duration. Thiourea was added to the solution containing chloride residue and 1.5 M $Fe_2(SO_4)_3$ in 5 : 1, heated to 70 °C, for a leach duration of 210 minutes. After chloride and cyanide leaching, adsorption of the precious metals on activated carbon was done in a glass container at room temperature for 30 minutes, and finally the solution was filtered using vacuum filtration. Figure 4.4 proposes the process flowsheet for the hybrid process (Quinet et al. 2005).

The studies showed that activated carbon has potential be used for extraction of metals from e-waste. Using a hybrid process using first stage cyanide solution and second stage adsorption on activated carbon (8.3 g/l), expensive metals such as gold, silver, and palladium were extracted with a yield of 99%, 95%, and 100%,

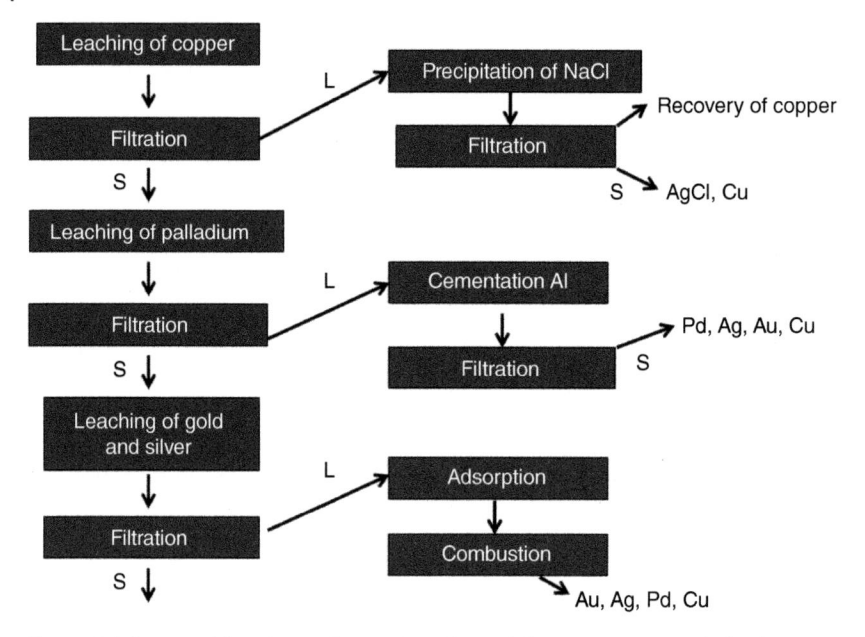

Figure 4.4 Proposed flowsheet of extraction of metals from e-waste using hybrid process.

respectively. However, the yield was only 30% for copper. Further modifications in the process may give better copper leaching.

Despite the encouraging bench scale results, efforts to scale up the technology are not reported elsewhere.

4.10 Summary and Outlook

Adsorption process provides a wide range of opportunities to improve the nature and properties of adsorbent particles used. Cost is one of the important parameters to scale up the process and therefore should be considered as a primary criterion for comparing the sorbent materials. Expense of individual sorbents varies depending on the degree of processing required and local availability (Nguyen et al. 2013); however, this information is generally ignored in the literature. Applicability of an adsorbent at large scale can be primarily measured on the basis of processing requirement, availability, and sorption capacity. Therefore, feasibility of agro-based "cost-effective green" adsorbents should be analyzed for the removal of heavy metals in detail based on certain aforementioned criterion. Secondly, adsorption technique using green adsorbents is somehow still limited to laboratory experimentation, mostly without pilot studies, or commercialization. Limited attempts for detailed economic and market analyses of these green adsorbents are available in literature. It will be more challenging to demonstrate the process at a pilot scale, but to really increase it to a large scale would call for a significant financial and technological effort.

Questions

1 Explain the different types of adsorption processes.

2 What makes adsorption a clean process as compared to other technologies for metal extraction from waste?

3 Name the industrial wastes that are potential source of adsorbents for metals from waste.

4 Discuss the benefits of using agricultural and industrial wastes used as green adsorbents.

5 Comment on the suitability of hybrid process for metal extraction. What is a better alternative?

5

Emerging Technologies for Extraction of Metals from Waste II. Bioleaching

"We need a new system of values, a system of the organic unity between humankind and nature and the ethic of global responsibility."
Mikhail Gorbachev

5.1 Introduction

Chapters 1 and 3 have highlighted the need for a paradigm shift toward a process that is not only green but also requires minimum fixed and operational cost. Bioleaching is one of most prominent emerging techniques based on the ability of microorganisms to convert insoluble form of waste to soluble form that can be used to extract metal out of it. It can be considered an effective technology for the recovery of precious metals due to the fact that the properties of certain types of inactive or dead microbial biomass materials (relatively inexpensive and available in abundance) allow them to bind and concentrate metal ions from industrial effluents and aqueous solutions (Dobson and Burgess 2007).

The process is defined as the "solubilization of metals from solid substitute either directly by metabolism of leaching bacteria or indirectly by the products of metabolism." In general, bioleaching is a process described as "the dissolution of metals from their mineral sources by certain naturally occurring microorganisms" or the use of microorganisms to transform elements so that the elements can be extracted from a material when water is filtered through it (Mishra et al. 2005). The process uses the catalytic effect produced by the metabolic activities of iron-oxidizing and sulfur-oxidizing microorganisms resulting in acceleration of the chemical degradation of the sulfides. Waste material is sent to a bioreactor, where it is bioleached using microorganisms. The resulting product is then sent to a solid–liquid separation tank, where metal is recovered from the sludge (Figure 5.1).

Commercial scale application of the process for the treatment of ores from mining industry is already gaining moment. Currently, approximately 20–25% of the world's total copper metal is extracted from its ore using bioleaching process only. A few industrial units have started using this technique to successfully extract metals from ores. For instance, a bioleaching plant of capacity 230 000 tonnes/yr in Morenci Mine, USA, demonstrates the success of the technique for large scale

Sustainable Metal Extraction from Waste Streams, First Edition.
Garima Chauhan, Perminder Jit Kaur, K.K. Pant, and K.D.P. Nigam.
© 2020 Wiley-VCH Verlag GmbH & Co. KGaA. Published 2020 by Wiley-VCH Verlag GmbH & Co. KGaA.

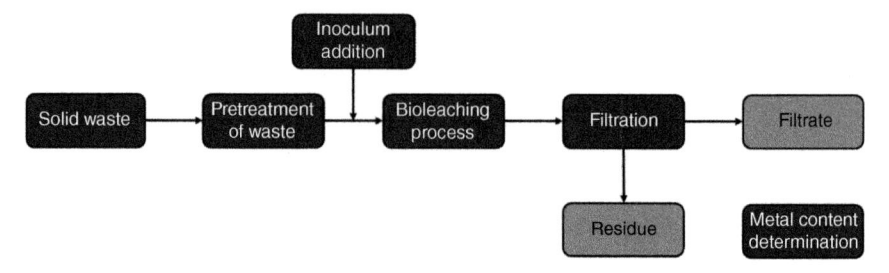

Figure 5.1 Major steps involved in the recovery of metals from waste using bioleaching process.

extraction of copper metal. Although the process is operational for mining industry and its waste treatment, the study of reaction conditions and scale-up of the technology for the recovery of metals from industrial waste or waste electrical and electronic equipment (WEEE) are in the nascent stage.

This chapter focuses on bioleaching as the emerging green approach for the recovery of metal from WEEE, industrial waste, and other types of waste. Fundamentals of bioleaching process including the principles and mechanism have been discussed in detail. Various types of reactors to operate the process in batch or continuous mode are described in brief. The chapter evaluates different technoeconomic issues associated with the process that need to be overcome to make it an attractive industrial process.

5.2 Bioleaching Process Description

In a typical bioleaching process, a bioreactor is required to allow the waste stream containing metals to interact with microorganisms. Residence time is optimized under laboratory conditions to allow maximum leaching of metals. Initially microorganisms need time to adjust to the conditions inside the reactor. This is known as lag phase. The growth of microorganisms begins after this initial adjustment period. Bioleaching process is performed by using microorganisms in bioreactors, which can be either batch or continuous. The process is generally performed by iron and sulfide oxidizing bacteria, or acid producing fungus. Some of the commonly used bacteria in this process are *Leptospirillum ferrooxidans*, *Thiobacillus ferrooxidans*, and some species of *Sulfolobus*, *Acidianus*, and *Sulfobacillus*. On completion of the bioleaching process, it is crucial to separate the two outlet streams, one containing the extracted metal and the other without metal.

During chemical treatments, high process investment, chemical cost, and effluent disposal problem have incited researchers toward the development of cost-effective eco-friendly technologies such as bioleaching. Efforts are directed toward maintenance of conditions that result in maximum bacterial growth and hence increased metal solubilization. Interestingly, it has been reported that use of the bioleaching process can help reduce the cost of process by 80% as compared to chemical treatment methods (Tyagi et al. 1988).

Traditionally, it is believed that bioleaching has certain associated advantages:

- The process is based on the use of simple basic equipment and is thus inexpensive to start.
- It is a clean green and eco-friendly process with less active waste stream residues.
- Moderate capital investment with low operating costs results in high economic value of the process.
- The process is accompanied with appropriate recovery of metals.

5.3 Factors Affecting the Process Efficiency

Maximum leaching occurs when conditions are optimum for bacterial growth and metal solubilization. Fundamental understanding of the effectiveness of bioleaching process is based on these factors, acting alone or in combination with each other. Experimental investigations have proved that factors such as microorganisms and metal to be leached, physicochemical factors such as temperature and pH, and operational variables such as reaction conditions, reactor design, and type of reaction, whether batch or continuous, have a significant influence on process efficiency as discussed below.

5.3.1 Types of Microorganisms

Bioleaching of metals is based on biological processes, in which microorganisms play a fundamental role. The driving factors for maximum metal recovery are the use of efficient microorganism and their sufficiently strong adhesion to the metal surface. The attached microorganisms should be capable of catalyzing the bioleaching process and release of metal ions to the surface.

In general, iron, sulfur, ammonia, phosphate, and magnesium salts are the required nutrients for the growth of microorganisms. As most of the time, chemolithoautotrophic bacteria are involved in bioleaching process, only inorganic compounds are required as nutrient. Environment and materials to be leached provide the minerals for their growth (Bosecker 1997).

Microorganisms have the ability to accommodate a variety of pollutants, both organic and inorganic. It is important to appreciate from the outset that microorganisms cannot destroy metals; however, they can influence the metals mobility in the environment by modifying their chemical and/or physical characteristics. Different types of microorganisms play an important role in the bioleaching process. Generally, the microorganisms exploited for bioleaching of metals can be classified broadly based on the temperature range for their growth as mesophiles and thermophiles (Pathak et al. 2009).

5.3.1.1 Mesophiles

Mesophiles are those microorganisms that grow at prevailing room temperature, i.e. 28–37 °C. The most dominant mesophiles used in bioleaching of metals from contaminated sites are sulfur-oxidizing bacteria (*Acidithiobacillus thiooxidans*) and iron-oxidizing bacteria (*Acidithiobacillus ferrooxidans*) (Wong et al. 2004).

These are chemolithotrophic bacteria that get their energy by oxidation of ferrous iron or reduced sulfur compounds. Among major groups of bacteria, the most commonly used are acidophilic and chemolithotrophic microbial consortia of *A. ferrooxidans, A. thiooxidans, and L. ferrooxidans*, and heterotrophs, for example, *Sulfolobus* sp. In addition, microscopic fungi such as *Penicillium* sp. and *Aspergillus niger* are examples of some eukaryotic microorganisms used in bioleaching during metal recovery from industrial waste (Willner et al. 2015). The optimum temperature required for growth of *A. ferrooxidans* is around 33 °C, although it can grow at any temperature in the range 20–40 °C. The growth occurs at a pH range 1.0–4.5 with an optimum value between 2.0 and 2.3 (Ruamsap et al. 2003; Mousavi et al. 2006). *A. ferrooxidans*, being a lithotroph, derives energy for its growth by oxidizing Fe(II) to Fe(III) and sulfur, sulfide, and different oxyanions of sulfur to sulfate. Besides *Acidithiobacillus* species, there are other species that are known to take part in various bioleaching environments. These species include *L. ferrooxidans* (iron-oxidizing bacterium; Markosyan 1972), *Acidithiobacillus albertis* (sulfur-oxidizing bacteria; Bryant et al. 1983), and *Ferroplasma* (iron-oxidizing archean; Edwards and Rutenberg 2001). However, they are not widely studied since they grow at a higher pH range at which efficient solubilization of metals does not take place.

5.3.1.2 Thermophiles

Moderate thermophiles are able to grow at a temperature of around 50 °C. There are a number of thermophilic strains isolated from different geothermal environments and mine sites (Kinnunen et al. 2003). An important moderate thermophile is *Sulfobacillus thermosulfidooxidans*, which has the ability to oxidize sulfur and iron (Xia et al. 2010). The bioleaching kinetics by moderate thermophiles is more than that with mesophiles as the leaching experiments are carried out at higher temperature. As the core temperature is 10–15 °C higher than the ambient, the moderate thermophiles are usually available inside the core of the dump.

Extreme thermophiles are those that can grow actively even at a temperature as high as 80 °C. The most important extreme thermophiles belong to the genus *Sulfolobus* (Jordan et al. 2006). A number of *Sulfolobus* species have been isolated such as *S. acidocaldarius, S. solfataricus, S. brierleyi, and S. ambioalous* (Norris 1997). They show the following properties:

(i) Anaerobic growth coupled with reduction of elemental sulfur
(ii) Aerobic growth coupled with oxidation of sulfur
(iii) Optimum growth temperature of 65–70 °C
(iv) Ability to oxidize both Fe(II) to Fe(III) and sulfur to sulfate.

Since the extreme thermophiles can grow at higher temperatures, the oxidation kinetics is more than that of mesophiles and moderate thermophiles (Norris 1997). The bioleaching kinetics in the presence of extreme thermophiles is higher than that of mesophiles and moderate thermophiles. The bioleaching dissolution reaction is exothermic; therefore, the temperature increases during the reaction. So if extreme thermophiles are used then a heat exchanger may not be required to control the leaching temperature.

At a higher temperature, archeans are the dominant species in the bioleaching environments. *S. thermosulfidooxidans* and other closely related species are moderate thermophiles that permit the use of higher temperature for a faster bioleaching rate. The extreme thermophiles that grow at 70 °C and use sulfur or thiosulfate as an energy source mainly include genus *Sulfolobus* viz *S. ambivalens* (Kletzin et al. 2004), *S. brierleyi* (Konishi et al. 1998), and *Thiobacter subterraneus* (Hirayama et al. 2005).

5.3.1.3 Heterotrophic Microbes

Heterotrophic microbes produce certain organic acids (oxalic acid, citric acid, and malic acid), which can supply both protons and metal complexing anions resulting in the recovery of metals from waste (Gadd 1999). Various species of bacteria such as *Acetobacter, Acidophilum, Arthrobactor, Pseudomonas*, and *Trichoderma* and fungi such as the genus *Penicillium* (Valix et al. 2001), *Aspergillus* (Mulligan and Cloutier 2003), and *Fusarium* (Burgstaller and Schinner 1993) have been reported to carry out bioleaching operation from ores and minerals.

5.3.2 Affinity Between Microorganisms and Metal Surfaces

Bacterial attachment to the metal surface plays an active role in the dissolution of substrate. The capacity of attachment and bioleaching action depends on the types of bacteria involved and the kinds of metals to be extracted. There is a difference in the level of attachments of different bacteria to multiple metals. Adhesion with surface properties of a few bacteria such as *T. ferrooxidans* is reported to be a function of electrophoretic mobility, zeta potential, cell surface hydrophobicity, and contact angle (Porro et al. 1997). However, Mozes et al. (1987) observed that there is no single cell surface property correlated with the number of attached bacteria or that appears dominant in attachment interaction.

A direct contact between bacterial cells and solid surfaces is a necessary condition for effective metal mobilization (Ostrowski and Sklodowska 1993). There are two levels of interactions between microorganisms and the mineral surface. The first level of interaction is a physical sorption because of electrostatic forces. Microbial cell envelopes are positively charged because of the low pH usually occurring in the leaching environments, causing electrostatic interactions with the mineral phase.

The second level of interaction is chemical sorption. In this process chemical bonds between cells and minerals are expected to be established (e.g. disulfide bridges). Extracellular metabolites are also formed and excreted closer to the attachment site. Sulfur oxidizers produce low molecular weight metabolites such as acids originating from the tricarboxylic acid (TCA) cycle, amino acids, or ethanolamine. High molecular weight compounds produced include lipids and phospholipids. Sulfur-oxidizing microorganisms from sewage sludge form a filamentous matrix similar to a bacterial glycocalyx in the presence of elemental sulfur. Thus, these extracellular substances are essential in the colonization of solid particles (Blais et al. 1993).

Although bacterial attachment to a surface is related to the type of substrate and composition of culture medium yet no linear correlation was observed.

Hydrophobic substrates such as sulfur yield more hydrophobic population of cells, which have a partial high adhesion to hydrophobic substrates such as molybdenite and sulfur as compared to less hydrophobic substrates such as ferrous ions. For instance, *T. ferrooxidans* grown on ferrous ion medium showed the greatest adhesion on covellite (which is less hydrophobic than sulfur).

The size of a micro-topographic feature could be responsible for the attachment of *Acidithiobacillus caldus* grown on pyrite. Even small changes in local surface alterations could strongly affect local adhesion parameters and affect the bacterial adhesion on mineral surfaces (Edwards and Rutenberg 2001). Surface roughness, chemistry of the surface, medium, and bacterial cell surface charges affect bacterial attachment to the metal surface and thus efficiency of the bioleaching process.

The populations of cells obtained by growth on a very hydrophobic substrate such as sulfur were more hydrophobic (more cells are transferred to the *n*-decane phase) than cells grown on ferrous ion. This greater hydrophobicity partially aids high adhesion of cells to hydrophobic substrates such as molybdenite and sulfur. *T. ferrooxidans* grown on ferrous ion medium showed the greatest adhesion on covellite (which is less hydrophobic than sulfur). On the other hand, a linear relationship between the percentage of attachment and the efficiency of bioleaching was not observed. In fact, molybdenite dissolution was negligible in all cases, despite the different degrees of attachment. The dissolution of covellite occurred in systems containing *T. thiooxidans*, although the attachment of these cells to covellite was great. However, there is a correlation between the degree of attachment to sulfur and bacterial action on it. Furthermore, it has been observed that *T. ferrooxidans* lost its ability to attack covellite if it was previously grown on sulfur.

5.3.3 Physicochemical Factors

5.3.3.1 Surface Properties

Corrosion and manufacturing of metallic waste result in difference in surface properties. Higher corrugation on the surface provides larger number of contact points and thus more bacterial attachment. Milling, sorting, and corrosion in industrial waste material are expected to have higher irregularities and thus would promote higher bacterial attachment.

5.3.3.2 Oxygen and Carbon Dioxide Content

Sufficient supply of oxygen is required for the growth and survival of microorganisms. Continuous supply of oxygen is also required for oxidation of Fe(II) and sulfur. Solubility of oxygen in water (8 g/m^3 at $35\,°C$) decreases with increase in the ionic concentration of the solution. It is reported that 0.07 g of oxygen is required by bacterium for each gram of Fe(II) oxidized, which needs to be supplemented to the solution regularly using aeration, stirring, and shaking in small scale leaching process. Bacterium metabolic activity can be retained at a dissolved oxygen concentration of 0.2 g/dm^3. For large scale leaching, supply of oxygen may cause difficulty. It is recommended that the total demand for oxygen should be evaluated using reaction kinetics in terms of metal dissolution and iron or sulfur oxidation. Design of reactor, impeller, and oxygen distribution are imperative for efficient oxygen distribution (Das et al. 1999). The assimilation of carbon dioxide by the microorganism is through Calvin–Benson cycle catalyzed by the ribulose bisphosphate carboxylase enzyme. Carbondioxide need not be added additionally.

5.3.3.3 pH Value of Solution

Adjustment of pH of the leaching medium is very critical to attain the optimum extraction efficiency as it directly affects the growth of bacteria and solubilization of metals in the medium. A suitable pH is important to enhance the bacterial activity by adaptation of the bacterial strain to the growth medium. Generally, very low pH is highly suitable for the growth of microorganisms. For ferrous ions and sulfide, a pH range of 2–2.5 is required. Inhibition of *T. ferrooxidans* will occur at pH below 2. Thus, the pH needs to be adjusted with addition of acid. *T. ferrooxidans* are unable to initiate growth on Fe(II) at a pH greater than 3.0. *A. ferrooxidans* and *A. thiooxidans* being acidophilic survive well under low pH and high acidic conditions. In case of large heaps of metal wastes to be treated, change in pH level may results in metal precipitation, which may severely affect the liquid and gas phase hydrodynamics, resulting in poor extraction performance.

5.3.3.4 Temperature

Maintenance of suitable temperature is of primary importance as this will affect the extraction efficiency of metal ions. It depends on the type of metals to be extracted and microorganisms involved in the process. For ferrous iron and sulfide oxidation temperature between 28 and 30 °C is essential to be maintained. For thermophilic bacteria high range of temperature between 50 and 80 °C is to be involved. For the extraction of copper, cobalt, nickel, and zinc, a low temperature range is required, may be as low as 4 °C.

5.3.3.5 Mineral Substrate

At high carbonate content the pH activity will increase, which will inhibit the growth of microorganisms. Addition of acid to maintain pH will not only result in increased cost but also precipitation of gypsum. Particle size may be decreased to improve the particle surface area. Pulp density can also be increased to improve the surface area. However, this may result in dissolution of certain substances that may have inhibitory or toxic effect on the leaching bacteria. Metal salts such as copper sulfides, chalcocite, and bornite are relatively easier to bioleach as compared to covellite and chalcopyrite. Quantitative research on the relation between metals to be extracted, reaction chemistry, and leach residue mineralogy has been reported by Watling (2006). Mineralogical changes also occur during acid leaching of oxide copper ore. Márquez et al. (2006) noted that the mineralogy in an ore and its transformation influence the design and operation of any industrial system.

5.3.3.6 Surface Chemistry of Metals

Surface chemistry of metals is an important parameter affecting the overall efficiency of bioleaching process. Certain metals are known to be toxic to bacteria. Metals such as copper, silver, nickel, magnesium, and chromium are known to be resistant to bacterial attachments. The capacity of an organism to sustain heavy metals is of varying extents. *Thiobacillus* bacteria have the capacity to tolerate a very high amount of heavy metals such as zinc, may be as high as 112 g/l. By gradually increasing the dose of metal to an individual species, it can be made to tolerate higher heavy metal concentration. The adaptation also helps to reduce the lag period, thus enhancing the overall leaching kinetics.

5.3.3.7 Surfactant and Organic Extractants

Surfactant concentration is an important factor influencing the microbial growth. Addition of surfactant can alter the surface properties of sulfur particles formed on the mineral surface. The surface tension between sulfur and sulfate solution may get reduced because of this interaction. This will enhance the bacterial attachment to sulfur formed. Thus, the dissolution rate of minerals will improve, resulting in improved recovery of metals. However, research studies show varying results depending on the type of microorganism used and the metal to be extracted.

Research findings show that it affects the bacterial growth rate and leaching efficiency. There are contradictory results reported by different investigators. Studies performed by Peng et al. (2012) showed that the addition of the surfactant Tween-80 reduced the lag phase of bacterial growth, enhanced the sulfur oxidation, and improved the copper extraction yield of chalcopyrite. This research finding showed that the addition of a certain optimum amount of surfactants may be an effective way to improve the bioleaching rate of chalcopyrite.

In another investigation, it was found that the amount of surfactant is negatively related to the bioleaching rate of cobalt ore. When the surfactant concentration increases, bacterial growth rate decreases as bacteria find it difficult to survive under high surfactant condition. In the presence of low surfactant concentration (polyoxyethylenesorbitan monolaurate, i.e. Tween-20), there was no significant change in the oxidation rate of ferrous ions. However, when the surfactant concentration was increased, the oxidation rate of ferrous ions was found to be decreased. Under high surfactant conditions, the growth of bacteria was inhibited. Surfactants and organic extractants have inhibitory effect on leaching bacteria. This might be due to decrease in surface tension and reduction in the mass transfer rate of oxygen (Liu et al. 2015).

When fungi are used as the microbial population, higher surfactant concentration acts as a growth promoter. It was found to be beneficial for its growth. The presence of surfactant also increased the bio-acid secretion by the fungi through elevated carbohydrate transformation, cumulatively effecting higher nickel recovery from pretreated chromite over-burden (COB), that is, about 39% nickel was extracted by *A. niger* in the presence of the surfactant whereas only 24% nickel extraction was achieved without surfactant (Behera and Sukla 2012).

5.3.4 Reactor Design

Bioreactor is an equipment where biologically assisted chemical transformation of product takes place. Biological reactors are quite similar in features to conventional chemical reactors. Biochemical reactors are designed taking care of all

Figure 5.2 Process factors affecting the efficiency of bioleaching process.

components as are required in the chemical reactor. In addition to this, care is taken to ensure that chemical reaction proceeds at appropriate rate, along with providing facilities for the growth, reproduction, and metabolic activity of the biological components (cells or whole organisms).

The selection of reactor depends on various parameters such as reaction kinetics, types of waste to be treated, cost, and extraction efficiency desired. Material of construction of reactor should be such that it is able to withstand not only oxidative environment but also should be nontoxic to microorganisms. Details about the various types of reactors used at various scales for various types of wastes are given in Section 5.5.

Figure 5.2 shows that the efficiency of a bioleaching process is an outcome of combinations of various factors. An optimum selection of all these factors can give the highest efficiency in bioleaching of treated waste.

5.4 Mechanism of Bioleaching Process

There are two different pathways to understand the dissolution of metals by microorganisms as explained below.

5.4.1 Biochemical Reaction (Direct vs. Indirect) Mechanism

Metal leaching is now recognized as mainly a chemical process, in which ferric iron and protons are responsible to carry out the leaching reactions. The main steps involved in the leaching process are presented in Figure 5.2. The role of the microorganisms is to generate the leaching chemicals and to create the space in which the leaching reactions take place. Microorganisms typically form an exopolysaccharide (EPS) layer when they adhere to the surface of a mineral but not when growing as planktonic cells (Sand et al. 1995). It is within this EPS layer rather than in the bulk solution that the bio-oxidation reactions take place most rapidly and efficiently and therefore the EPS serves as the reaction space (Sand and Gehrke 2006; Ilyas et al. 2007).

It has been widely mentioned in the literature that the metal dissolution reaction is not identical for all metal sulfides and the oxidation of different metal sulfides proceeds via different intermediates (Schippers et al. 1999). Microorganisms can oxidize metal sulfides via a direct mechanism where electrons are obtained directly from the reduced minerals. In this case, cells have to be attached to the mineral surface and a close contact is needed. The adsorption of cells to suspended mineral particles takes place within some minutes or hours. In another mechanism, called the "indirect" mechanism, the oxidation of reduced metals is mediated by ferric(III) ion and this ferric is formed by microbial oxidation of the ferrous iron present in the minerals. Ferric iron acts as an oxidant and can oxidize metal sulfides and is reduced to ferrous iron, which, in turn, can be microbially oxidized. In this case, iron acts as an electron carrier. It was proposed that no direct physical contact is needed for the oxidation of iron. In many cases it was concluded that the "direct" mechanism dominates the "indirect" one, mostly because the direct mechanism involves direct physical contact of bacteria with the mineral surfaces.

Equations (5.1)–(5.5) describe the direct and indirect mechanisms for the oxidation of pyrite.

Direct:

$$2FeS_2 + 3.5O_2 + H_2O \rightarrow Fe^{2+} + 2H^+ + 2SO_4^{2-} \tag{5.1}$$

$$2Fe^{2+} + 1/2O_2 + 2H^+ \rightarrow 2Fe^{3+} + H_2O \tag{5.2}$$

Indirect:

$$FeS_2 + 14Fe^{3+} + 8H_2O \rightarrow 15Fe^{2+} + 16H^+ + 2SO_4^{2-} \tag{5.3}$$

$$[M][S] + 2Fe^{3+} \rightarrow M^{2+} + S^0 + 2Fe^{2+} \tag{5.4}$$

$$S^0 + 1.5O_2 + H_2O \rightarrow 2H^+ + SO_4^{2-} \tag{5.5}$$

However, the model of direct and indirect metal leaching is still under discussion. Recently, this model has been revised and replaced by a model that is not dependent on the differentiation between direct and indirect leaching mechanisms. The key aspects of the leaching have been integrated and a mechanism has been developed, which is characterized by the following features:

(1) Cells have to be attached to the minerals and in physical contact with the surface.
(2) Cells form and excrete exopolymers.
(3) These exopolymeric cell envelopes contain ferric iron compounds that are complexed to glucuronic residues. These are regarded as part of the primary attack.
(4) Thiosulfate is formed as intermediate during the oxidation of sulfur compounds.
(5) Sulfur or polythionate granules are formed in the periplasmic space or in the cell envelope.

Briefly, a thiosulfate mechanism has been proposed for the oxidation of acid insoluble metal sulfides such as pyrite (FeS_2) and molybdenite (MoS_2), and a polysulfide mechanism for acid soluble metal sulfides such as sphalerite (ZnS), chalcopyrite ($CuFeS_2$), or galena (PbS). Since the industrial waste samples are mostly oxidic in nature, for action of acidophiles, substrates are added externally.

5.4.2 Mechanism of Metal Sulfide Dissolution (Polysulfide Pathway)

Depending on the solubility of metals in acid solution, there are two proposed pathways for the dissolution of metal as shown in Figure 5.3. Metals that are insoluble in acids follow thiosulfate mechanism, while soluble metals dissolve through polysulfate mechanism (Figure 5.3).

A combination of proton attack and oxidation processes results in the dissolution process, which is controlled by the mineral species. In bioleaching process, bacteria are used not only for biological generation of protons but also to maintain the oxidizing state of iron ions (as Fe(III) ions) for an oxidative attack.

Monosulfide or disulfide structures do not control the pathway of dissolution. Acid-nonsoluble metal sulfides such as pyrite, molybdenite, and tungstenite

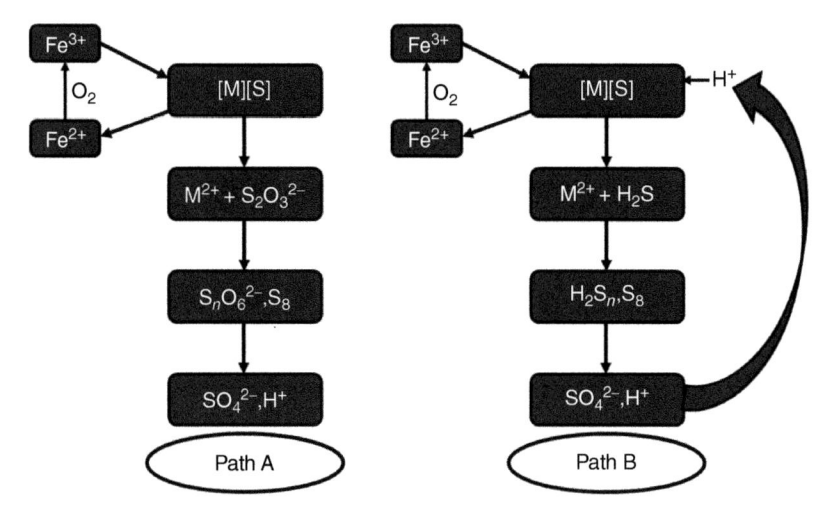

Figure 5.3 Two indirect pathways of bioleaching process. Source: Mishra et al. (2005). Reproduced with permission of Springer Nature.

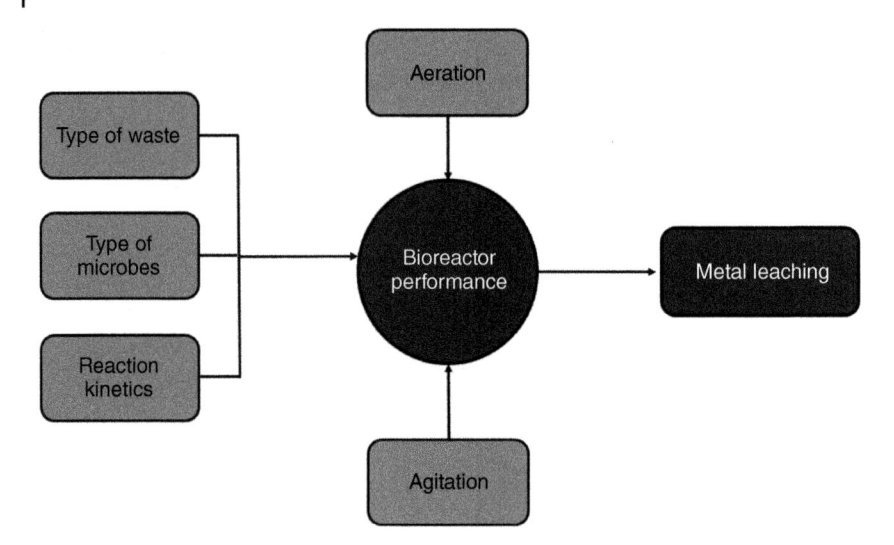

Figure 5.4 Factors affecting the efficiency of biological CSTR.

(FeS$_2$, MoS$_2$, and WS$_2$ respectively) dissolve through thiosulfate pathway. Metals are exclusively oxidized via electron extractions by ferric (Fe(III)) ions. The chemical bonds between sulfur and metal moiety remain intact till the evalution of thiosulfate.

Dissolution of acid-soluble metal sulfides such as ZnS, CuFeS$_2$, and PbS follows polysulfate pathway. In this mechanism, proton attack combined with electron extraction by Fe(III) ions helps in the dissolution process (Figure 5.4). The chemical bond between metal and sulfur moiety is broken by proton attack. As a result of binding of two protons hydrogen sulfide is emitted. It is expected that a sulfide cation (H$_2$S$^+$) is first liberated. It is then dimerized to free disulfide (H$_2$S$_2$) and is oxidized through higher polysulfides and polysulfide radicals to elemental sulfur. Thus, the mechanism was named as "polysulfide pathway" (Mishra et al. 2005).

$$FeS_2 + 6Fe^{3+} + 3H_2O \rightarrow S_2O_3^{2-} + 7Fe^{2+} + 6H^+ \tag{5.6}$$

$$S_2O_3^{2-} + 8Fe^{3+} + 5H_2O \rightarrow 2SO_4^{2-} + 8Fe^{2+} + 10H^+ \tag{5.7}$$

Polysulfide mechanism

$$[M][S] + Fe^{3+} + H^+ \rightarrow M^{2+} + 0.5H_2S_n + Fe^{2+} + H^+ \quad (n \geq 2) \tag{5.8}$$

$$0.125\,H_2S_n + Fe^{3+} \rightarrow 0.125S_8 + Fe^{2+} + H^+ \tag{5.9}$$

$$0.125S_8 + 1.5O_2 + H_2O \rightarrow SO_4^{2-} + 2H^+ \tag{5.10}$$

5.5 Engineering Practices in Bioleaching Process

A chemical reactor is defined as a well-defined space where chemical or biological conversion reactions take place. For bioleaching process, microbes interact with waste material under controlled conditions, in the specially designed reactors known as biochemical reactors. Their mode of operation, i.e. batch or continuous, quality and quantity of waste materials and microbes, and their proportions in the mixture are essential to be considered before designing a reactor.

A good bioreactor should be designed for better productivity and consistent and higher quality products in an economically feasible manner. Production of organism, optimum conditions required for the desired product formation, product value, and its scale of production affect operations of a bioreactor. An effectively designed bioreactor should positively control the occurrence of simultaneous side chain biological reactions and prevent any foreign contamination. The capital investment and operating cost are essential components to be considered while designing a bioreactor design. Depending on the applications, different shapes, sizes, and forms of reactors are designed. The sizes of the bioreactor can vary from a few cubic millimeters to shake flask (100–1000 ml) to pilot level (0.3–10 m^3) to plant scale (2–500 m^3) for large volume industrial applications. The rate of reaction, rate of cell growth and process stability, gas (i.e. air, oxygen, nitrogen, carbon dioxide), flow rates, temperature, pH and dissolved oxygen levels, and agitation speed/circulation rate, foam production, etc. should be closely monitored and controlled.

Particularly for bioleaching processes, reactors are designed to handle large volumes of sludge and industrial waste.

The reaction takes place in three phases:

1) *Aqueous phase*: Solution of salts that act as nutrients for microflora exists in aqueous phase. It is the suspended medium where all reactions take place. Reactions involved in the process are as follows:
   (a) Growth of microorganisms
   (b) Encounter of solid phase with microorganisms
   (c) Encounter of solid with chemical solutions added
   (d) Release of metal ions
   (e) Distribution of gases.
2) *Solid phase*: Waste material from which metal is to be extracted is in solid phase. This phase acts as food for microorganisms.
3) *Gaseous phase*: Atmospheric gases, oxygen, and carbon dioxide required for the process to take place exist in gas phase. The phase supplies gases for biosynthesis of microorganisms.

There are two main types of reactors in use for bioleaching process, namely batch and continuously stirred tank reactors (CSTRs). In addition to this, there are certain mixed modes of reactors under the stage of development.

5.5.1 Batch Process

Batch reactors are the simplest form of reactors with a tank to hold the reactants, i.e. microbes and waste materials to be leached. There is a provision for input of reactants and output of products. It may contain a cooling or heating jacket outside. Following the reaction, products are poured; the tank is cleaned for the next batch of reactions to be carried out. Most of the laboratory scale studies of bioreactors have been performed on batch mode of operations only.

Ease of operation makes batch reactors a preferable choice for researchers in laboratories. They are convenient to operate and can generate large amount of database for scale-up studies. Since reducing conditions of anaerobic sludge can be maintained easily in batch bioreactors, they are generally used for the same.

Various process parameters such as rate of aeration, pH, and temperature of solution affect the metal removal efficiency. There are a series of phases under which typical bacterial conversion takes place as follows:

(a) *Lag phase*: Microflora needs some time to adjust to the new environment of inoculated medium. In this phase, no growth of cells takes place.
(b) *Logarithmic growth phase*: After an initial period of adjustment in the lag phase, microorganisms start growing exponentially. This period is also known as the balanced period. If substrate and nutrients are available, balanced bacterial growth takes place under this period.

In one of the studies, batch experiments were conducted in a pulsed plate bioreactor for bioleaching of Cu from e-waste. After the first 48 hours of batch run, the bioleaching media with the leached metal was drained, and the reactor was filled with a fresh bioleaching media. The process was found to be less energy intensive and controllable, and resulted in 63.5% copper removal efficiency on five batch runs. It was reported that extraction efficiency can be improved by increasing the number of runs of batches.

There are certain challenges related to the economic viability of batch reactor as reported by Tyagi and Couillard (1987, 1989) and Couillard and Mercier (1991). Higher leaching duration (6–12 days) with efficiency of only 80% of Cu, Zn, Cd, and Ni solubilization makes the reactors less attractive.

Example 5.1
Problem: If we start bioleaching experiments with a bacterial culture containing 10 000 cells, cell generation time is two hours. Find the number of cells in the culture after a period of 1 day and 2 days.

Microbial cells also follow the batch process for generation as given by the equation

$$B = 2^n B_0$$

B_0 = Initial concentration of the cells
N = Number of generations
B = Final number of cells generated

Solution: Number of cell generations in 1 day (24 hours) = 12.
Thus $B = 2^{12} \times 10^3$.

5.5.2 Continuous Process

The CSTR, also known as vat- or back-mix reactor, is a common ideal reactor type in chemical engineering. In a perfectly mixed reactor, the output composition is identical to composition of the material inside the reactor, which is a function of residence time and rate of reaction. In a CSTR, the reaction vessel is equipped with an impeller (stirrer) to stir the reagents. This ensures good mixing and uniform product composition throughout. Aeration coil and mechanical stirrer are essential parts of the bioreactor design. Figure 5.4 illustrates the major factors that may affect the performance of a bioreactor and that should be kept in mind while designing a reactor.

The difference between a chemical CSTR and a biological CSTR is that in chemical processes the amount of catalyst remains the same, while in biological CSTR the biocatalyst process "pseudo autocatalytic process" performed by growing microorganisms varies with time.

Bioleaching in stirred tank reactors can provide more homogeneous reacting masses and allow close control of the main process variables, compared with the heap and column leaching operation. In order for the bioleaching process in stirred tank reactors to be economical, the pulp density needs to be high. The balance equation for CSTR in generalized form is

Rate of accumulation = Net rate of transport + rate of transformation

where net rate of transport = Transport in − Transport out

In addition to this, the microbial cell consumes energy for various reasons: for cell division, for cell growth, cell motility, for the formation of cellular products, and maintenance of cell functions.

Besides maintaining the cell structure, maintenance energy is required for integrity of cells. This type of energy is not available for either growth of cells or their biosynthesis. The maintenance energy required for an active microbial cell is negligible as compared to the amount of energy required for their growth or motility. Cell decay occurs when a microbial cell is depleted of substrate. Cell decay results in decrease in the size of microbial population and should be taken into account in the mass balance equation. Thus, the new balance equation becomes

Rate of accumulation = Net rate of transport + rate of transformation − rate of microbial decay

Characteristic parameters affecting the efficiency of CSTR are the following:

1) *Pulp density and residence time*: In the operation of stirred tank reactors the quantity of solids (pulp density) that can be maintained in suspension is limited to 10% (w/v). When pulp densities >10% (w/v) both the physical and microbial efficiencies decrease. The slurry becomes very thick with an inefficient gas transfer rate, which is essential for the bioleaching process to take place and for microorganisms to grow. At increased pulp densities >10% (w/v) physical damage of microbial cells by the shear force of the impellers is also reported in the literature.
2) *Tank design*: The number of tanks required and the volume of each tank affect the overall reaction performance. Optimum tank geometry (size and aspect ratio) results in better interaction between metal and microorganism.

3) *Impeller design*: The impeller plays three important roles. Solid suspension, mixing, and dissolution of required amount of atmospheric gases into aqueous phase are done with the help of the impeller. The impeller design includes its type, quantity, diameter, speed, baffle number, and geometry.
4) *Aeration design*: A compressor is used to force air into the reactor. It is recommended that air should be dry and oil free to avoid contamination to the medium. Care should be taken to sterilize the air. Air sparger provides thorough mixing and high mass transfer rates through bulk liquid and bubble boundary layers. It breaks the air into small bubbles. Generally, ring shaped spargers consisting of hollow tube with small holes are used. It is assembled below the agitator allowing bubbles to rise below the impeller blade.
5) *Kinetics of reaction*: Faster the growth kinetics of microorganisms faster is the oxidation process.

CSTR can treat larger volumes of sludge in a shorter period of time. They are economically more suitable for larger volumes of sludge. The conditions where microbial population have to be in metabolically active state are most suitable for CSTR bioreactors. Operational simplicity, lower capital and operating costs, environment friendliness, and suitability for the treatment of complex and low grade ores make CSTR an attractive option for leaching of metals.

However, there are certain challenges associated with conventional CSTRs. Stirred tank reactors have some common drawbacks such as severe collision and friction between solid particles, high shear force, and no more than 20% of pulp density in bioleaching systems (Rossi 2001), which can lead to low efficiency bio-oxidation of sulfide minerals. It has been found that the collision and friction between solid particles were the main factors to impair Fe^{2+} oxidation by *A. ferrooxidans*. Therefore, the reactors used in sulfide bioleaching should be designed to reduce collision and friction between solid particles for avoiding loss in the microorganism's bioactivity. The Kasese bioleach plant, which treats a cobaltiferous pyrite concentrate, comprises six bioleaching reactors, each with a volume of $1400\,m^3$. At the time of construction, these were the largest individual reactor volumes for a plant of this type.

There are rotating drum type reactors also used for the reaction purpose. Compared with stirred tank reactors, rotating drum reactors have a different mechanism of slurry mixing. Herrera et al. (1998) found that the total concentration of iron in a rotating drum reactor was 2.3 times higher than that in a stirred tank reactor at high pulp densities in the bioleaching of refractory gold concentrate, but its gas mass transfer performance was not satisfactory for the lack of an efficient gas-sparger (Barrera-Cortés et al. 2006). Jin et al. (2010) improved the rotating drum reactor by fitting a gas-sparger of microfiltration ceramic membranes to optimize its gas mass transfer performance. However, the application potential of the improved rotating drum reactor still needs to be verified by the bioleaching of sulfide minerals. Table 5.1 lists some of the research trends in the field of bioreactor design.

Table 5.1 Some significant latest research on the developments in bioreactors.

Type of reactor	Metal bioleached and type of waste	Microbes involved	Reaction conditions	Metal removal efficiency	References
CSTR and column reactor	Uranium from mineral ores	*Acidithiobacillus ferrooxidans*	100 l/h, pH 1.8, initial Fe^{2+} concentration 4 g/l and temperature 35 °C		Tavakoli et al. (2017)
CSTR	Printed circuit board (e-waste)	*Acidithiobacillus caldus*, *Leptospirillum ferriphilum*, and uncultured *Thermoplasmatales archaeon*	3 l glass cylindrical reactor, temperature: 45 °C, agitator speed: 400 rpm, pH 2.0	Cu: 77%; Al: 85%; Zn ~100%; Pb ~10%; and Sn <10%	Xia et al. (2017)
Rotating drum type reactor	Arsenopyrite from mineral ores	*A. ferrooxidans*	Pulp density 30% and 45% Drum rotation speed 5 rpm		Jin et al. (2012)
	Mo, Ni, V, and Al from spent refinery catalyst	*A. thiooxidans* *A. ferrooxidans*		Ni: 100%, Al: 55%, Mo: 81%, and V: 100% Ni: 94%, Al: 55%, Mo: 77%, and V: 99%	Srichandan et al. (2014)
Hybrid reactor (combination of two batch and a CSTR) with inclusion of continuously sequencing reactor (CSR) and bioleaching	Au, Ag from printed circuit board (e-waste)	*C. violaceum*, *P. chlororaphis*	pH 7.0, temp: 22.5 °C, glycine (4.4 g/l), methionine (2 g/l), rotation speed 80 rpm	Maximum recovery rates of Cu and Au were 88.1 wt% and 76.6 wt%, respectively	Jujun et al. (2015)
Sequence of batch reactors	Zn, Cu, and Pb from sewage sludge	*A. thiooxidans* and *A. ferrooxidans* mixed in a ratio of 4 : 1	6 d of bioleaching at 15 °C, pH of 6, a 15% (v/v) inoculum concentration	15 °C, 89.6% of Zn, 72.8% of Cu, and 39.4% of Pb	Zhou et al. (2017)

5.5.3 Hybrid Processes

In one the latest research studies, hybrid technology of corona electrostatic separation (CES) was used as step one to separate the nonmetallic part of WEEE from the metallic part. In bioleaching process, combination of three reactors was used. The first tank A was used to grow cyanogenic strain with no other microbes with operation in an aseptic box, set for 12 hours. An aerator was used to provide the dissolved oxygen regularly. Cyanogenic strain was then mixed with metallic particles from crushed waste printed circuit boards (PCBs) in the second tank (Tank B equipped with sliding baffle at the bottom).

Example 5.2
Problem: From a wastewater plant, the aerobically digested sludges (20 g/l of total solids) were obtained. Microbial leaching tests were performed using iron-oxidizing and sulfur-oxidizing microflora. Each of 5% the corresponding adapted sludge (sulfur or iron-oxidizing bacteria) was treated with 0.5% substrate. The time required to solubilize metals (60%) at different temperatures is presented in the table below. Find the optimum temperature required to achieve the above solubilization of metal.

Leaching process	Metal	Temperature (°C)					
		10	15	20	25	30	35
With ferrous ions	Cd	240	160	130	80	70	65
	Cu	200	110	65	40	35	30
With elemental sulfur	Cd	320	230	160	106	100	95
	Cu	260	210	140	100	95	90

Answer: 25 °C.

Tank B was covered by a compact filter layer. During the interaction of bacterial strains with metal, the active metal was dissolved first leaving behind the inert metal. After a limited bioleaching time, the lixivium of active metal was pumped from Tank B to Tank C. A real time monitoring system detected the inert metal in Tank C. The presence of inert metal ion in Tank C indicates the complete extraction of the active metal. For the separation of active metal from inert metal solutions, replacement reaction could be used. The experimental setup was used to extract Cu (88.1 wt%) and Au (76.6 wt%) from the waste metal stream (Jujun et al. 2015).

5.6 Application of Bioleaching in Extracting Metals from Waste

Bioleaching has been considered as the key technology for the treatment of solid waste such as fly ash, spent catalyst, electronic scraps, spent batteries, and many more for the last few decades. Several research studies have been reported to

Table 5.2 Comparison of various technologies for extraction of metals from e-waste.

Technology	Advantages	Drawbacks	References
Chemical technology	• High separation rates for metals	• Environmental non-friendly methods • Incidences of pollution • Chemical reagents cause security risk to labor and ecosystem	Khaliq et al. (2014)
Bioleaching	• Eco-friendly	• Low extraction efficiency • Nonmetallic parts are difficult to be separated • Nonmetallic parts may destroy the process	Mishra et al. (2010)
Corona electrostatic separation (CES)	• Preferred method for e-waste separation • Less pollution	• Unsuitable to separate precious metals	Li et al. (2008)
Hybrid technology (combination of CES and bioleaching)	• First CES separates nonmetals of crushed waste PCB • Better efficiency than simple bioleaching • Eco-friendly technique	• In the nascent stage of development	Awasthi and Li (2017)

illustrate the applicability of the environment friendly and economical approach for extraction of metals from a variety of waste. Table 5.2 lists the possible advantages and disadvantages of this emerging bioleaching process over other conventional processes for the removal of metals from WEEE.

Efforts are constantly being made to establish the process as a sustainable approach on industrial scale. This section reviews the literature available on bioleaching process to recover base and precious metals from solid waste.

5.6.1 Extraction of Metals from WEEE

Although the process has been used extensively for treatment of mining industry waste, yet its application for WEEE treatment as a potential alternative to conventional processes (Chapter 3) for resource optimization is still in its early stage. Table 5.3 shows the effectiveness of various microorganisms to treat

Table 5.3 Use of microbes to extract metals from variety of WEEE.

WEEE	Microorganisms involved	Metal (%)	References
Electronic scrap	*Aspergillus niger* *Penicillium simplicissimum*	Cu, Sn 65% Al, Ni, Pb, Zn >95%	Brandl et al. (2001)
PCB	*S. thermosulfidooxidans*	Ni: 81%; Cu: 89%; Al: 79%; Zn: 83%	Ilyas et al. (2007)
PCB	*Chromobacterium violaceum*	Au: 68.5%	Brandl et al. (2008)
PCB	*A. ferrooxidans*	96.8% Cu, 83.8% Zn, 75.4 Al%	Yang et al. (2009)
WPCB	*A. ferrooxidans*, *A. thiooxidans* *A. ferrooxidans + A. thiooxidans*	Cu: 99% Cu: 74.9% Cu: 99.9%	Wang et al. (2009)
Electronic scrap	*T. sulfobacillus + T. acidophilum*	Cu: 86%; Zn: 80%; Al: 64%; Ni: 74%	Ilyas et al. (2010)
WPCB	*T. ferrooxidans*	70% Cu	Gao et al. (2010)
Ground WEEEs (copper-rich)	*Acidithiobacillus thiooxidans* (ATCC8085)	60% Cu	Hong and Valix (2014)
PCB	*A. ferrooxidans*	Cu: 98–99%;	Willner and Fornalczyk (2013)
WPCB	*Pseudomonas chlororaphis*	52.3% Cu	Ruan et al. (2014)
PCB	*A. ferrooxidans, A. thiooxidans/albertensis, A. caldus, L. ferrooxidans, S. thermosulfidooxidans, S. thermotolerans*	Cu: 99%	Makinen et al. (2015)
WPCB	*Acidithiobacillus ferrivorans* and *Acidithiobacillus thiooxidans*	98.4%	Isildar et al. (2015)
WPCB	*Aspergillus niger*	80% Cu	Jadhav and Hocheng (2015)
WPCB	*Acidithiobacillus ferrooxidans* strain Z1	96% Cu	Nie et al. (2015)
Mobile phone PCB	*Acidithiobacillus ferrooxidans*	99% Cu and Ni	Arshadi and Mousavi (2015)
WPCB	*Leptospirillum ferriphilum*	84.2% Cu	Shah et al. (2015)

(Continued)

Table 5.3 (Continued)

WEEE	Microorganisms involved	Metal (%)	References
WPCB	*Acidithiobacillus ferrooxidans* and *Sulfobacillus thermosulfidooxidans*	94% and 99% respectively	Rodrigues et al. (2015)
Solar cells	*Cellulosimicrobium funkei* isolated from contaminated soil	Indirect leaching: 92% Ga; direct leaching: 78%	Maneesuwannarat et al. (2016)
Lithium-ion mobile phone batteries	*Aspergillus niger*	100% for Cu, 95% for Li, 70% for Mn, 65% for Al, 45% for Co, and 38% for Ni in spent medium	Horeh et al. (2016)
Spent zinc–manganese WEEE	Sulfur oxidising bacteria (SOB): *Alicyclobacillus* sp. under accession	10.09 ± 0.71 g/l for Zn	Niu et al. (2016)
WPCB	*A. ferrooxidans*	Cu: 4%, Zn: 92%, Pb: 64%, Ni: 81%	Priya and Hait (2018)
Lapton battery	*A. thiooxidans*	Li: 99%, Co: 50%, Ni: 89%	Heydarian et al. (2018)

different types of WEEE. PCB itself is known to be toxic for the growth of various microorganisms. Thus, low concentration of solution is used in studies to bioleach e-waste.

Hong and Valix (2014) investigated the direct vs. indirect leaching mechanism for copper solubilization from WEEE using *A. thiooxidans*. It was reported that 60% Cu was recovered using biogenic acid leaching and direct leaching mechanism; on the other hand, 98% recovery of Cu was attained through abiotic leaching process. Poor metal recovery (in case of biogenic acid leaching and direct leaching) may correspond to copper passivation and galvanic leaching due to the presence of incompletely oxidized sulfide compounds and the sulfates, which led to copper monosulfide (CuS) and copper sulfate ($CuSO_4$) precipitates (Chauhan et al. 2018).

Use of thermophiles (Ilyas et al. 2010; Vestola et al. 2010) and fungi such as *A. niger and Penicillium simplicissimum* (Jadhav et al. 2016; Kolenčík et al. 2013) is widely known for recovery of heavy and precious metals from WEEE. Moderate thermophiles have been reported to be better than mesophiles and extreme thermophiles in terms of rate of metal leaching from metal ores. Cyanogenic bacteria such as *Chromobacterium violaceum, Pseudomonas fluorescens, and Pseudomonas plecoglossicida* (Brandl et al. 2008; Chi et al. 2010) have also been employed for the extraction of gold from the WEEE. Pradhan and Kumar (2012) employed a two-step bioleaching process to investigate the leaching efficiency of single and mixed culture of three different cyanogenic bacteria strains (*C. violaceum, Pseudomonas aeruginosa,* and *P. fluorescens*). *C. violaceum* strain exhibited the maximum bioleachability of all the metals (Cu [79.3% w/w]; Au [69.3% w/w]; Zn [46.12% w/w]; Fe [9.86% w/w]; and Ag [7.08% w/w]) among single cultures. A mixture of *P. aeruginosa* and *C. violaceum* leached maximum 83.46% Cu followed by 73.17% Au, 49.11% Zn, 13.98% Fe, and 8.42% Ag.

Recently, Sinha et al. (2018) employed a hybrid combination of bioleaching as the recovery step and biosorption as the purification step followed by electro treatment to recover copper from waste printed circuit board (WPCB) in its reusable form. Leaching process parameters (reaction pH, substrate concentration, inoculum size, pulp density, agitation speed) were optimized in the presence of an isolated strain *USCT-R010* to maximize metal solubilization. The leach liquor containing mobilized copper was subjected to biosorption using dead biomass of *Aspergillus oryzae* and Baker's Yeast under optimized reaction conditions. The recovery of more than 86% copper was reported during desorption using 0.1 N HCl from dead biomass. Further, electrowinning was carried out at 2 A current for 150 minutes to recover 92.7% Cu from the eluate. The process flow sheet for the hybrid recovery of copper from WEEE using the two-stage biorecovery process followed by electrochemical treatment is given in Figure 5.5.

Jadhav et al. (2016) also investigated the fungal assisted bioleaching of metals in a two-step process where fungus was grown for organic acid production in the first step. The culture supernatant without fungal cells was used for metal leaching during the second step of experiments. Rasoulnia and Mousavi (2016)

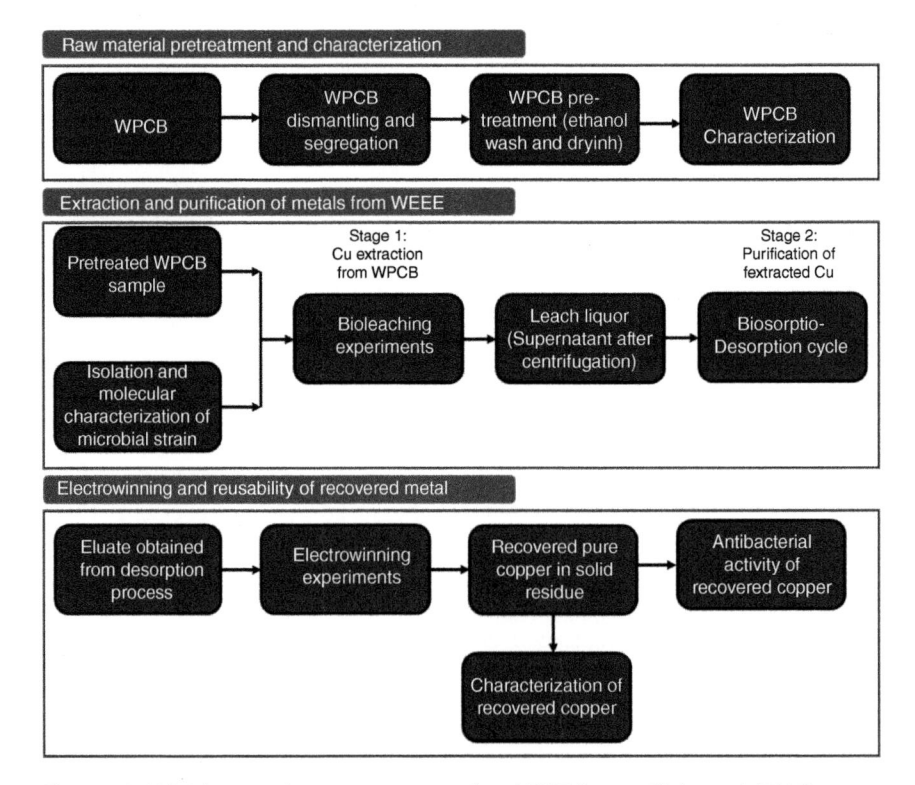

Figure 5.5 Hybrid approach to recover copper from WEEE. Source: Sinha et al. (2018). Reproduced with permission of Elsevier.

suggested that the possibility of WEEE contamination being treated by the microbial biomass and metal toxicity toward fungus can be minimized using the supernatant from fungal growth.

Studies on bioleaching process of WEEE show that it is a complex process. Various factors such as the type of microorganisms, pH, concentration of Fe^{2+} in the system, qualitative and quantitative composition of waste, toxicity of ingredients and fineness of the material, temperature, and reaction affect the efficiency of metal extraction. Practical difficulties in the separation of nonmetallic parts from metallic part may deteriorate the quality of leaching by causing negative effect on growth of microbes and make the process cumbersome. Studies on designing of reactors for large scale operations are scanty in the literature.

5.6.2 Extraction of Metals from Industrial Waste

Most of the studies on bioleaching of spent catalysts have focused on the use of mesophilic microorganisms (growing at 15–40 °C) including chemolithotrophic bacteria such as *A. ferrooxidans and A. thiooxidans* or heterotrophic eukaryotic fungi such as *Penicillium* and *Aspergillus*. However, slow rate of leaching is the major constraint associated with the process using these microbes. Although thermophiles have the potential for increasing the metal solubilization rate as they survive at higher range of temperature and can thus break the metal bonds in spent catalyst, yet there is no significant research in this field. Attempts have been made by various researchers to compare chemical leaching with bioleaching. Possible pros and cons associated with both processes are listed in Table 5.2. Higher extraction efficiency of bioleaching as compared to chemical agents is attributed to the fact that biological agents may mobilize the heavy metal from bulk to the surface of spent industrial catalyst. In order to remove carbonaceous deposits and other organic impurities, spent catalyst can be pretreated by decoking it. Spent catalyst can be bioleached by using combinations of various stages such as one-step, two-step, and spent medium leaching. Bharadwaj and Ting (2013) investigated the bioleaching of spent hydrotreating catalyst by treating with thermophile *Acidianus brierleyi* using pretreatment (i.e. decoking) of the catalyst resulting in extraction of Al: 22.9%; Fe: 53.2%; Ni: 2.3%; Mo: 9.9%. Table 5.4 shows that bioleaching technique was found to be more effective than or comparable to chemical leaching of spent catalyst.

Chemical reactions involved during metal bioleaching from exhaust catalysts can be simplified as follows (Eqs. (5.11)–(5.15)) for a bivalent metal, M^{2+}, present as a metal sulfide, [M][S], in the solid matrix (Vegilo et al. 2000).

$$[M][S] + \tfrac{1}{2}O_2 + H + \text{Bacteria} \rightarrow M^{2+} + S^0 + H_2O$$
(direct mechanism of bioleaching) (5.11)

$$[M][S] + Fe^{3+} + H_2O \rightarrow M^{2+} + Fe^{2+} + S_2O_3^{2-} + H^+$$
("indirect mechanism" of bioleaching) (5.12)

$$4Fe^{2+} + O^{2+} + 4H + \text{bacteria} \rightarrow 4Fe^{3+} + 2H_2O$$

Table 5.4 Metal leaching efficiency (%) in chemical leaching and bioleaching for coked and decoked spent catalyst.

Metal	Bioleaching (%)		Chemical leaching (%)	
	Coked catalyst	Decoked catalyst	Coked catalyst	Decoked catalyst
pH	2.4	2.1	1.5	1.2
Al	10.8	15.4	57.5	26.6
Fe	31.8	0.7	53.9	5
Ni	95.7	44	65.3	27.9
Mo	2.4	47.4	10.1	47.0

Source: Adapted from Bharadwaj and Ting (2013).

(ferric iron generated by bio-oxidation) (5.13)

$$S^0 + 3/2 O_2 + H_2O + \text{bacteria} \rightarrow 2H^+ + SO_4^{2-}$$
(elemental sulfur bio-oxidation) (5.14)

$$S_2O_3^{2-} + H_2O + 2O_2 + \text{bacteria} \rightarrow 2SO_4^{2-} + 2H^+$$
(thiosulfate bio-oxidation) (5.15)

Aung and Ting (2005) employed *As. niger* for the bioleaching of spent fluid catalytic cracking (FCC) catalyst and showed that the optimum pulp density for metal extraction was only 1%. The high metal concentration at higher pulp density gave rise to heavy metal toxicity from the leached liquor that imposes a constraint on the bioleaching process. Similar results have been reported by Santhiya and Ting (2005) for spent hydroprocessing catalyst using one-step and two-step bioleaching processes and the highest values of extraction of metal from the spent catalyst at 1% w/v pulp density were found to be 54.5% Al, 58.2% Ni, and 82.3% Mo in 60 days of bioleaching. An increase in tolerance of the fungi to heavy metals may be achieved through isolation of microbial strains from a suitable environment, and adaptation of the microorganism. Olson and Clark (2008) tried to bioleach molybdenum from molybdenite-containing mine waste samples and from high purity molybdenite in solutions amended with ferrous sulfate and indicated that the high redox potential required for decomposition of molybdenite is virtually impossible to reach using bioleaching processes. Mo extraction from finely ground (10–13 μm), decopperized waste stream molybdenite was 50% after 1 month, but slowed after this time, and after 6 months Mo extraction was 84–88%. Table 5.5 lists the literature related to bioleaching of industrial waste(s) (specifically spent catalyst) for the recovery of metals.

Beolchini et al. (2010) recovered metals from hazardous spent hydroprocessing catalysts involving the bioleaching abilities of Fe/S oxidizing bacteria (*A. ferrooxidans*, *A. thiooxidans*, and *L. ferrooxidans*). The authors concluded that nickel and vanadium dissolution kinetics are significantly faster than molybdenum dissolution ones. Results showed that Fe/S oxidizing bacteria are very effective in Ni and V extraction especially in the presence of ferrous iron. The presence of Fe^{2+}

Table 5.5 Bioleaching process for the recovery of metals from spent catalyst.

Catalyst	Metals	Microorganisms	Metal recovery	References
Spent hydroprocessing catalysts	Ni, V, Mo	A. ferrooxidans, A. thiooxidans, L. ferrooxidans	83% Ni, 90% V, 30–40% Mo	Beolchini et al. (2010)
Mine waste samples	Mo	Fe/S oxidizing microorganisms	84–88% after 6 months	Olson and Clark (2008)
Spent catalyst	Mo	A. Ferrooxidans	(<1%) after one month of bioleaching	Askari et al. (2005)
Spent refinery catalyst	Ni, V, Mo	Sulfur (SOB) and iron oxidizing bacteria (IOB)	90% Ni, 80% V, and 54% Mo by IOB and 88% Ni, 94% V, and 46% Mo using SOB after 7 d	Kim et al. (2010)
Spent hydro processing catalyst	Al, Ni, Mo	Unadapted Aspergillus niger	54.5% Al, 58.2% Ni, and 82.3% Mo in 60 d of bioleaching	Santhiya and Ting (2005)
Spent refinery processing	Ni, Mo, Al	Adapted Aspergillus niger to various metals	78.5% Ni, 82.3% Mo, and 65.2% Al over 30 d	Santhiya and Ting (2006)
Spent petroleum catalyst	Ni, V, and Mo	A. ferrooxidans (IOB) and A. thiooxidans (SOB)	—	Pradhan et al. (2009)
Spent refinery catalysts	Ni, V, Mo, and Al	Adapted and unadapted bacterial cultures	—	Kim et al. (2010)
Spent hydrotreatment catalyst	Fe, Ni, Mo, Al	Acidophilic thermophile, Acidianus brierleyi	Nearly 100% Fe, Ni, and Mo, 67% Al	Bharadwaj and Ting (2013)
Spent refinery catalysts	Al, Mo, Ni	Acidianus brierleyi	35% Al, 83% Mo, 69% Ni	Gerayeli et al. (2013)
Spent refinery catalyst	Al, Mo, Ni	Aspergillus niger	$99.5 \pm 0.4\%$ Mo, $45.8 \pm 1.2\%$ Ni, $13.9 \pm 0.1\%$ Al	Amiri et al. (2012)
Tungsten rich spent hydro-cracking catalyst	Ni, Mo, Fe, W, Al	Penicillium simplicissimum	100% W, 100% Fe, 92.7% Mo, 66.43% Ni, 25% Al	Amiri et al. (2011)

can be responsible for a cycle triggered by Fe/S oxidizing bacteria metabolism: ferrous iron favors bacteria adaptation to the high metal content, they oxidize Fe^{2+} to Fe^{3+}, which dissolves Ni and V sulfides on the spent catalyst by means of an oxidative attack producing other ferrous iron for the bacterial metabolism, while molybdenum showed both extraction yields and process kinetics significantly lower than nickel and vanadium, even in the presence of Fe^{2+} and S^0. This may suggest that molybdenum follows a dissolution pattern not associated to sulfides oxidation.

5.6.3 Extraction of Metals from Mineral Waste

Naturally occurring ores with high concentrations of metals are preferred industrially while the leaner deposits are usually ignored due to the lack of economic viability of the traditional extraction processes. Limited availability of natural mineral sources calls for efficient and eco-friendly techniques for resource recycling (Das and Mishra 2010). Lower energy consumption coupled with green methods is considered as the most efficient metal extraction process for the mining industry. It uses the inherent ability of microorganisms that enables them to convert insoluble conjugated minerals into their soluble and extractable forms. The process is an economic option for recovering minerals from low to medium grade ores and waste residues using microorganisms.

Mining wastes include low grade ores, mine tailings, and sediments from lagoons or abandoned sites. Reactors such as Pachuca tanks, rollings reactor, or in propeller vessels have been used. Heap leaching is more common since it allows large volume wastes to be treated in place. To enhance this process, aeration can be forced through the pile or hydrophilic sulfur compounds can be added. Thiobacilli bacteria are responsible for the oxidation of inorganic sulfur compounds. Applications include metal dissolution. For slurry processes, oxidation rate per reactor volume, pH, temperature, particle size, bacterial strain, slurry density, and ferric and ferrous iron concentrations need to be optimized. Bioleaching is very effective for recovery of gold from refractory gold pyrite and copper from chalcopyrite. One of the most significant applications of biohydrometallurgy is in low grade ore recovery. Economical recovery of gold has been the main focus of some bioleaching processes.

Mulligan et al. (2004) obtained maximum copper dissolution of 68%, 46% for zinc, 34% for nickel, and 7% for iron. The medium with sawdust and sucrose showed also significant copper solubilization. Maximum zinc dissolution was 46% with sulfuric acid pretreatment. Similarly, low grade apatite ore of uranium mines was used to extract uranium. With the shaking flask at 10% (w/v) pulp density, pH 1.7, and 35 °C with fine particles of <45 µm size, 96% uranium biorecovery was achieved in 40 days (Abhilash and Pandey 2011). In one of the recent studies, Ghosh and Das (2017) used bioleaching of manganese from mining waste deposits using indigenous bacterial strain *Acinetobacter* sp. with recovery of 76% in 20 days period at initial pH 6.5, 2% w/v inoculums, and 2% pulp density at 30 °C with an agitation speed of 200 rpm.

5.6.4 Extraction of Metals from Municipal Sewage Sludge

The treatment of municipal wastewater leads to production of sewage sludge. Generally, the sludge is disposed of as landfill. Sewage sludge consists of N, P, other micronutrients (copper, zinc, molybdenum, boron, iron, magnesium, and calcium), and organic matter. It is found to be beneficial to forestry, vegetation production, and landscaping. It results in improved texture and water holding capacity of soil (USEPA 2000). Besides being a good fertilizer, sludge consists of heavy metals (~0.5–2% on dry weight basis, may be as high as 6%), and may cause release of heavy metals to soil and in turn to water bodies. The whole food chain may get polluted with heavy metals, causing metabolic disorder and chronic diseases in humans. Therefore, removal of heavy metals from sewage sludge is of great importance. In recent years, several attempts have been made for cleaning of sewage sludge of all the contaminants.

The heavy metal content of sewage sludge is about 0.5–2% on a dry weight basis (Wong and Henry 1984). However, in some cases, extremely high concentrations of 4% w/w of chromium, copper, lead, and zinc have been reported (Lester et al. 1983). The discharge of heavy metals can be controlled either by source control of discharge or by extraction of metals from the sludge. Identification of source of metal discharge to sewage sludge is an arduous task.

Most of the bioleaching studies for heavy metal removal from sewage sludge have been reported using laboratory scale batch reactors. Ease and convenience of operation helps to generate data required for the development of the process for large-scale applications. The drawbacks of longer treatment time using batch bioleaching process can be overcome by using a continuous process. This can treat a larger volume of sludge in shorter time period and hence appears to be more suitable for applying on a larger scale. However, very few studies have been carried out in continuous flow system for bioleaching of metals. Most of the studies on bioleaching in continuous mode have been carried out in CSTR having provision for aeration and agitation to mix up the reactor content.

5.7 Technoeconomic Opportunities and Challenges

The commercialization of any technology is associated with many risks and challenges. The stages of development from conceptualization to commercialization are critical and affect the overall effectiveness of the operation. The economics of any process technology must be assessed before commercialization of the process. With the success of industrial scale bioleaching application for copper and gold, scale-up studies for the extraction of other precious metals have gained momentum as well.

Economic analysis of the plant involves both calculations of capital cost and operational cost and profitability investigations. The capital costs of bioleaching process encompass construction, provision services, operating costs, chemicals, maintenance, and services. Generally, the comparative investment for capital for the plants based on bioleaching process is reported to be less than that for conventional treatment techniques such as smelting and roasting (Mishra et al. 2005). Besides, bioleaching processes are less expensive in terms of chemical cost. However, bioleaching plants work at very low pH, thus the associated maintenance cost is high. It has been found that bioleaching process (sulfur-oxidizing process) is attractive only at low plant capacity and at high solids content (Sreekrishnan and Tyagi 1995).

Bioleaching process is considered as an efficient and economical process compared with traditional methods of metal removal. It was reported that bioleaching process requires only 1/5th of the cost of the chemicals required for leaching and recovery of metals compared to the traditional chemical methods. Generally, the bioleaching process requires 16–20 days and sufficient aeration and mixing for the entire duration. In addition to the costs of chemicals, the costs of mixing, aeration, construction of holding tank, and operational maintenance will have to be added to the total cost for carrying out a satisfactory cost analysis (Tyagi et al. 1988). Moreover, the total cost of metal removal should also include the cost of sludge conditioning, dewatering, and the cost associated with metal recovery from acid sludge filtrate.

Therefore, more in-depth cost analysis of the overall bioleaching process needs to be performed. Unfortunately, not much information is available in the literature on the economical aspect of the bioleaching process, which limits the scope of detailed analysis.

Besides the factors discussed above, there are certain other factors that cause problems in commercial scale operation of the process. The material of construction is of prime importance as it should withstand highly acidic conditions. Removing heat of exothermic mineral oxidation process should be taken care of. Biological population of bacteria or fungi should not undergo shear process. One of the designing challenges includes large residence time of slurry to be treated. Generally, stainless steel is a material of choice for the construction of aerators, agitators, and cooling system. Mesophilic and thermophilic conditions are maintained in commercial reactors with temperature conditions in the range of 35–50 °C, which are easier to maintain. Commercial plants using high range of temperature are not generally preferred.

BHP-Billiton is actively developing proprietary technologies for stirred tank bioleaching of chalcopyrite concentrates. Gericke et al. (2009) mentioned about a stirred tank bioleach pilot plant (170 m^3 capacity) in Monterrey, Mexico, using moderately thermophilic microorganisms. It has been reported that the objective of this plant is to recover metals from polymetallic (chalcopyrite, sphalerite, galena) concentrate. The report suggests that 96–97% Cu, 99% Zn, 98–99% Au, and 40% Ag were extracted daily (feed rate of 2.7 t/d).

Stirred tank bioleaching process can be used for impure concentrates as they have associated high cost for smelter penalty.

In addition to stirred reactors, bioleaching of copper from chalcopyrite ore in heaps using thermophiles will likely become a reality within the next few years. The GEOCOAT process, developed by Geobiotics, Lakewood, Colorado, is a unique heap leach system for bio-oxidation pretreatment of refractory precious metal concentrates and bioleaching copper, zinc, or nickel sulfide concentrates. Electrochemical interactions in the bioleaching of complex sulfides could take an advance seat in the metal extraction process. Mobilization of metals from electronic waste materials through the bioleaching process robustly helps waste

Table 5.6 Commercial copper bioheap leach plant.

Plant and location	Size (t/d)	Years in operation	Metals
Gunpowder's Mammoth Mine, Australia	1.2 million tonnes	1991–present	Copper
Mt. Leyshon, Australia	1 370	1992–1997	Copper
Girilambone, Australia	2 000	1993–present	Copper
Cerro Verde, Peru	15 000	1996–present	Copper
S&K Copper Project, Myanmar	15 000	1998–present	Copper
Fairview, South Africa	40	1986–present	Gold
Sao Benso, Brazil	150	1990–present	Gold
Laizhou, China	100	2001–present	Gold

Source: Mishra et al. (2005). Reproduced with permission of Springer Nature.

management in the electronic and galvanic industry. There are many commercial metal extraction plants, some of which are mentioned in Table 5.6.

The challenges involved in the commercial development of metal extraction plants include the following:

- Slow leach kinetics and need for large reactors
- High power consumption for oxygen supply
- Confidence that high plant availability will be achieved
- Economic recovery of precious metals.

5.8 Summary and Outlook

A reflection of the increasing interest in the green process and sustainability can be seen in the increasing number of publications in the field of bioleaching processes. Various hybrid approaches are being employed these days in order to improve the leaching efficiency and process selectivity. Along with several laboratory and pilot scale studies, the process is gaining the attention of industrialists for commercial applications. Mining industries also recognize that biotechnology offers another tool for economic recovery of metals of value. Literature indicates the significant potential of microbes to solubilize metal ions, which leads to efficient recovery of base and precious metals from solid waste. However, there are still unresolved issues in this research area such as the high sensitivity of microorganism to reaction parameters and waste contaminants during the process, slow leaching rate, and long processing time that demand further research in this area to establish it as a sustainable method for resource optimization. It is believed that future development will expand the scope of this emerging stream in waste treatment and the process will be reckoned as a successful demonstration of "greening the waste" concept in order to generate substantial economic, environmental, and social benefits from solid waste.

Questions

1. What do you understand by the term "bioleaching?" How is it different from "biosorption?"

2. Discuss how affinity between microbes and metals affects the leaching efficiency.

3. Describe the effect of various process parameters on the bioleaching process.

4. Write a short note on types of microorganisms.

5. Discuss the mechanism of bioleaching process.

6. List five recent research activities on the extraction of metals from WEEE using microorganisms.

7. What could be the possible limitations that hamper the scope of bioleaching process on industrial scale? Write your ideas to minimize these limitations.

8. "Bioleaching will be reckoned as a green process in future." Do you agree? Please give reasons for your opinion.

6

Future Technology for Metal Extraction from Waste: I. Chelation Technology

Every great advance in science has issued from a new audacity of imagination.

John Dewey, The Quest for Certainty, 1929

Abbreviations

DTPA	diethylenetriaminepentaacetic acid
ED3A	ethylenediaminetriacetic acid
EDDA	ethylenediaminediacetic acid
EDDG	ethylenediamine-N,N-diglutaric acid
EDDM	ethylenediamine-N,N-dimalonic acid
EDDS	ethylenediamine-N,N'-disuccinic acid
EDTA	ethylenediaminetetraacetic acid
EGTA	ethylene glycol-bis(β-aminoethyl ether)-N,N,N',N'-tetraacetic acid
GCG	L-5-glutamyl-L-cysteinylglycine
GLDA	N,N-Dicarboxymethyl glutamic acid tetrasodium salt
HEDTA	hydroxyethylethylenediaminetriacetic acid
HEIDA	hydroxyethyliminodiacetic acid
HSAB	hard and soft (Lewis) acids and bases
IDA	iminodiacetate
IDSA	iminodisuccinic acid
KPDA	2-ketopiperazine-1,4-diacetic acid
KPMA	2-ketopiperazine monoacetic acid
MGDA	methylglycine diacetic acid
NTTA	nitrilotris-(methylene)triphosphonic acid
PDA	pyridine-2,6-dicarboxilic acid
TMDTA	trimethylene-dinitrilotetraacetic acid

6.1 Introduction

Chelating agents have been used widely in the field of medicine for several decades. Yet, the application of these reagents is unexplored for resource

Sustainable Metal Extraction from Waste Streams, First Edition.
Garima Chauhan, Perminder Jit Kaur, K.K. Pant, and K.D.P. Nigam.
© 2020 Wiley-VCH Verlag GmbH & Co. KGaA. Published 2020 by Wiley-VCH Verlag GmbH & Co. KGaA.

recovery from solid waste materials. Limited studies have been carried out at laboratory scale to investigate the potential of the chelation process to recover metals from soil, spent catalyst, and other contaminated sites. It is postulated on the basis of laboratory scale studies that chelating agents have the tendency to bind with metals and form metal–ligand complexes. Understanding the mechanism of this complexation between metals and chelating ligands may open the doors to find a sustainable way to extract metals from waste streams, especially industrial wastes and WEEE(s). This chapter covers the basics of chelation technology ranging from definition, chemistry associated, factors affecting metal–ligand interaction, stability of the metal–ligand complexes, environmental fate of chelating agents, and novel applications of the process in the field of solid waste management.

6.2 Defining "Chelation"

"Chelation" is, in general, defined as the formation of stable metal–ligand complex in aqueous environment. The stability of the metal–ligand complex strongly depends upon the denticity of the ligands. "Denticity" refers to the number of donor groups in a ligand that bind to the central metal ion during the complexation process. If only one donor group from the ligand binds to the metal ions, the ligand is monodentate with denticity one. If there are more than one donor groups available in the ligand to bind to the metal ions, the ligand is called a "polydentate ligand." Chelating agents can be monodentate (e.g. water molecule) or polydentate ligands (e.g. ethylenediaminetetraacetic acid). Complexation between metal and chelating agent is primarily the replacement of a monodentate ligand (H_2O, in case of aqueous media) by polydentate ligands in the metal–ligand complex on the basis of ligand substitution reaction. The metal ions get hydrated in the presence of aqueous media and form metal–H_2O complex. The solvation sphere is further substituted by chelating agents (polydentate ligands) to form a metal–ligand complex. Chemical activity of the metal ions present in a complex is always different from that of un-complexed metal ions.

6.3 Classification of Ligands

Ligands can be classified in many ways depending upon their charge, identity of coordinating atom(s), and denticity (monodentate, bidentate, and polydentate ligands). Table 6.1 lists some common ligands with their structures, charge, and most common denticity. The polydentate ligand produces one or more rings by linking two or more of its donor atoms to the single metal ion simultaneously. Ring formation increases the stability of metal–chelate complex from one to several orders of magnitude greater than those of the monodentate complexes, and thus is considered as one of the major defining characteristics of chelation chemistry. Formation of each additional ring by the same ligand is believed to increase

6.3 Classification of Ligands | 125

Table 6.1 Ligands with their structures, IUPAC names, and most common denticity.

Ligands	Structures	Charge	Denticity
Fluoride ion	:F:⁻	Monoanionic	Monodentate
Water	H–O–H	Neutral	Monodentate
Ammonia	H–N(H)–H	Neutral	Monodentate
Hydroxide ion	⁻:O–H	Monoanionic	Monodentate
Thiocyanate ion	⁻:S–C≡N:	Monoanionic	Monodentate
Acetylacetonate ion	H₃C–C(=O)–CH=C(O⁻)–CH₃	Monoanionic	Bidentate
Oxalate ion	⁻O–C(=O)–C(=O)–O⁻	Dianionic	Bidentate
Phenanthroline	(phenanthroline structure)	Neutral	Bidentate
Triazacyclononane	(1,4,7-triazacyclononane ring)	Neutral	Tridentate
Corrole	(corrole macrocycle with three phenyl substituents)		Tetradentate
Pyrazine	(pyrazine ring)	Neutral	Ditopic
Ethyleneglycol-bis(oxyethylene-nitrilo)tetraacetate	(EGTA structure)	Tetraanionic	Octodentate

Table 6.2 Classification of chelating agents based on structural classes.

Structural classes	Chelating agents	Molecular formula	Abbreviation
Aminopoly-carboxylic acid	Ethylenediamine tetra acetic acid	$C_{10}H_{16}N_2O_8$	EDTA
	Hydroxyethylethylenediaminetriacetic acid	$C_{10}H_{18}N_2O_7$	HEDTA
	N-Dihydroxyethylglycine	$C_6H_{13}NO_4$	2-H_xG
	Diethylenetriaminepentaacetic acid		DTPA
	Ethylenebis(hydroxyphenylglycine)	$C_{18}H_{20}N_2O_6$	EHPG
Hydroxy carboxylic acid	Gluconic acid	$C_6H_{12}O_7$	
	Tartaric acid	$C_4H_6O_6$	
	Citric acid	$C_6H_8O_7$	Cit
	5-Sulfosalicylic acid	$C_7H_6O_6S$	5-SSA
Polyphosphate	Sodiumtripolyphosphates	$Na_5P_3O_{10}$	STPP
	Hexametaphosphoric acid	$H_6O_{18}P_6$	
Polyamines	Ethylenediamine	$C_2H_8N_2$	en
	Diethylene triamine	$C_4H_{13}N_3$	dien
	Triethylenetetramine	$C_6H_{18}N_4$	trien
	Triaminotriethylamine	$C_6H_{18}N_4$	tren
Amino alcohols	Triethanolamine	$C_6H_{15}NO_3$	TEA
	N-Hydroxyethylethylenediamine	$C_6H_{12}N_2O$	Hen
Phenols	Salicylaldehyde	$C_7H_6O_2$	
	Disulfopyrocatechol	$C_6H_6O_8S_2$	Tiron, PDS
	Chromotropic acid	$C_{10}H_8O_8S_2$	DNS
Phosphonic acids	Nitrilotrimethylenephosphonic acid	$C_3H_{12}NO_9P_3$	NTP, ATMP
	Ethylenediaminetetra (methylenephosphonic acid)	$C_6H_{20}N_2O_{12}P_4$	EDTMP
	Hydroxyethylidenediphosphonic acid	$C_2H_8O_7P_2$	HEDP
Polymers	Polyethyleneimines	$(C_2H_5N)_x$	PEI
	Polymethacryloylacetone	$(C_7H_{10}O_2)_x$	
	poly(p-vinylbenzyliminodiacetic acid)	$(C_{13}H_{15}NO_4)_x$	
Sulfur compounds	Toluenedithiol (Dithiol)	$C_7H_8S_2$	Tdth
	Dimercaptopropanol	$C_3H_8OS_2$	BAL
	Sodium diethyldithiocarbamate	$C_5H_{11}NS_2$	ANA
	Diethyl dithiophosphoric acid	$C_4H_{11}O_2PS_2$	
	Dithizone	$C_{13}H_{12}N_4S$	dz
Aromatic heterocyclic bases	Dipyridyl	$C_{10}H_8N_2$	Bipy
	o-Phenanthroline	$C_{12}H_8N_2$	Phen
Amino phenols	Oxine, 8-hydroxyquinoline	C_9H_7NO	Ox
	Oxinesulfonic acid	$C_9H_7NO_4S$	
Oximes	Dimethylglyoxime	$C_4H_8N_2O_2$	
	Salicylaldoxime	$C_7H_7NO_2$	

the stability of the complex due to the entropy effect of displacing a coordinated solvent molecule (Chauhan et al. 2015a).

Chelating agents can also be characterized based on their structural classes. Polyphosphates are the best known inorganic chelating agents that are used in fertilizer industry, potable and wastewater treatment, mineral processing, commercial detergents, flame retardants, etc. However, these are hydrolytically unstable at high temperature and pH. Aminopolycarboxylates (e.g. EDTA, DTPA, NTA) are the most popular organic chelating agents and are used in various applications ranging from detergent, pulp and paper, wastewater treatment, agrochemical industry, household cleaning, and many others. Hydroxy-carboxylate chelating agents such as citric acid and gluconic acid are widely used in electrochemical remediation process (Yeung and Gu 2011), in medicine, textile manufacturing process, and many more.

Out of these three widely known chelating agents, aminopolycarboxylates bind metal ions more strongly than polyphosphates and maintain their sequestering ability over a wider pH range than hydro-carboxylates (Hong et al. 2002). A survey report by Glauser et al. (2010) suggested that amino-polycarboxylate contributes the largest share (46%) in the worldwide consumption of chelating agent while the consumption of hydroxycarboxylic acid and organophosphate was 28% and 16% respectively in year 2010. Phosphonic acids have also been gaining the researcher's attention in recent years in the group of organic chelating agents due to complexing properties similar to those of inorganic polyphosphates such as threshold-scale inhibition, but unlike the polyphosphates, the phosphonates are stable in water at high temperature and pH. In addition, this class of chelating agents are reported to be effective even if present in less than the stoichiometric ratios of chelate to metal ion. Polyamines, amino phenols, amino alcohols, and sulfur based chelating agents are also used in various industrial and household activities. Table 6.2 lists a few chelating agents, grouped according to recognized structural classes.

Strict environmental legislations and broadening applications of chelating agents are the major key factors to drive the market growth of chelating agents. In recent years, considering the increase in demand of biodegradable chelating agents, several new chelating agents (EDDS, GLDA, IDSA) have emerged as the potential replacement for conventional chelating agents. These are expected to significantly share the market of chelating agents in the forthcoming years to avoid environmental risks associated with nonbiodegradable chelators.

6.4 Chemistry Associated with Chelation

6.4.1 Theories Derived for Metal–Ligand Complexation

As mentioned before, the chelation process works based on incorporation of metal ions into a heterocyclic ring in aqueous media. Alfred Werner proposed the concept of metal complexation in the year 1893 by illustrating the octahedral and square planar model for transition metal complexes (Jackson et al. 2004). He unriddled the puzzling aspect of having higher valences than requisite for transition metal compounds, which provides the foundation for transition metal coordination chemistry. The theory explains the existence of secondary valence for transition metal compounds in addition to a "primary" valence of appropriately charged counter ions (Bowman-James 2005). The primary valence meets the principles of neutrality, whereas the secondary valence provides the coordination number to the complex. Werner's theory explains the coordination environment around the metal ions according to which the metal ions are expected to be coordinated by neutral or negatively charged ligands whose number and arrangement around the center depend on the oxidation state of the metal (Telpoukhovskaia and Orvig 2013).

An 18-electron rule was formulated by Irving Langmuir in 1921 to investigate the stability of certain transition metal compounds (Jensen 2005). Langmuir derived an equation relating the covalence (v_c) of a given atom in complex ion to the difference between the number of valence electrons (e) in the isolated atom and the number of electrons (s) required for completion of its valence shell as shown in Eq. (6.1):

$$v_c = s - e \tag{6.1}$$

Bonding of metal–ligand complexes can be explained based on ligand field theory [1] (LFT), which incorporates all levels of covalent interactions into the model. This theory uses molecular orbital theory to describe the bonding, orbital arrangement, and other characteristics of coordination complexes. In octahedral complexes, ligands approach along the x, y, and z axes, so their σ-symmetry orbitals form bonding and antibonding combinations with the d_{z^2} and $d_{x^2-y^2}$ orbitals. The nonbonding metal orbitals (d_{xy}, d_{xz}, d_{yz}) have T_{2g} symmetry, whereas the bonding metal orbital (s) has A_{1g} symmetry; p_x, p_y, p_z have T_{1u} symmetry and $d_{x^2-y^2}$, d_{z^2} have E_g symmetry. Some weak bonding (and antibonding) interactions with the s and p orbitals of the metal also occur, to make a total of six bonding (and six antibonding) molecular orbitals. Interactions occur between frontier metal orbitals and the pi orbitals of the ligand to make a pi (π) bond. The d^7-d^{10} configuration with filling of E_g levels are labile and tend to have large Jahn–Teller distortions and/or low crystal field stabilization energy. The d^8 has a $^3A_{2g}$ ground state, which is immune to Jahn–Teller distortion; therefore, with strong field ligands, d^8 may be square planar and inert.

Morgan and Drew (1920) introduced the word "chelate" for caliper-like groups that function as two associating units and fasten to the central atom so as to produce heterocyclic rings. Aminopolycarboxylates and phosphonates were the major chelating agents in the nineteenth century and a lot of research work has been done on these two types of chelating agents. The first phosphonate was synthesized in 1897 in the form of a biphosphonate. Phosphonic acids were considered as an effective complexing agent in 1949 whereas the first chelating agent "nitrilotriacetic acid" (NTA) was synthesized by Heintz (1862). Complex forming ability of EDTA was first described by Pfeiffer and Offermann (1942).

---

1 LFT is not discussed here in detail. Students can find many relevant articles and textbooks to illustrate the LFT.

6.4.2 Attributes of Metal Ions for Complexation

Chelate formation is affected by various properties of the metal ion such as electronic structure, coordination number of the metal ion, oxidation state, and size and character of the acceptor. The nature of metal–ligand bond may vary from essentially electrostatic to almost covalent based on the properties of metal ions. Metals can be classified into the "a/b" and the "hard/soft" classification according to their acceptor character toward ligands present in aqueous solution. In general, most of the metals in their common oxidation states are class "**a**" acceptors that can form their most stable complexes with electronegative ligands containing nitrogen, oxygen, and fluorine and thus prefer ligands with acidic functional groups such as carboxylates. Class "**b**" acceptors form their most stable complexes with elements such as phosphorus, sulfur, and chlorine. These class "**b**" metal ions bind with neutral ligands preferentially and form a covalent bond with the ligands due to higher polarizing power than class "**a**."

There are several borderline metal ions that can be a part of class "a" and "b" depending upon their electronic configurations. Metal ions with ns^2 or ns^2np^6 configuration in the outermost shells (such as Mg^{+2}, Ca^{+2}) belong to class "**a**," whereas when the number of electrons starts to increases in d-subshell (transition metals), the class "**a**" character changes to class "**b**." Furthermore, when d-subshell fills completely (Zn^{+2}, Cd^{+2}, Hg^{+2}), class "**b**" character changes back to class "**a**" with increasing charge of the ion. On the other hand, class "b" character dominates in the case of "inert" pair of s-electrons $(n-1)d^{10}ns^2$.

Since metal ions behave as Lewis acids (electron pair acceptor) and ligands as Lewis bases (electron pair donor), metals can be classified according to the hardness/softness of metal ions and ligands. A qualitative concept of "hard and soft acids and bases" (HSAB) theory was introduced by Ralph Pearson to explain the stability of metal complexes and the reaction mechanism. According to the HSAB theory, the Lewis acid and bases can be further divided into hard or soft or borderline types. A characterization of different Lewis acids and bases has been listed in Table 6.3.

It can be seen from Table 6.3 that metal ions such as Cu^{2+}, Ni^{2+}, Co^{2+} are borderline Lewis acids and thus can be considered of intermediate strength. Metal ions with intermediate strength are expected to have nitrogen and oxygen atoms in the coordination sphere. Interaction between hard acid and hard base or soft acid and soft base combinations results in the formation of stable complexes. Ligands containing highly electronegative donor atoms, which are difficult to polarize, are classified as hard bases (e.g. the carboxylic acid donors). Like class **a** ions, hard metal ions retain their valence electrons strongly and are not easily polarized. Ions that are small in size and possess high charge are classified as hard. Soft metal ions in turn, like class **b** ions, are relatively large, do not retain their valence electrons firmly, and are easily polarized. Hancock and Marsicano (1978) investigated the hardness order of metal ions ($Ca^{2+} > Mg^{2+} > La^{3+} > Fe^{3+} > Mn^{2+} > Pb^{2+} > Zn^{2+} > Cd^{2+} > Cu^{2+} > Hg^{2+}$) as relative ionicity vs. covalence in metal–ligand bonds.

Another important factor responsible for the formation of metal–chelate complex is the size of a metal ion that can be affected by several factors,

Table 6.3 Characteristics of Lewis acids and Lewis bases according to HSAB theory.

Acid/base	Characteristics	Example
Hard acids	Atomic center of small ionic radii (<90 pm), high positive charge, empty orbitals in the valency shell, low electronegativity and low electro affinity, likely to be strongly solvated, high energy LUMO	H^+, Li^+, Na^+, K^+, Be^{2+}, Mg^{2+}, Ca^{2+}, Sr^{2+}, Sn^{2+}, Al^{3+}, Ga^{3+}, In^{3+}, Cr^{3+}, Co^{3+}, Fe^{3+}, Ir^{3+}, La^{3+}, Si^{4+}, Ti^{4+}, Zr^{4+}, Th^{4+}, VO^{2+}, UO_2^{2+}, $BeMe_2$, BF_3, BCl_3, $B(OR)_3$, $AlMe_3$
Soft acids	Large radii (>90 pm), low or partial positive charge, completely filled orbitals in their valence shells, intermediate electronegativities (1.9–2.5), low energy LUMOs with large magnitude of LUMO coefficients	Cu^+, Ag^+, Au^+, Hg^+, Cs^+, Tl^+, Hg^{2+}, Pd^{2+}, Cd^{2+}, Pt^{2+} metal atoms in zero oxidation states
Borderline acids		Fe^{2+}, Co^{2+}, Ni^{2+}, Cu^{2+}, Zn^{2+}, Pb^{2+}, $B(CH_3)_3$, SO_2, NO^+
Hard bases	Small radii (around 120 pm) and highly solvated, electronegative atomic centers (3.0–4.0), weakly polarizable, difficult to be oxidized, high energy HOMO	H_2O, OH^-, F^-, Cl^-, $CH_3CO_2^-$, PO_4^{3-}, SO_4^{2-}, CO_3^{2-}, NO_3^-, ClO_4^-, ROH, RO^-, R_2O, NH_3, RNH_2, N_2H_4
Soft bases	Large atoms (>170 pm) with intermediate electronegativity (2.5–3.0), high polarizability, easily undergo oxidation, low energy HOMOs but large magnitude HOMO coefficients	RSH, RS^-, R_2S, I^-, CN^-, SCN^-, $S_2O_3^-$, R_3P, R_3As $(RO)_3P$, RNC, CO, C_2H_4, C_6H_6, R^-, H^-
Borderline bases		Aniline, pyridine, N_3^-, Br^-, NO_2^-, SO_3^{2-}, N_2

including coordination number, nature of the metal–ligand bonds, nature of linked molecules, and spin state. Moving to the right in periodic series corresponds to a decrease in the ionic size. If the ionic charge remains constant, the decrease in size is smooth and moderate; however, a precipitous drop in the ionic radii can be observed with the increase in ionic charge. Increase in the oxidation state causes a shrinkage in size for certain metal ions due to loss of electron density and increasing cationic charge that attracts the negatively charged ligands closer to the metal ions. Spin states predominantly affect the effective ionic radii of transition metals. Increasing coordination number has an increasing effect on the ionic radii because of repulsions among the coordinating ions.

6.4.3 Metal–Chelate Complex Formation

Formation of metal–ligand complex follows a sequence of reactions including two transition states where ligand substitution takes place to convert the reactants into products. The general reaction profile for the metal–ligand interaction can be given as Eq. (6.2):

$$M + L \leftrightarrow [ML]^* \leftrightarrow [ML] \leftrightarrow [ML]^{**} \leftrightarrow ML$$
$$\text{Collision} \quad \text{Trasition} \quad \text{Intermediate} \quad \text{Trasition} \quad \text{Final}$$
$$\text{of energy} \quad \text{state} \quad \quad \quad \text{state} \quad \text{product}$$

(6.2)

The schematic representation of the general reaction mechanism of metal–ligand interaction is given in Figure 6.1a. The complexation process starts due to energy collision between metal and ligand molecules. An intermediate can be seen at the local energy minima in Figure 6.1a whereas the transition state occurs at an energy maxima point and cannot be isolated (Chauhan et al. 2015a). Here, W defines the width related to entropy (narrow for $-\delta S$, wide for $+\delta S$).

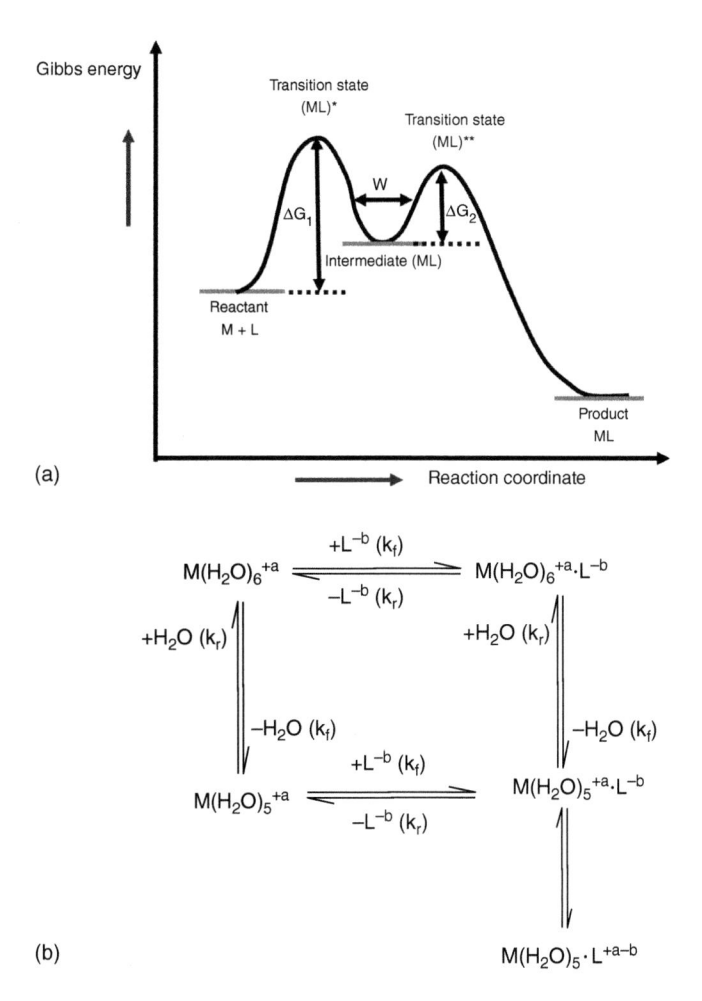

Figure 6.1 Metal–ligand interaction. (a) General mechanism; (b) reaction in aqueous phase. Source: Chauhan et al. (2015a). Reproduced with permission of Royal Society of Chemistry.

As mentioned before, the metal–ligand complexation process takes place in aqueous phase due to ligand substitution of water molecules in the coordination sphere of the metal by polydentate ligand. The reaction mechanism of metal–chelate complexation in aqueous system is demonstrated in Figure 6.1b. The process can be illustrated in terms of a rapid pre-equilibrium in which an outer sphere metal–water complex $[M(H_2O)_n]^{+m}$ forms and then the water molecules are replaced in the rate-determining step of the inner sphere metal–ligand complex [ML] formation.

6.4.4 The Chelate Effect

The term "chelate effect" refers to the enhanced stability of a metal–ligand complex during the chelation process. Chelating ligands are less prone to be replaced from a complex than are monodentate ligands of the same type (Kauffmann 1983). It can be explained by Eq. (6.3):

$$[M(H_2O)_6]^{2+} + 3en \rightarrow [M(en)_3]^{2+} + 6H_2O \tag{6.3}$$

Here, the chelation process releases six water molecules and thus increases the number of particles. Therefore, chelate effect can be related to the entropy change of the reaction due to release of more molecules of monodentate ligand in solution than molecules of chelating ligand used in the reaction. The formation constant for any chelation process is given by the thermodynamic relationship as given in Eq. (6.4):

$$\Delta G° = -RT \ln(K) = \Delta H° - T(\Delta S°) \tag{6.4}$$

Enthalpy contributes to the thermodynamics of chelation in the form of ligand repulsion, ligand distortion, and crystal field stabilization energy, and thus it is a measure of the binding energy whereas the entropy of formation (ΔS) includes all probability factors controlling the stability of the complex.

Schwarzenbach (1952) proposed the "chelate effect" as a combined effect of enthalpy and entropy while conducting studies on EDTA versus the corresponding IDA and linked the stability of chelating species with the favorable entropy change. Adamson (1954) investigated the effect of standard thermodynamic states and concluded that asymmetry in the choice of the standard thermodynamic states causes the chelate effect. The choice of standard state is primarily one of varying the translational entropy contributions. The use of hypothetical mole-fraction standard state for reactions of similar charge type tends to minimize the translational entropy effect.

Myres (1978) suggested enthalpy difference of ligands as a significant contributor to ΔG_0 in aqueous solution while working on the thermodynamic data of various chelation reactions using Born–Haber analysis. The significant contribution of the entropy of solution of ligands to the position of equilibrium and the entropy of solution of the coordinated metal ions was also demonstrated in his work. An increase in translational entropy was observed when molecules are released in solution during the chelation process; however, the entropies of solution of all species are generally not ideal. It was also concluded that the entropy of solution of the monodentate and chelated cations can differ by as much as 34 eu, whereas the enthalpy can easily deviate from ideal by more than 4.8 kcal.

The entropy change of a chelation reaction is considered a combination of various entropy forms as written in Eq. (6.5):

$$\Delta S^\circ = \Delta S_{solvation} + \Delta S_{transition} + \Delta S_{intrinsic\ rotation} + \Delta S_{symmetry} \\ + \Delta S_{isomer} + \Delta S_{vibration} + \Delta S_{internal\ rotation} \quad (6.5)$$

Here, the entropy terms based on solvation and internal rotation contribute negatively to the overall energy of the system. The negative magnitude of solvation entropy could be due to the difference in solvation entropies of the ligands and the different standard states employed. The solvation factor in organic solvents will be less significant than in aqueous solution; therefore, the entropy change of chelation in the gas phase or in an organic solvent will be larger than that in aqueous solution. In addition, internal rotation of free ligands and the complexes containing monodentate ligands are lost in the chelation reaction, and therefore, the entropy decreases with an increase in chelate ring. Vallet et al. (2003) demonstrated the contributions of rotation and vibration entropies as significant as the translation entropy contribution to the total entropy of the reaction. Thus, the high stability of a metal–ligand complex can be considered a result of the combined effect of enthalpy and various forms of entropies of the chelation process.

6.5 Chelation Process for Extraction of Metals

6.5.1 Framework for Chelating Agent Assisted Metal Extraction from Solid Waste

Recovery of metals from solid waste using a chelating agent is a solid–liquid extraction process due to the transportation metal–ligand complex from the solid surface to aqueous solutions. The final product of the process does not depend upon whether the reaction takes place at the interface or in the aqueous phase. Nevertheless, it depends on the electron count of the metal complex during the ligand substitution reaction.

Extraction of metals from solid waste using chelation reaction is postulated to take place in two stages. In stage 1, the metal and ligand form a complex through ligand substitution mechanism (as mentioned before) on the solid–liquid interface and then kinetic detachment of the complex takes place in stage II of the process to release the metals from the solid surface into bulk aqueous solution. Chauhan et al. (2015b) proposed a conceptual framework to define the overall chelating agent assisted extraction process (Figure 6.2). The authors categorized the process into primary (Eqs. (6.6)–(6.8)) and secondary reactions (Eqs. (6.9)–(6.11)) based on the possible interaction of the chelating agent (ligands) with target metals and other contaminants (impurities) present on the solid surfaces.

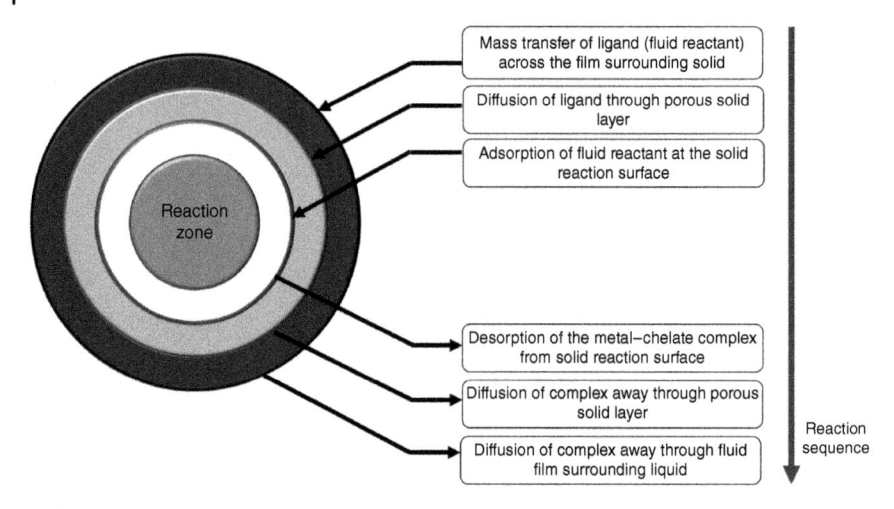

Figure 6.2 Conceptual framework for metal–ligand complex formation.

Primary Reactions:

$$([SW] - O) - M_i^{m+} + L^{n-} + H_2O \rightarrow [SW] - OH + [M_i - L]^{m+n-} + OH^- \quad (6.6)$$

$$M_i(OH)_m + L^{n-} \rightarrow [M_i - L]^{m+n-} + 2OH^- \quad (6.7)$$

$$[M_i - O]^{m+2-} + L^{n-} + H_2O \rightarrow [M_i - L]^{m+n-} + 2OH^- \quad (6.8)$$

Secondary Reactions:

$$[SW] - OH + [M_i - L]^{m+n-} \leftrightarrow [SW] - L^{n-} - M_i^{m+} + OH^- \quad (6.9)$$

$$([SW] - L - M_i)^{m+n-} + ([SW] - O - M_{ii})^{p+2-} + H^+ \leftrightarrow$$
$$([SW] - L - M_{ii})^{p+n-} + [SW] - OH + M_i^{m+} \quad (6.10)$$

$$([SW] - L - M_{ii})^{p+n-} + OH^- \leftrightarrow [SW] - OH + [M_{ii} - L]^{p+n-} \quad (6.11)$$

The target metals M_i^{m+} bound to surface hydroxyls (Eq. (6.6)) or present in metal hydroxides (Eq. (6.7)) and metal oxides (Eq. (6.8)) are extracted by surface complexation and dissociation into solution. Secondary reactions include the re-adsorption process and target metal substitution with other metals present on the solid surface, which depends upon the stability constant of metal–ligand complexes and approachability of other available metals (M_i^{p+}). The dissociated metals may also present as free ions in solution or may partially re-adsorb onto the catalyst surface sites.

It is worth noting that the rates of individual chemical stages of the process are considered in the region of "mixed kinetics" where the rates of mass transfer and of chemical reactions are comparable in magnitude. The process can be considered a "diffusion controlled" process at low rate of phase mixing, whereas a "kinetic region" can be achieved by increasing the mixing and lowering the particle size (Chauhan et al. 2015a).

6.5.2 Process Parameters Affecting the Metal Extraction Process

Several process parameters such as molar concentration of chelating agents, solution pH, reaction temperature, presence of competing ions in aqueous phase, reaction time, stirring speed, and particle size play an important role in the chelating agent assisted extraction process of metals from solid wastes. In order to understand the effect of mass transfer on the complexation process, stirring speed of the system and solid particle size are investigated whereas reaction time and reaction temperature are important parameters to study the reaction kinetics associated with the chelation process. This section briefly discusses the effect of various process parameters on the recovery of metals from industrial waste/WEEE using the chelation process.

6.5.2.1 Effect of Reaction pH

Solution pH may affect the stability of metal–ligand complex, solubility of chelating agents, trace metal sorption/desorption, ion exchange phenomena, as well as re-adsorption mechanisms and thus strongly affect the performance of the chelating agent for extracting metals from contaminated sites (Fangueiro et al. 2002). Extraction of metal–ligand complex from the solid surface into aqueous solution can be explained by Eq. (6.12):

$$M^{n+} + nHL \leftrightarrow ML_n + nH^+ \tag{6.12}$$

The extraction constant for the above metal–ligand complexation reaction is given as

$$K_{excons} = \frac{[ML_n][H^+]^n}{[M^{n+}][HA]^n} \tag{6.13}$$

Here, if the formation of intermediate complexes with the reagent and side product formation due to hydrolysis or competing reaction in aqueous phase are neglected for a certain pH range, Eq. (6.13) can be rearranged as shown below in Eq. (6.14):

$$K_{excons} = E\left(\frac{[H^+]^n}{[HA]^n}\right) \quad \text{or} \quad E = K_{excons}\left(\frac{[HA]^n}{[H^+]^n}\right) \tag{6.14}$$

where E = distribution coefficient.

The logarithmic of Eq. (6.14) can be represented as Eq. (6.15):

$$\log(E) = \log(K_{excons}) + npH + n\log[HA]^n \tag{6.15}$$

It is evident from Eq. (6.15) that molar concentration of the reagent and solution pH are the two major process parameters in the chelating agent assisted extraction process. If molar concentration of chelating agent is kept constant, then solution pH is the only variable and the curve between distribution coefficient ($\log E$) and solution pH should be a straight line with a slope of n.

Various biodegradable and conventional chelating agents have been employed to investigate the effect of solution pH on extraction of metals from contaminated sites, industrial waste, electronic waste, and soils. Tandy et al. (2004) studied the metal (Cu, Zn, Pb) extraction from non-calcareous soils by varying the solution

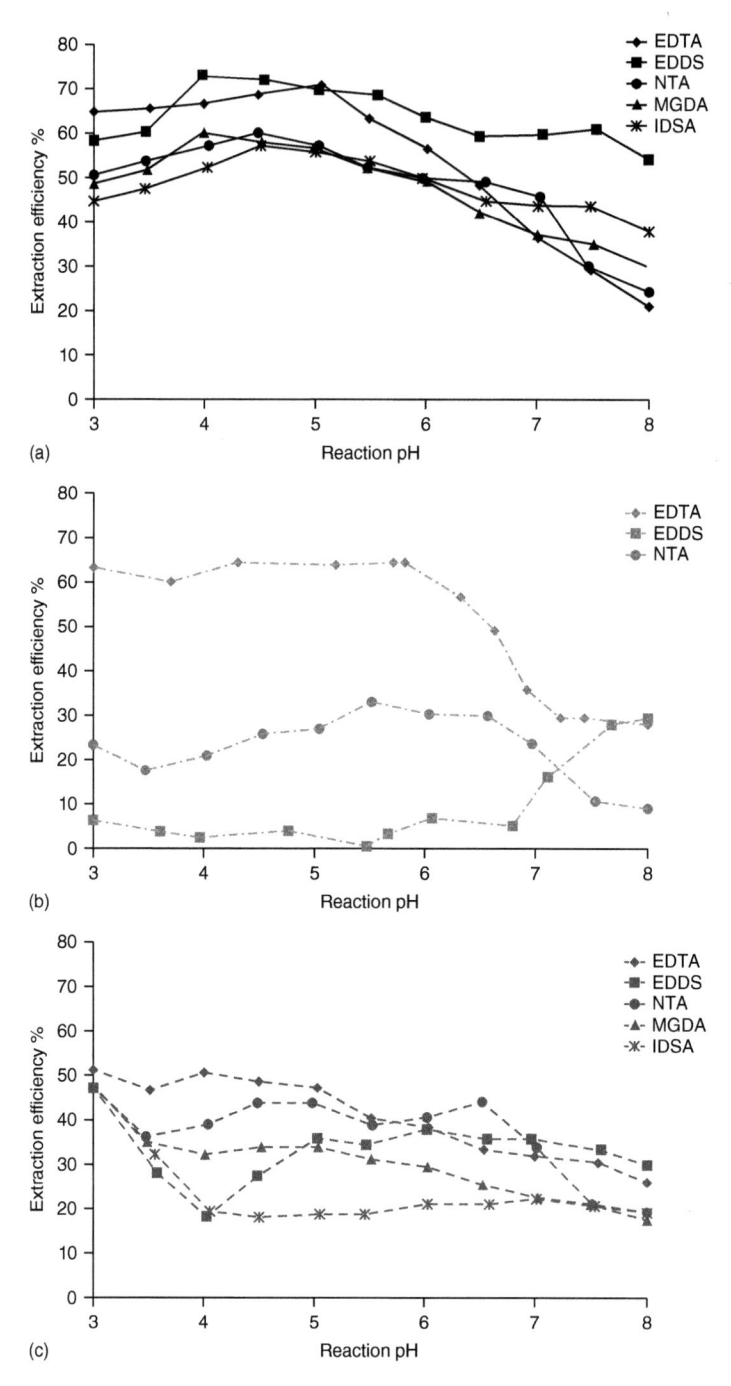

Figure 6.3 Effect of reaction pH on extraction efficiency of various chelating agents for (a) Cu (black markers), (b) Pb (light gray markers), and (c) Zn (gray markers) (Tandy et al. 2004).

pH in a wide range. A variety of chelating agents (EDTA, NTA, EDDS, IDSA, MGDA) were explored and the results are redrawn in Figure 6.3.

EDTA was able to extract metals efficiently over the entire pH range whereas the biodegradable chelating agent EDDS was found more effective for extraction of Cu and Zn than EDTA at pH >6 and low chelate to metal ratios. NTA also showed comparable extraction of Zn as EDDS at neutral pH but the ligand was not found competitive enough for the extraction of Cu. IDSA and MGDA performed better than EDTA at neutral pH and equimolar ratio of chelating agent to metal for Cu extraction. Pb metal was found to make the strongest complex Pb–EDTA with EDTA, than with other chelating agents; however, similar Pb extraction efficiency was noted for EDTA and EDDS at pH >7.

In another study, EDDS, GLDA, and EDTA were investigated for the extraction of Cu, Pb, Cd, and Zn from artificially contaminated soils by varying various process parameters such as solution pH, metal distribution, molar ratio of chelating agent, and stability constants. EDDS was reported to be the best extractant at neutral pH among all the ligands investigated. GLDA was found to be a better extractant at acidic pH for the extraction of Zn, Cd, and Cu, whereas EDTA showed the highest extraction of Pb among all the targeted metals. The average extraction efficiency at neutral pH is either lower or comparable to that of acidic or alkaline reaction conditions (Begum et al. 2013). Acidic pH conditions may increase the concurrent release of the other metal ions and an exchange of H^+ ion from functional groups present at the soil surface (Lim et al. 2004), whereas alkaline pH conditions may be responsible for increasing the reactive species L^{n-} in aqueous solution and higher formation rate of the soluble coordination compounds (Fischer and Bipp 2002), which could be the possible reasons for improved extraction efficiency at acidic and alkaline pH. The lower pH, sometimes, may also solubilize undesirable metal cations that may compete with the target metal for ligand binding sites.

Goel et al. (2009) performed extraction experiments using EDTA by varying the solution pH over a wide range to recover nickel from spent catalyst. Extraction efficiency was observed to increase with an increase in pH value up to 10; however, after pH of 10 a decreasing pattern was observed in the extraction efficiency. The possible reason behind this observation could be the presence of anionic ML_{n+1}^- complexes in the solution or hydrolysis of the metals. Chauhan et al. (2012) investigated the performance of biodegradable chelating agent EDDS to extract nickel from spent catalyst and concluded that EDDS requires a narrower pH range for the chelation–dechelation process than EDTA due to lower stability constants of Ni–EDDS complex. It has also been reported that EDDS can also extract higher amount of heavy metals at neutral pH than EDTA.

Effect of solution pH was also investigated for the extraction of Cu from WEEE. The highest extraction efficiency (~84%) was obtained at neutral pH, while with the increase in reaction pH extraction efficiency started to decrease. Approximately 15% reduction in Cu recovery was reported with the increase in pH from 7 to 11 (Jadhao et al. 2016). Presence of competing metals ions and accumulation of insoluble copper hydroxide at alkaline pH could be the possible reasons to affect the formation of Cu–EDTA complex and the consequent decrease in metal recovery at higher pH.

The effective pH range for extraction of a particular element may differ for different ligands, which can be explained on the basis of the percentage distribution of protonation stages of various ligands. Each ligand has a certain pH range for different protonation stages. This deviation in the percentage distribution of various protonation stages strongly affects the effective pH range for a ligand. It should also be kept in consideration that the favorable pH range for the efficient extraction of heavy metals may also correspond to the precipitation range of the corresponding chelate, which can be determined either by the stability constants or by the distribution constants of the complexes (Chauhan et al. 2015a).

6.5.2.2 Effect of Molar Concentration of Chelating Agent

Molar concentration of chelating agents has been a subject of several studies to investigate the effect on metal extraction from contaminated sites. This process parameter can also be reported as molar ratio of ligand (chelating agent) to metal ions (MR) in the literature. Kim et al. (2003) studied the chelation process for the extraction of metal from lead contaminated soils and observed significant dependence of extraction efficiency on molar concentration of the ligand (EDTA in reference study). Metal extraction was also found affected by the soil properties and therefore at a given stoichiometric ratio, extraction efficiencies were different for different soil samples as shown in Figure 6.4. However, the effect of soil properties could be minimized at high concentration of EDTA and most of the lead can be extracted from soil samples.

Jiang et al. (2011) varied the chelating agent concentration from 0.1 to 0.4 g/l to extract Cu and Ni from artificially contaminated soil. An increase in metal recovery was observed with the increase in molar concentration of the reagent; however, once the equilibrium concentration was attained when the reagent

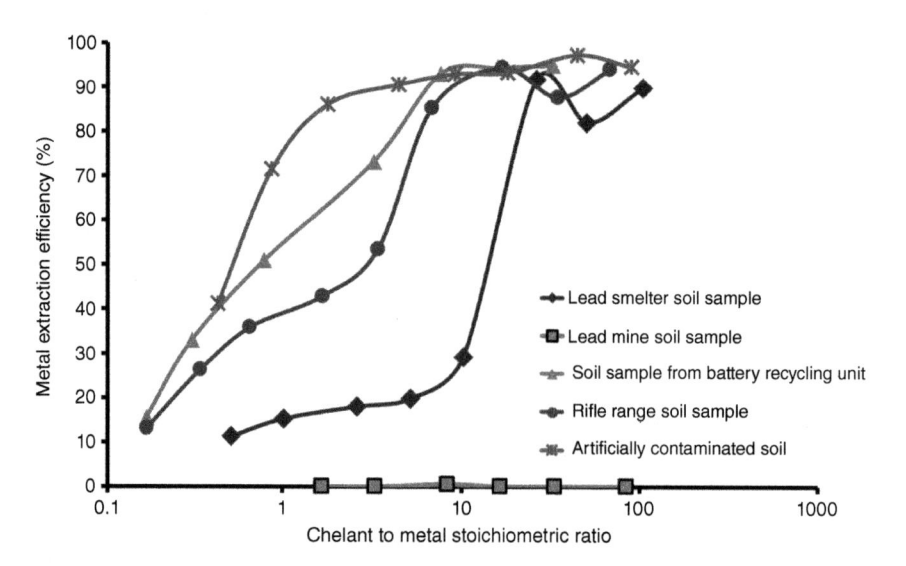

Figure 6.4 Effect of chelant (EDTA) to metal stoichiometric ratio (MR) on metal extraction for different soil samples. Source: Chauhan et al. (2015a). Reproduced with permission of Royal Society of Chemistry.

amount is enough for soil requirement, extraction became nearly constant. Chauhan et al. (2012) demonstrated that higher concentration of chelating agent pushes the reaction in the forward direction according to Le Chatelier's principle due to the reversible nature of the chelation reaction. Therefore, a gradual change in Ni recovery was seen with increase in MR from 1.2 to 4.8. The initial rate of extraction was calculated at different MR values and MR = 3.6 showed the highest initial rate of extraction, whereas beyond MR = 3.6 no significant change was observed in Ni recovery.

Molar concentration of chelating agent is associated with the solid (contaminated sites or solid waste) to liquid (volume of chelating agent solution) ratio (S/L) in chelation reaction. Effect of S/L was investigated for extraction of Co and Mo from multimetallic spent catalyst at constant MR and a significant increase in extraction efficiency was reported with the decrease in S/L ratio (Chauhan et al. 2013a) to a certain level and afterwards, it started to decrease. Decrease in S/L ratio corresponds to the lesser concentration of chelating agent in aqueous solution (at constant MR). However, if the S/L ratio is considerably high, then it would be difficult to provide efficient mixing of particles in order to minimize the external resistance; consequently, extraction efficiency will be less, and therefore, optimization of S/L is a prerequisite to achieve the required metal recovery.

In some studies, effect of S/L ratio is considered trivial. Kim et al. (2003) varied S/L from 1/10 to 1/3 for extraction of Pb from contaminated soils; however, the authors could not find any noteworthy change in extraction efficiency with the change in S/L ratio. The possible reason for this observation could be the presence of excess amount of chelating agent from which only a small fraction was effectively utilized to extract Pb from soil. The remaining amount of chelating agent may be freely available in the solution or may form complexes with other metal cations; therefore, no effective increase in extraction efficiency can be observed beyond a certain S/L value when the concentration of reagent is enough for the metal extraction. Manouchehri et al. (2006) also urged that L/S cannot be considered an indicator to ensure the maximum metal extraction efficiency; however, this ratio must be investigated with respect to all the extractable metal ions present in the contaminated sites.

It is recommended that the molar concentration of chelating agents should be greater than that of the target metal species present in the contaminated site to minimize the competing effect of other undesirable ions; however, some reports also suggest that after a certain extent, extraction efficiency either becomes unaltered or starts decreasing with increase in concentration, which could be related with the higher dose of reagent concentration than required at the contaminated site for metal extraction. Some other limitations may also arise due to higher concentration of chelating agent such as the reagent being difficult to remove, and its residual traces reducing the accuracy of the determination. Using a lower concentration of ligand and high S/L ratio may prove beneficial in order to prevent clogging of the solid particles during leaching and to provide better mixing. However, generation of large amount of wastewater after the extraction process may enhance the treatment cost and pollution possibility, which is not desirable. Therefore, an optimized value of the concentration of chelating agent and S/L ratio are necessary for effective extraction.

6.5.2.3 Effect of Reaction Temperature

Reaction temperature always plays a significant role in extraction kinetics, and therefore, study of the thermodynamics aspects associated with chelate assisted metal extraction process is necessary in order to optimize the process parameters. Several studies have been carried out to investigate the effect of reaction temperature on metal extraction efficiency. Chauhan et al. (2013a) varied the reaction temperature from 100–140 °C for extraction of Co and Mo from spent catalyst using EDTA and reported that extraction efficiency increased with increase in reaction temperature. The increase in (%) metal extraction with increase in temperature may be explained on the basis of the possibility of enhanced reaction kinetics due to Arrhenius behavior of the surface reaction that dissolves the metal. An asymptotic extraction (Co 80.4% and Mo 84.9%) was reported at a reaction temperature of 120 °C within four hours of reaction time.

Figure 6.5 demonstrates the extraction of Ni reported at different reaction temperatures under atmospheric and autogenous reaction conditions using EDTA and EDDS as chelating agents. It can be depicted from Figure 6.5 that 95% Ni recovery was achieved under hydrothermal condition in an autoclave at reaction temperatures of 150 °C, over a four hours reaction time (Vuyyuru et al. 2010). High temperature may also cause an autogenous pressure buildup in the reactor. Chauhan et al. (2012) also performed the extraction experiments at a reaction temperature of 90 °C under atmospheric reflux conditions and a significant difference in extraction efficiency was observed when results were compared with those of autogenous reaction condition at similar reaction conditions (Vuyyuru et al. 2010). This could be due to the autogenous pressure in autoclave and better mixing of particles, which is necessary for effective extraction. Chauhan et al. (2013a) also reported increase in vessel pressure up to

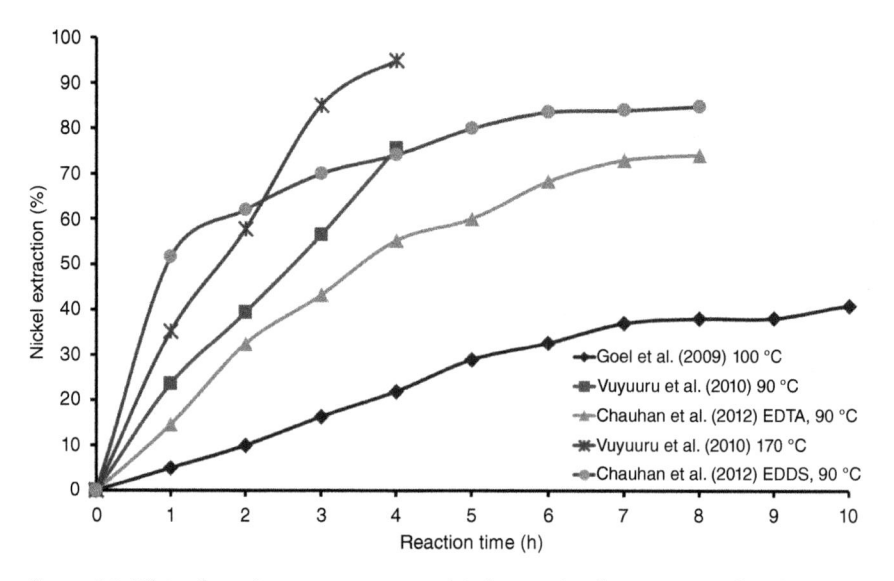

Figure 6.5 Effect of reaction temperature on nickel extraction from spent catalyst. Source: Chauhan et al. (2015a). Reproduced with permission of Royal Society of Chemistry.

2.7 atm at temperature 140 °C during metal extraction using EDTA, which favors the rate of the forward reaction; hence, metal extraction efficiency improves. Extraction process at similar reaction conditions was performed under atmospheric reflux conditions at a reaction temperature of 100 °C, and about 58% Mo was extracted, which was nearly 12% less than the extraction efficiency obtained under autogenous conditions. Hong et al. (2008) also investigated pressure assisted lead extraction from soil samples and concluded that soil particles may break into smaller fractions due to high pressure, and therefore, lower particle size enhances the metal extraction efficiency.

A synergistic relation between MR and reaction temperature was established through statistical optimization of EDTA assisted extraction of Ni and Cu from industrial waste (Chauhan et al. 2013b) and waste printed circuit boards (Sharma et al. 2017) respectively.

Chelate assisted extraction process under autogenous reaction conditions is considered a feasible and efficient way of metal extraction on industrial platform based on its high extraction efficiency (Figure 6.5); nevertheless, high reaction temperatures (170 °C) can adversely affect process economics in terms of consumption of steam utilities and equipment cost.

6.5.2.4 Presence of Competing Ions in Reaction Zone

Species in aqueous solution exist in formation–dissociation equilibrium, and consequently, displacement reactions of one element (metal ion) or organic reagent (ligand) by another may occur. Metal–ligand complexation depends on the displacement equilibrium constants when a chelating agent is added to the solution containing two or more metal cations. It is suggested that a ligand must overcome competing metal precipitation, surface complexation, and precipitation of solid particles to achieve an effective extraction of metals from the contaminated site. Therefore, enough molar concentration of chelating agent in the solution is also desirable to combine with the target metal and other ions, which can displace the target metal. The chelating agent should have different stability constants for two metals if selective complexation of one metal is required in the presence of the other.

As mentioned in Section 6.5.1, the presence of competing ions in the reaction zone may interfere with the extraction of target metals. Interaction of ligands with metals other than the target metals, polymerization, hydrolysis, and complex formation with extraneous complexing agents are the major examples of competing reactions in aqueous phase. Literature suggests that if all the compounds are mononuclear and ML_n is the only target species, then the distribution coefficient is calculated according to Eq. (6.16) where the effect of all other competing reactions is also considered.

$$E = \frac{[ML_n]_0}{\left([M] + \sum_{i=1}[ML_i] + \sum_{j=1}[M(OH)_j] + \sum_{q=1}[MX_q]\right)} \quad (6.16)$$

where $[ML_i]$ is the concentration of the complexes of the metal with the chelating agent including $[ML_n]$; $[M(OH)_j]$ is the concentration of all hydroxylated components; and $[MX_q]$ is the concentration of complexes formed with the extraneous complexing agents. In et al. (2008) employed the solvent extraction process with chelating agent salen-(NEt$_2$)$_2$ to extract Cu^{+2}, Mn^{+2}, and Zn^{+2} from water samples and investigated the effect of the concomitant ions on the metal extraction process. Mg^{2+} was observed to interfere with Mn^{+2} extraction; however, no significant effect was observed on the extraction of Cu^{+2} and Zn^{+2}. Extraction efficiency of Mn^{+2} increased with an increase in the concentration of Mg^{2+}, which could be explained by the competing reaction of Mg with Mn^{+2} in the complex formation. Therefore, if Mg^{2+} coexists in the solution at a very high concentration (>150 000 times the Mn^{+2}), then Mn^{+2} cannot be determined due to strong interference. Manouchehri et al. (2006) illustrated the simultaneous occurrence of various competitive reactions for the major elements (Al, Ca, Fe, and Mg) toward EDTA depending on the calcium content of soil. Calcareous soils preferred the formation of Ca–EDTA due to abundance of Ca in soil, while Al– and Fe–EDTA complex formations dominate in non-calcareous soil. The relatively weak bonding strength of Ca to certain soil fractions during the concurrent extraction process in calcareous soils may cause preferable Ca dissolution and displacement of other metals at higher reaction times. Tandy et al. (2004) observed a strong competitive reaction between target metals and Ca as an important factor for extraction using EDTA but not with EDDS, which results in a decrease in extraction efficiency for EDTA compared to EDDS. Vandevivere et al. (2001) illustrated that Fe may be neglected during speciation of EDDS in the soil sample. On the contrary, Fe–EDTA was observed to decrease with increase in pH due to Ca interference while Fe–EDDS is a relevant species at neutral pH.

6.5.3 Factors Affecting Stability of Metal–Ligand Complex

The complexation reaction between the metal ion and ligand is given in Eq. (6.17):

$$M^{m+} + nL^- \rightarrow ML_n; \quad \beta_{comp} = \frac{[ML_n]}{[M^{m+}][L^-]} \tag{6.17}$$

$$[ML_n] = ML_{n(0)}; \quad E_{ML} = \frac{[ML_n]_0}{[ML_n]} \tag{6.18}$$

Here, β_n and E_{ML} refer to the overall stability constants and the distribution constant of the metal–ligand complex respectively.

By combining Eqs. (6.17) and (6.18) with Eq. (6.13), it can be rewritten as

$$K_{excons} = \frac{(\beta_n E_{ML} K_{HL}^n)}{E_{HL}^n} \tag{6.19}$$

where K_{HL}^n and E_{HL}^n are the dissociation constant and distribution constant respectively of the molecular form of the ligand.

$$\log K_{excons} = \log(\beta_n E_{ML}) - n \log\left(\frac{E_{HL}}{K_{HL}}\right) \tag{6.20}$$

It can be said from Eq. (6.20) that the stability constant (K_{excons}) of the metal–ligand complex and dissociation constant (K_{HL}) of the ligand are interlinked. Thus, stability of any metal–ligand complex depends primarily on ligand properties.

The presence of long chains or bulky ring structures on any of the donor groups may affect the stability of the metal–ligand complex due to steric hindrance. In general, 5- and 6-membered rings are considered the most stable in chelation reaction, although steric hindrance plays a dominant role in defining the stability of the complex. For example, the biodegradable chelating agent EDDS has lesser stability than the conventional chelating agent EDTA complex due to the presence of two six membered rings in EDDS, which causes steric hindrance and weakens the metal–ligand bond. DTPA is a more stable chelating agent than EDTA due to the presence of more donor atoms than in EDTA and EDDS; on the other hand, NTA is considered relatively biodegradable and forms a less stable complex due to the availability of only four binding sites for complexation. The larger the complex formation constant, the more stable is the species. In addition, ligands with relatively low or moderate stability constants (EDDS and NTA) are less resistant to biodegradation than the ligands that form metal–ligand complexes with high stability constants such as EDTA and DTPA.

Steric hindrance may also affect the formation of the metal–ligand bond. When sizes of the metal ion and donor atom are disparate, steric effects increase. Sovago and Gergely (1976) studied the effect of steric factors on the equilibrium and thermodynamic conditions of mixed ligand complexes of Cu^{+2} ion with diamines and concluded that steric hindrance in CuA_2^{2+} complexes containing substituted diamines and histamine leads to the formation of protonated complexes CuA_2H^{3+} or mixed hydroxo complexes. Gothard et al. (2012) investigated the steric hindrance caused by attaching bulky *tert*-butyl groups in bis(2,9-di-*tert*-butyl-1,10-phenanthroline)Cu(I), [Cu(I)(dtbp)(2)](+). The two bulky *tert*-butyl groups on the dtbp ligand lock the excited state into the pseudotetrahedral coordination geometry and completely block solvent access to the copper center in the metal-to-ligand charge transfer excited state of [Cu(I)(dtbp)(2)](+). Steric hindrance effect was also observed in the adsorption process of metal–organic complex onto the chelating agents. Deepatana and Valix (2008) investigated the adsorption of Ni and Co complexed with citrate, malate, and lactate on chelating aminophosphonate Purolite S950 and observed that the larger metal–organic complexes predominate the effect of steric hindrance to the metal uptake to the resins under acidic conditions. The adsorption of lactate complex was found to be higher than that of smaller hydrous metal ions. Chauhan et al. (2015c) investigated the stability of transition metal complexes with various aminopolycarboxylates using density functional theory calculations and inferred that the metal–chelate ligand complexes follow the stability trend (EDTA) > (EDDS) > (NTA). The high stability of $EDTA^{4-}$ ligand with a favorable minimal strain was reported as one of the major reasons for its inherent resistance to degradation.

6.6 Novel Applications of Chelating Agents

Chelating agents have been used in medicine and household activities for several decades. More than 70% of the global demand for chelates comes from household and industrial cleaners, water treatment, and pulp and paper industries

(Chauhan et al. 2015a). Expanding at a rapid growth rate from domestic end-use markets, such as food, photography, paper, and surfactants, applicability of chelating agents is now being explored in soil remediation and metallurgical processes. Several research studies at batch scale are being carried out to investigate the economic and environmental feasibility of chelation process for extraction of metals from industrial wastes (Chauhan et al. 2012, 2013a) and electronic wastes (Sharma et al. 2017; Jadhao et al. 2016). This section gives a brief review of the application of chelating agents in metal extraction from industrial waste and WEEE and in soil remediation.

6.6.1 Chelating Agents Used for Metal Extraction from Metal-Contaminated Soil

6.6.1.1 Hydrometallurgical Route of Chelation Process (Direct Use)

The increasing intensity of industrial activity is responsible for polluting the environment and posing an irrecoverable damage to the ecosystem. It was mentioned earlier that heavy metals may leach into the soil and water during the landfilling or recycling of solid wastes. A significant amount of metals has been reported in the soil of nearby areas to WEEE recycling units (Chen et al. 2009; Deng et al. 2014; Leung et al. 2010; Wang et al. 2015). These trace metals present in the soil affect the crop yields, soil productivity, and may cause harmful health effects by entering into the food chains. Sorption of these metals to the soil particulates can limit the metal mobility and therefore heavy metals tend to bioaccumulate for long time into the environment. In addition, the possibility of surface polymerization increases with the increase in concentration of metals in the soil matrix that may cause precipitation of metals in the solid phase.

Various soil remediation techniques have been investigated in the literature such as pneumatic fracturing, solidification/stabilization, vitrification, excavation and removal of contaminated soil layer, and phytoremediation. These methods work basically on two fundamental principles, i.e. either completely remove contaminants from polluted sites or transform pollutants into nontoxic forms. However, certain limitations are associated with these treatment methods such as long time duration, generation of secondary wastes, plant predation, and high cost, and therefore, cannot be considered as long-term solution for metal contamination. Soil washing process is also drawing attention these days due to the rapid kinetics, economic efficiency, and high metal extraction efficiency using chelating agents, surfactant, acids, alkalis, and complexing agents. Davis and Singh (1995) employed chlorination, hydrochloric acid leaching, and chelation process using organic ligands EDTA and DTPA to extract Zn^{2+} from artificially contaminated soil column. The highest removal efficiency was obtained with the chelating agent at optimized flow rate. Kociałkowski et al. (1999) evaluated the performance of different chelating agents (HEDTA, EDTA, EGTA, and NTA) for extraction of heavy metals Cu^{2+}, Zn^{2+}, Pb^{2+}, and Mn^{2+} from arable soil samples and concluded that EDTA was the best chelating agent for extraction of all metals. Lim et al. (2005) proposed a cost effective closed loop utilization of EDTA ligands by recovering and reusing the EDTA in multiple cycles for the extraction of Cd^{2+}, Pb^{2+}, and Ni^{2+} from soil. Palma et al. (2011) investigated the performance of

Table 6.4 Chelating agents used for metal extraction from soil in literature.

Type of soil	Chelating agent	Target metals	References
Superfund site soil	EDTA	Pb	Ellis et al. (1986)
Soil	GCG, EDTA, NTTA, TMDTA	Pb, Cu, Cd, Zn, Ni, and Hg	Tandy et al. (2004)
Soil from battery recycling facility	EDTA, NTA	Pb	Elliott and Brown (1989)
Quartz-rich sediment and soil samples	NTA	Pb, Ni, Zn	Howard and Shu (1996)
Wetland soil	Acetic acid, EDTA, DTPA	Zn, Fe, Mn, Ca, Cd, Mg, K, Al, Na, Cu, Pb	Sistani et al. (1995)
Industrially contaminated soil, agriculture soil	EDTA, EDDS, IDSA, MGDA, NTA	Zn, Cu, Pb, Cd, and Ni	Tandy et al. (2004)
Soil from 25–50 cm below ground surface near a bay	EDDS	Cu, Zn, Pb	Yip et al. (2009)

EDTA, its structural isomer EDDS, rhamnolipids, and citric acid for the removal of metals from a contaminated harbor sediment by varying reagent concentration and washing time. Nearly 70% metal extraction was attained using both the chelating agents whereas rhamnolipids and citric acids were not found suitable for metal removal process. Yip et al. (2009) employed EDDS for the extraction of Cu^{2+}, Zn^{2+}, and Pb^{2+} from artificially and field contaminated soil samples and observed that higher extraction rates can be achieved in multimetal contaminated soil samples than mono-metallic contaminated soils due to the presence of a large proportion of carbonated and exchangeable fraction of heavy metals. Table 6.4 gives a list of chelating agents that have been used for metal extraction from soil in the literature.

6.6.1.2 Phyto-remediation of Soils in Presence of Chelating Agents

Phyto-remediation is a greener way of extracting metals from contaminated soil in order to improve the soil quality. It is also known as "green remediation" technique. The process can be categorized into phytoextraction (accumulate metals in shoots), phytomining (use of plant biomass), phytovolatilization (use of microbes to convert soil elements into volatile form), and phytostabilization (conversion of highly toxic, nonbiodegradable materials into less toxic materials). In recent years, chelating agents are also being employed in the phytoextraction process to enhance the uptake of metals by plant shoots and biomass. It is argued that phyotextraction follows either the split uptake mechanism (i.e. selective uptake of only free metal ions) or uptake of purportedly intact metal–ligand complexes; however, it is still an unresolved argument. Table 6.5 lists some of the studies illustrating the effect of the addition of various synthetic aminopolycarboxylates on the mobilization and uptake of heavy metals.

Table 6.5 Effect of various chelating agents on the phytoextraction of heavy metals.

Chelating agents	Heavy metals	Metal concentration in soil	Heavy metal uptake	Crop	References
EDTA	Cd	42-fold	11-fold	Rainbow pink (*Dianthus chinensis*)	Lai and Chen (2005)
	Zn	6.3-fold	Not significant		
	Pb	50-fold (H_2O-extractable)	15-fold		
EGTA	Pb	Several 100-fold (H_2O-extractable)	Several 1000-Fold	Indian mustard (*B. juncea*)	Blaylock et al. (1997)
DTPA	Pb	60-fold (H_2O-extractable)	7.1-fold	Corn (*Zea mays*)	Huang et al. (1997)
			23.8-fold	Pea (*Pisum sativum*)	
NTA	Cd	8-fold	2-fold	Indian mustard (*B. juncea*)	Quartacci et al. (2005)
EDDS	Cu	190-fold	45–fold	Corn (*Z. mays*)	Luo et al. (2005)
	Pb	20-fold	9-fold		
	Zn	7-fold	2.3-fold		
	Cd	3-fold	1.5-fold		
	Cu		135-fold	White bean (*Proteus vulgaris*)	
	Pb		42-fold		
	Zn		4.5-fold		
	Cd		1.5-fold		
HEIDA	Cu	300-fold (H_2O extractable)	3–4-fold	Corn (*Z. mays*), Vetiver grass (*Vetiveria zizanoide*)	Chiu et al. (2005)

Blaylock et al. (1997) investigated the application of *Brassica juncea* (Indian mustard) as a hyper-accumulator with EDTA alone and EDTA in combination with acetic acid to extract PB^{2+} from soil. The combined effect of the chelating agent and acid resulted in 2X higher accumulation of metals in mustard shoots than the extraction with EDTA alone. Vassil et al. (1998) suggested that coordination of Pb^{2+} transport by EDTA enhances the mobility of insoluble Pb^{+2} ion within the plants whereas high concentrations of EDTA may cause reductions in both the transpiration rate and the shoot water content due to the presence of free protonated EDTA (H-EDTA) in the hydroponic solution. Meers et al. (2005) also compared the performance of EDDS, NTA, and citric acid with the synthetic chelator EDTA to enhance shoot accumulation of heavy metal ions (Cd^{2+}, Cu^{2+}, Cr^{3+}, Ni^{2+}, Pb^{2+}, and Zn^{2+}) in *Helianthus annuus* and confirmed the applicability of EDDS for phytoextraction of heavy metals.

It is evident that chelating agents can be successfully employed in phytoextraction to improve the metal uptake from contaminated soils. Still, all the research work had been carried out at laboratory scale only. Field studies are needed to ensure the successful implication of chelate induced phytoextraction in real conditions. An effective prevention of leaching into groundwater, need for more field studies to have a better outlook of phytoextraction possibilities, and justified reasons for different observations by various researchers are the most important factors before chelating agent enhanced phytoextraction can be considered as a promising solution.

6.6.2 Chelating Agents Used for Metal Extraction from Industrial Waste

Industrial wastes are one of the most hazardous waste streams and certainly need treatment prior to their disposal in the environment. Spent catalyst is the primary solid waste generated from refinery and fertilizer industries. These catalysts lose their activity probably due to poisoning of active sites, sintering at high temperature, deposition of coke and other impurities on the surface, loss of mechanical strength, etc. as mentioned in Chapter 2. Generation of spent catalysts, in recent years, has increased significantly because of a steady increase in the processing of heavy feedstock, with high contents of sulfur, nitrogen, and metallic heteroatoms. Disposal of spent catalysts requires compliance with stringent environmental regulations. In addition, the significant fraction of valuable metals present in spent catalysts calls for new research goals to explore various new methods for metal extraction from these industrial wastes. It has been reported that recycling of metal scraps is an eco-efficient way to reduce energy consumption by 33% and pollutant generation by 60% than the production of fresh fuel (Marafi and Stanislaus 2008). This section briefly discusses various research efforts to extract heavy metals from spent catalyst using chelation process.

Metal–ligand complexation does not require high temperature for the chelation; therefore, the reaction can be carried out at ambient to low temperature range. Secondly, chelating agents can be recovered after the reaction cycle and reused in the subsequent cycle after washing, which makes it more economic than other conventional hydrometallurgical routes. In addition, use of less hazardous and less corrosive chemicals eliminates the need for specially designed equipment and material of construction. No/less hazardous by-products are liberated during the process. High efficiency of metal extraction, high thermodynamic stabilities of the metal complexes, low adsorption of the chelating agents to a catalyst, and minor impact on the physical and chemical properties of the solid matrix as compared to acids make this technology more favorable than any other technology for metal recovery.

Nickel (Ni^{2+}) was recovered from spent catalyst generated in the fertilizer industry using EDTA as a chelating agent. Nearly 96% recovery of Ni^{2+} was reported in the form of nickel sulfate ($NiSO_4$) at the molar concentration of fresh EDTA = 0.8 M, solid to liquid ratio (S/L) = 1/50, reaction time = 10 hours, particle size = 100 μm, stirring speed = 700 rpm, temperature = 100 °C, and pH 10. Dechelation experiments were performed to recover the EDTA from the

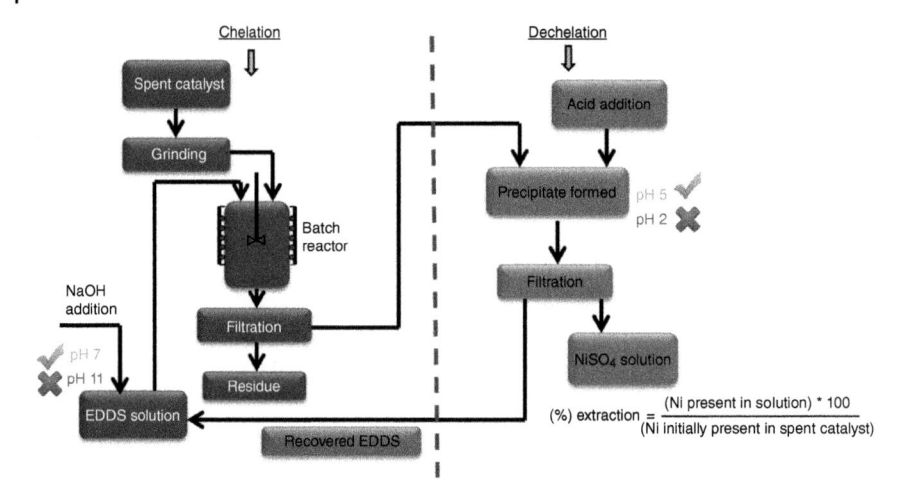

Figure 6.6 Process flow chart for extraction of nickel from spent catalyst using [S,S]-EDDS.

first cycle. Significant recovery of EDTA was reported, which could be utilized in the successive cycle with considerable extraction of Ni, although at an extraction efficiency less than that of fresh EDTA (Goel et al. 2009). This slight decrease in metal extraction could be due to the interference of competing cations such as sulfur in the recycled extracted EDTA.

Despite having high extraction efficiency, EDTA is not considered as a suitable chelating agent for metal extraction processes due to the nonbiodegradability issue associated with it. Chauhan et al. (2012) instigated the performance of a biodegradable chelating agent [S,S]-EDDS (a structural isomer of EDTA) for the extraction of metals from spent catalyst in batch mode under atmospheric reflux condition and recovered 84% nickel at optimum reaction conditions in one cycle. It was observed that the reagent [S,S]-EDDS works at a narrower pH (pH 4–7) range than EDTA (pH 3–10) for chelation–dechelation process. The complete process flow chart is shown in Figure 6.6.

Goel and Gautam (2010) studied the exchange behavior of Ni toward both competitive chelators EDTA and DTPA from the primary reforming waste catalyst used in the fertilizer industry. It was concluded that extraction of metal depends on the complexing affinity of a chelate for a metal and on the affinity between solid and metal. DTPA illustrated higher Ni^{2+} recovery than EDTA; however, the former chelating agent could not be recovered in the dechelation process. Chauhan et al. (2015c) suggested that EDTA has six binding sites, four acid sites and two amine sites, to make a ring structure with metal ions while DTPA has eight sites, three amine and five acid sites. The higher number of available binding sites provides stronger binding capacity, and thus higher extraction capacity could be attained than EDTA. However, the recovery of DTPA is relatively difficult due to the higher energy required to break the coordination bond.

Chauhan et al. (2013a) employed the combination of chelation and selective precipitation for the extraction of cobalt and molybdenum from multimetallic spent catalyst. Nearly 80% Co^{2+} and 85% Mo^{6+} were recovered in the first chelation cycle using fresh EDTA. Selective precipitation was performed based on the maximum solubility of metal oxides at different pH ranges. If Co is present in higher concentration in the solution, the possibility of coprecipitation of Co and Mo increases at acidic pH. Therefore, Co was precipitated first to avoid any coprecipitation of Co with Mo. The complete flow sheet for the proposed chelation–dechelation selective precipitation cycle is redrawn in Figure 6.7. The recovered EDTA was used in successive chelation cycles at optimum reaction conditions; however, nearly 20% loss in extraction efficiency was observed for EDTA recovered from the fifth cycle than for the fresh EDTA. The loss in extraction efficiency could be related with the loss in the number of metal-binding sites due to repetitive precipitation of EDTA and added number of impurities during each cycle. The recovered metals were reused for synthesizing new Co–Mo based catalyst. Structural analysis of the spent catalyst before and after the chelation cycles confirms the successful removal of metals from the solid surface of the catalyst.

Dzulkefly et al. (2002) used the chelating resin Amberlite IRC-718 for the extraction of nickel from spent catalyst samples obtained from hydrogenation of palm and palm kernel oils. Oil content was removed from the catalyst by soxhlet extraction using *n*-hexane. Amberlite resin showed more than 90% extraction efficiency for Ni^{+2} with purity of higher than 90% with regard to other metals Mg^{2+} and Ca^{2+} in the catalyst samples. Natural bentonite, orange peel, chitosan, and anaerobic granular sludges for heavy metal removal have also been investigated for metal extraction but these materials have very low adsorption capacities in as-received forms; therefore, chemical modification of these materials is required to improve the performance. Repo et al. (2010) investigated the adsorption properties of EDTA and DTPA modified chitosans in the aqueous solutions of Co^{2+} and Ni^{2+} ions in batch technique. The effect of pH for metal recovery proved that Ni^{2+} recovery occurred at lower pH than Co^{2+}. Results indicated that adsorption efficiency of EDTA modified chitosan was 99.2% for Co^{2+} and 99.5% for Ni^{2+} while DTPA modified chitosan removed 96.7% of Co^{2+} and 93.6% of Ni^{2+} under similar operational conditions, which could be explained on the basis of the cross-linked structure of DTPA-chitosan and lower surface coverage of DTPA compared to EDTA. In the similar manner, Wang et al. (2010a,b) applied the adsorption process to extract Pb^{2+} and Cu^{2+} in micro-polluted water source using PS-EDTA resin. It was illustrated that at optimum reaction conditions, the PS-EDTA surface adsorbs Cu^{2+} much better than Pb^{2+} and seems to be a better adsorbent for removal of both ions off the aqueous single solutions.

6.6.3 Chelating Agents Used for Metal Extraction from WEEE

Research work on the extraction of metals from WEEE is in its infancy. Very few research studies have been carried out so far in this particular area. Use of chelation technology for the recovery of metals from electronic scrap was initiated by our research group. Jadhao et al. (2016) performed metal–ligand complexation reaction using EDTA ligand to extract Cu^{2+} from waste printed circuit boards. The chelation performance was compared with the conventional acid leaching

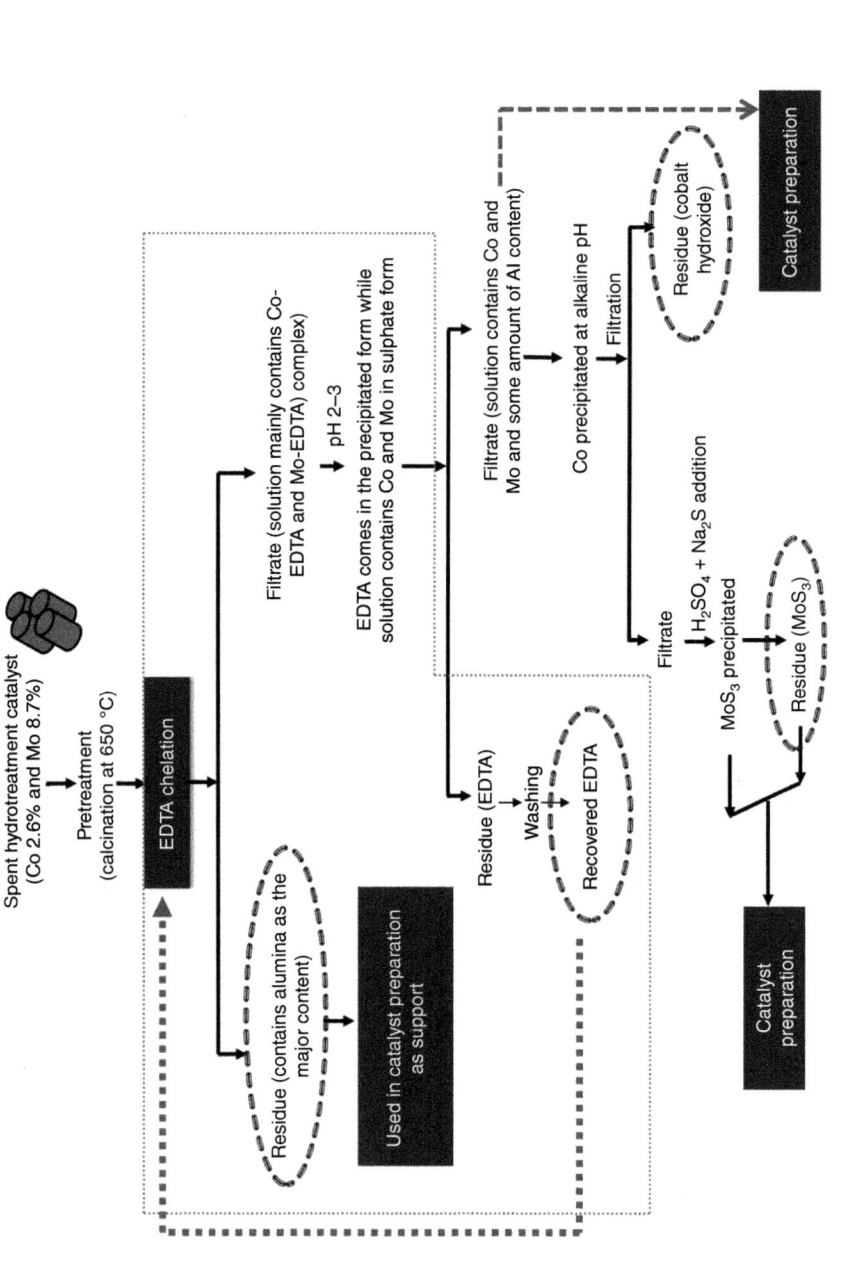

Figure 6.7 Process flow sheet for chelation–dechelation process in combination with selective precipitation of metals. Source: Chauhan et al. (2015a). Reproduced with permission of Royal Society of Chemistry.

method using H_2SO_4 in the absence of any oxidant. EDTA assisted extraction of metals resulted in 50% higher extraction than H_2SO_4 assisted leaching process. Recently, Sharma et al. (2017) used a similar chelation–dechelation concept for the extraction of Co^{2+} from mobile batteries and performed statistical optimization by response surface methodology. More than 94% Co^{2+} was extracted from mobile batteries while experiments were performed under optimum range of process parameters, i.e. molar concentration of EDTA (X1: 0.33–0.41 M); S/L ratio (X2: 1/25–1/28 g/ml), reaction time (X3: 163–178 minutes), reaction temperature (X4: 87–92 °C), and reaction pH (X5: 8.3–8.8).

Bench scale studies suggest the wide applicability of chelating agents in the extraction of metals from industrial waste, electronic waste, and contaminated soils; however, this research area can still be considered unexplored. Experimental studies are at their early stage and more research efforts are needed to progress in a sustainable green era.

6.7 Ecotoxicological Concerns and Biodegradability

Although the wide acceptability of chelating agents in the area of metallurgical processing opens a new research direction, the environmental concerns associated with these reagents cannot be ignored. Various processes such as chelate adsorption, reactive transport, photo and chemical degradation, metal mobilization, and mineral dissolution should be kept in consideration when assessing the reactions of metal–chelate complex in the environment. It is important to account for ecotoxicological concerns and biodegradability issues associated with various chelating agents to understand the environmental fate of chelation technology as a green alternative of traditional hydro- and pyro metallurgical methodologies.

Complexing agents suppress the short-term toxicity of toxic heavy metals when used either as a sodium salt or in a fully protonated form and therefore, chelation therapy is effectively employed in medical applications for many years. However, several studies are available where environmental and medical effects are contradictory. NTA, being a readily biodegradable chelating agent, can be employed in metal extraction; however, the possible carcinogenicity of NTA and kidney toxicity prohibit the use of NTA on a wide scale. Complexing agents are expected to have harmful long-term effects in concentrations at mg/l level only.

Ligands may be responsible for the enhanced mobilization of heavy metals and sometimes, these chelating agents may increase the environmental mobility of radioactive materials. Chelates may affect the eutrophication process too due to the presence of nearly 10% nitrogen in the molecule. Metal–ligand complexes enhance the bioavailability of hazardous heavy metals, although the isolated chelating agent molecule should not be considered a risk for bioaccumulation. Dissolution of metal phosphates and consequently phosphorus content in water may also be enhanced due to the presence of chelating agent.

Biodegradable chelating agents are likely to capture a significant share of the chelating agent's market in the coming years, owing to the potential health and environmental hazards associated with the use of nonbiodegradable organic chelating compounds. High resistance to biodegradability is always a desired factor for the stability of metal–chelate complex in industrial processes; therefore, EDTA and DTPA are employed in technical processes predominantly. These chelating agents cannot be eliminated from the aquatic systems by the conventional physical and biochemical methods, and therefore, the xenobiotic content in drinking water is increasing day by day. In addition, aminopolycarboxylate chelating agents can enhance the migration of heavy metals or radionuclides from soils undesirably. The biodegradation of these conventional chelating agents (EDTA, DTPA, organophosphonates) can only be performed under very specific process conditions using isolated bacterial strains (Nortemann 2005); still, these compounds do not satisfy the criteria for biodegradability (OECD 1992).

Sykora et al. (2001) suggested that biodegradability of a complexing agent depends on the functional group available on the ligand and the number of nitrogen atoms present in the ligands. NTA with a single nitrogen atom in its molecule is a readily biodegradable chelating agent. Resistance to biodegradability also increases with the number of tertiary amino groups (EDTA, DTPA). Optical isomerization is another important factor for biodegradability, e.g. $[S,S]$-EDDS stereoisomer is subjected to easy degradation whereas the $[R,R]$ isomer remains undegraded and the $[R,S]$ isomer degrades very slowly and incompletely (Chauhan et al. 2015a). OECD guidelines for testing of chemicals described internationally used standard test methods in the year 1993 for investigating biodegradability of chemicals (Painter et al. 1993). A brief description of biodegradability tests has been adapted from Chauhan et al. (2015a) and listed in Table 6.6.

Gerike and Fischer (1979) confirmed the nonbiodegradability of EDTA by performing several biodegradation tests (coupled unit test, Zahn–Wellens test, MITI test, AFNOR test, Sturm test, OECD screening test, and closed bottle test) on EDTA. The European Union risk assessment report also confirms that EDTA is not readily biodegradable. Biotechnologically produced EDDS is a structural isomer of EDTA and is drawing the attention of researchers these days due to its biodegradable properties. MGDA showed good stability in a wide pH range and satisfied the OECD (1992) criteria. Borowiec et al. (2009) illustrated GLDA as a chelator that is biodegradable and nontoxic for the ecosystem. EDDG and EDDM were also recognized as biodegradable chelators. PDA also showed the efficiency to chelate metals in a ligand to metal ratio of 2 : 1.

There are a few processes available to degrade these chelating agents that are resistant to biodegradation. Advanced oxidation processes (AOP) apply combinations of radiation, oxidants, and catalysts for degrading the target compounds. The iron and copper complexes of NTA, EDTA, DTPA, and EDDS can be easily degraded using UV radiation. Another major approach for the degradation of complexing agents is focused on pure bacterial cultures capable of biodegradation of organic pollutants.

Aerobic Gram-negative bacterial strains DSM 9103 and DSM 6780 (BNC1) were mentioned as the most promising ones in the degradation of EDTA; however, no bacterial strain has still been found in the literature to be able to degrade

6.7 Ecotoxicological Concerns and Biodegradability

Table 6.6 Standard test methods for investigating biodegradability of chemicals (Chauhan et al. 2015a)

Major categories	Sub-categories	Basis of testing	Test duration
Ultimate biodegradability screening tests	Modified Sturm test/OECD 301B	By measuring the analytical parameters of mineralization (CO_2 evolution or O_2 consumption)	4 wk (batch tests)
	Modified MITI test/OECD 301C		
	Closed bottle test/OECD 301D		
	Manometric respirometry test/OECD 301F		
	BODIS test (BOD test for Insoluble substances, ISO 10708)		
	DOC die-away test/OECD 301A	By determining the extent of ultimate biodegradation by DOC removal	
	Modified OECD screening test/OECD 301E		
Inherent biodegradability screening tests	Modified Zahn–Wellens test/OECD 302B	By measuring carbon removal, high sludge inoculums are used to obtain high bacterial concentration	28 d
Continuous activated sludge (CAS) test	Coupled unit test/OECD 303A	By carbon removal (DOC) measurement using activated sludge	3 h, continuous mode of experiment

DTPA. It was also illustrated that slow metal exchange kinetics plays a significant role in the biodegradation of EDTA because of the strict speciation requirements. Biological degradation of complexing agents by finding their degradation products and degradation pathway is increasing its scope nowadays to investigate the degradability aspects of the chelating agent. The degradation pathway can also be estimated by studying the bond dissociation energy associated with the cleavage of C—N bond using computer simulations. Many of the primary and secondary degradation products of EDTA and DTPA may form soluble complexes with metal ions, such as ED3A, EDDA, and IDA, whereas some of the degradation products are expected to be even more recalcitrant to degradation than EDTA or DTPA themselves, such as KPDA, KPMA, and ketopiperazine. Various intermediate products of EDTA degradation obtained by different chemical and biological methods have also been listed in Table 6.7.

Chauhan et al. (2015c) proposed the biodegradation pathway for $HEDDS^{3-}$ and $HNTA^{2-}$. The proposed degradation pathway and the bond dissociation energy of all investigated C—N bonds are shown in Figure 6.8. It can be seen in Figure 6.8 that NTA can be split into IDA and glyoxalate from which IDA can be further

Table 6.7 Various intermediate products of EDTA degradation.

Mode of degradation	Intermediates	References
Bacterial degradation (mixed culture of BNC1–BNC2)	ED3A, N,N'-EDDA, ED, IDA,	Yuan and VanBriesen (2008)
Photocatalytic degradation over TiO_2	(1) ED3A, U-EDDA, S-EDDA, EDMA, EDA, glyoxylic acid, oxalic acid, formic acid (2) IMDA, glycine, oxamic acid, glyoxylic acid, oxalic acid, formic acid (3) Glyoxylic acid, oxalic acid, formic acid	Babay et al. (2001)
Bacterial degradation (whole cells and cell-free extracts of strain BNC1)	ED3A, N,N'-EDDA ED3A, IAA, IDA	Kluner et al. (1998)
EDTA-degrading bacterium BNC1	ED3A, EDDA, EDMA, ED	Liu et al. (2001)
Cellular degradation	(1) ED3A, (N,N-EDDA and N,N'-EDDA), EDMA, ED, Tricarboxylic acid cycle (2) ED3A, IAA, IDA, glycine, tricarboxylic acid cycle	Nörtemann (1999)
EDTA-degrading bacterium BNC1	ED3A and glyoxylates using EDTA monooxygenase	Payne et al. (1998)

EDDS degradation pathway

$E_{BDE, 1} = 46.35$
$E_{BDE, 2} = 49.86$
$E_{BDE, 3} = 50.37$
$E_{BDE, 4} = 49.68$

Ethylenediamine disuccinic acid

$E_{BDE, 1} = 30.16$
$E_{BDE, 2} = 42.62$
$E_{BDE, 3} = 38.24$

N-(2-Aminoethyl)aspartic acid

Ethylenediamine

NTA degradation pathway

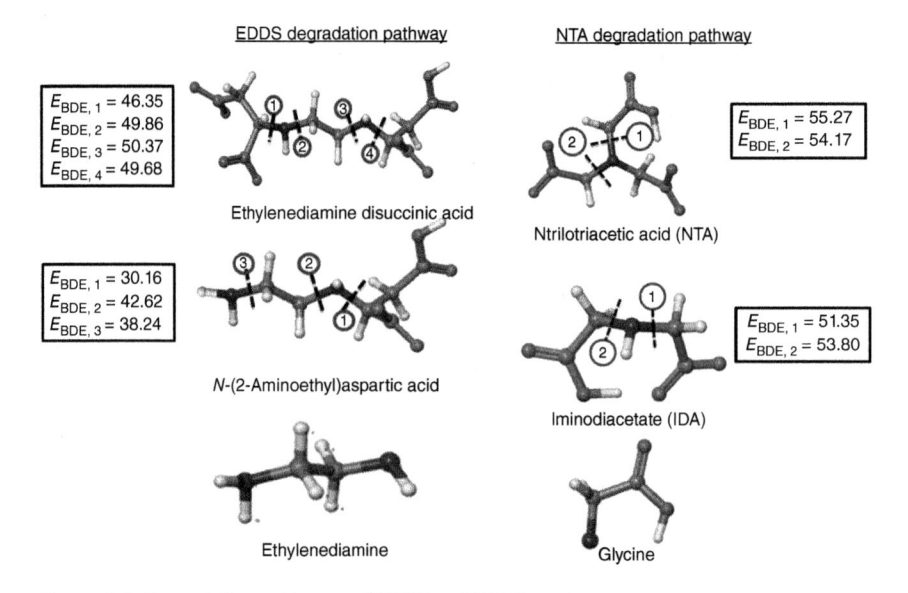

$E_{BDE, 1} = 55.27$
$E_{BDE, 2} = 54.17$

Ntrilotriacetic acid (NTA)

$E_{BDE, 1} = 51.35$
$E_{BDE, 2} = 53.80$

Iminodiacetate (IDA)

Glycine

Figure 6.8 Degradation pathways of EDDS and NTA ligands.

degraded to yield glycine and glyoxalate as the final products. The preferential cleavage sites of the ligand species are determined on the basis of lower bond dissociation energy. The theoretical results were found to be in agreement with the literature (Uetz and Egli 1992) in which aerobic *Chelatobacter heintzii* ATCC 29600 enzyme complex cleaved the C—N bond and isolated IDA and glyoxylate as the products of microbial degradation of NTA (Chauhan et al. 2015c).

It can be said that stability of a metal–ligand complex is interlinked with the degradability of the ligands. Methods are available to determine the biodegradability of the reagents; nevertheless, there are very few studies to degrade the resistant reagents. Theoretically, the biodegradation pathways can be determined, although the approach still needs more research efforts to find practical solutions to the biodegradability issues.

6.8 Summary and Outlook

Chelation technology for extraction of heavy metals from contaminated sites (soil, water, industrial waste) is drawing sincere concern these days in order to develop new promising green methods, which should be efficient on the ecological and economical platform. Various R&D studies are currently underway to identify process parameters that may affect the formation and stability of the metal–ligand complex. Higher extraction efficiency, reusability of chelating agents, and diverse applicability make it a convincing technology; however, still a number of challenges need to be addressed. Synthesis of new biodegradable mobilizing agents and identification of their degradation pathways by means of molecular simulation or biological methods are open research topics with immense opportunities for further development in the field of chelation technology. Different possible methods to recover and recycle the chelating agents should also be explored to bring about an efficient "closed loop" chelation process. Successful application of the chelation technology for metal extraction from contaminated sites is a very elegant example of ligand substitution mechanism, although the industrial application of the process is still restricted by the lack of adequate knowledge about competing reactions, which may affect the metal–ligand complexation. More experimental studies will be helpful to have a better perceptivity of the ligand substitution mechanism and to explore the diverse applicability of chelation technology. It is conceived that modern computational tools may suggest an ongoing renaissance of research activities in chelation technology to carve out the new territory of metal–ligand complexation. Thus, future attempts should focus on sustainability, economics, and environmental impact of the process to fulfill the growing industrial demand.

Questions

1 Discuss the "metal–ligand complexation" process.

2 Classify the ligands based on their denticity, structure class, and HSAB theory.

3   List the major factors to define the stability of metal–ligand complex.

4   What are the major process parameters that can affect the chelating agent assisted metal extraction process? Discuss any two.

5   "Stability of the complex and degradability of the ligands are interlinked". Explain.

6   Discuss the applicability of chelation technology for the recovery of metals from spent catalyst generated from a fertilizer industry.

7   Discuss the major advantages and limitations associated with the chelation process.

8   "EDDS, a structural isomer of EDTA, is a biodegradable chelating agent; but EDTA is not degradable at all." Explain.

9   Write a short note on the applicability of chelating agent in the phytoextraction process.

10 Give an example of chelating agent assisted extraction of metals from WEEE.

11 "Chelation technology is a future technology for waste management in a sustainable way." Do you agree with the statement? If Yes/No, discuss why.

7

Future Technology for Metal Extraction from Waste: II. Ionic Liquids

Projects we have completed demonstrate what we know – future projects decide what we will learn.

Dr. Mohsin Tiwana

Abbreviation

[3-BuPyr][NTf$_2$]	3-butylpyridinium bis(trifluoromethanesulfonyl)imide
[A336][DEHP]	tri-n-octylmethylammonium di(2-ethylhexyl)phosphate
[BMIM]BF$_4$	1-butyl-3-methyl-imidazolium tetrafluoroborate
[BMIM]HSO$_4$	1-butyl-3-methyl-imidazoliumhydrogen sulfate
BMIM	1-n-butyl-3-methyl imidazolium
[Cmim][CF$_3$SO$_3$]	1-alkyl-3-methylimidazolium trifluoromethylsulfonate
DESs	deep eutectic solvents
DODGAA	N,N-dioctyldiglycol amic acid
EMIM	1-ethyl-3-methylimidazolium
EPR	electro-paramagnetic resonance
[Hmim]HSO$_4$	1-methylimidazolium hydrogen sulfate
ILs	ionic liquids
[mimSBu][NTf$_2$]	2-butylthiolonium bis(trifluoromethanesulfonyl)imide
MMIM	1,3-dimethylimidazolium
NATO	the North Atlantic Treaty Organization
[P$_{666,14}$][Cl$_3$]	trihexyl(tetradecyl)phosphonium trichloride
PILs	protic ionic liquids
RTILs	room temperature ionic liquids
SPS	"switch" polarizable solvents
SILMs	supported ionic liquid membranes
[TBA][DEHP]	tetrabutylammonium di(2-ethylhexyl)phosphate
THF	tetrahydrofuran
[THTP][DEHP]	trihexyl(tetradecyl)phosphonium
TSILs	task specific ionic liquids
TODGA	N,N,N',N'-tetra(n-octyl)diglycolamide

Sustainable Metal Extraction from Waste Streams, First Edition.
Garima Chauhan, Perminder Jit Kaur, K.K. Pant, and K.D.P. Nigam.
© 2020 Wiley-VCH Verlag GmbH & Co. KGaA. Published 2020 by Wiley-VCH Verlag GmbH & Co. KGaA.

7.1 Introduction

"Ionic liquids (ILs) might be 'neoteric solutions' which would change the face of organic chemistry": the given concept was popularized by Kenneth R. Seddon (Holbrey and Welton 2018) nearly two decades ago by performing an intense research in the field of these "not intrinsically green" solutions and creating a broader research landscape. Research activity started to flourish in this area after the first international meeting "the North Atlantic Treaty Organization (NATO) advanced research workshop" on ionic liquids, which was held in Heraklion, Crete, in April 2000. Today, over 75 000 research articles and 12 000 patents have been published involving ionic liquids, and thus it is evident that ILs have the potential to revolutionize all activities where molecular solvents can be used (Shiflett and Scurto 2017).

While talking about ILs, "diversity" is the key word. Diversity of anion–cation combinations, diversity of modes of preparation, modes of purification, and nature of impurities (quality), diversity of properties, diversity of mode of use, and diversity of applications make it difficult to generalize the properties or applications of ILs. This chapter covers the different aspects of ILs, from the fundamental properties of ILs to a diverse range of their applications. The evolution and milestones achieved in the development of ILs and the synthesis and purification methods available for the design of new generations of ILs are discussed in detail. Ionic liquids are offered as a potentially "green" alternative to volatile organic compounds (VOCs) in spite of being "non–intrinsically green" solutions. In this regard, it is important to investigate the effect of this class of solvents on the environment to confidently label them as "green." Therefore, environmental fate of ionic liquids has been discussed and, in the end, applicability of ionic liquids in diverse industrial applications is summarized.

7.2 What Are Ionic Liquids?

In the simplest way, ionic liquids (ILs) are defined as solvents composed of only ions, with "no molecular species present" in them. This unique property differentiates these solvents from molecular solvents (only neutral molecules) and salt solutions (ions + molecules). Several other terminologies such as fused salts, nonaqueous ionic liquids, neoteric solvents, etc. are also used in the literature to describe ILs. Ions are believed to be poorly coordinated in ILs, which results in the low melting point of these solvents. Most of the ILs are liquids at temperatures below 100 °C, or even at room temperature (room temperature ionic liquids, RTILs). However, it should not be considered as the strict constraint to define ionic liquids. As Welton (2018) said, "Also, there is no scientific justification for believing that a salt with a melting point of 90 °C is in any necessary way different from one with 110 °C," considering ionic liquids as salts melting below 100 °C is always not recommended, because the choice of

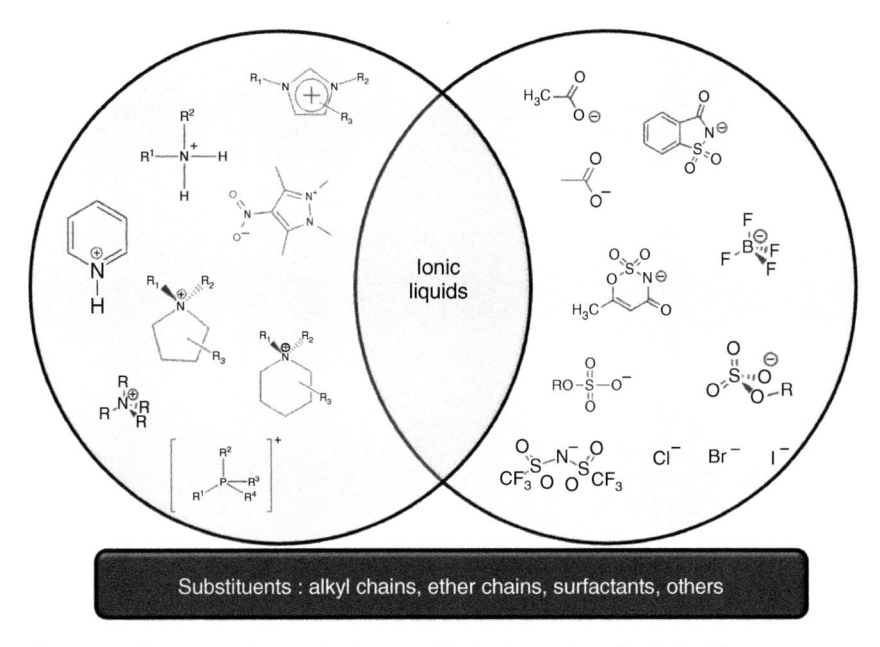

Figure 7.1 Common cations and anions used in the formation of ionic liquids.

this upper temperature is very arbitrary and salts melting at higher temperatures can still have the typical properties of ionic liquids.

ILs being solely composed of ions, the interaction in these solvents arises mainly from the coulomb forces. Other forces such as hydrogen bonds, dispersion forces, and Π–Π interactions depend upon the structures and chemistries of the constituent ions, and hence may lead to strong difference in IL behaviors. In order to form a hydrogen bond, the solvent must essentially have a strong hydrogen bond donor (a cation) and a suitably strong hydrogen bond acceptor (an anion). The widely used cations are ammonium, sulfonium, phosphonium, imidazolium, pyridinium, etc. with different alkyl/aromatic substituents. $[BF_4]^-$, $[PF_6]^-$, $[(CF_3SO_2)_2N]^-$, $[NTF_2]^-$, etc. are listed into frequently employed anion in the ILs structures. Figure 7.1 gives a general illustration of ionic liquids with common cations and anions.

Several research studies are being carried out to fine-tune the physical and chemical properties of both the cations and anions by varying the composition of ILs mixture (Villar-Garcia et al. 2014; Serra Morenoa et al. 2015; Ribot et al. 2012). Based on this unique feature of having tuneable properties with regard to polarity, solvent miscibility, and hydrophobicity through the appropriate modification of cation and anion, these solvents are regarded as "Designer Solvents." There are estimated to be hundreds of thousands of simple ion combinations to make ionic liquids and a near endless (10^{18}) number of potential ionic liquid mixtures. In addition, this ability to design specific solvents offers a wide range of applications

Table 7.1 Evolution of ionic liquids as a novel liquid: milestones.

Milestone	Ionic liquid	References
First ionic liquid	Ethylammonium nitrate	Gabriel and Weiner (1888)
First room temperature IL	Ethylammonium nitrate	Walden (1914)
Fist IL with bromochloro aluminate species in the melt	1-Ethylpyridinium bromide-aluminum chloride	Hurley and Wier (1951)
The term "ionic liquid" was coined		Bloom (1961)
Quantitative determination of ionization strength of ionic medium	Tetra-n-hexylammonium benzoate	Swain et al. (1967)
First series of RTILs with chloroaluminate ions	1-Butylpyridinium chloride-aluminum chloride ([C_4py]-$AlCl_3$)	Chum et al. (1975) and Robinson and Osteryoung (1979) (Osteryoung's Research Group)
Thermodynamics and "hydrogen bonding" behavior in controlling IL's solvent properties	Ethylammonium nitrate	Evans et al. (1981)
Introduction of 1,3-dialkylimidazolium cations into ILs	1-Alkyl-3-methyl-imidazolium chloride aluminum chloride ionic liquids ([C_nC1im]Cl-$AlCl_3$, where $n = 1$–4	Wilkes et al. (1982)
ILs as stationary phase in gas chromatography	Ethylammonium nitrate	Pacholec et al. (1982)
Electrochemical aspects of relevant transition metal complex	Chloroaluminate as nonaqueous polar solvent	Scheffler et al. (1983)
ILs as nonaqueous solvent for biochemical systems	Ethylammonium nitrate	Magnuson et al. (1984)
Spectroscopic and complex chemistry experiments	Chloroaluminate as nonaqueous polar solvent	Appleby et al. (1986)
New reaction media for Friedel–Crafts catalysts	Acidic ILs with chloroaluminate ions	Boon et al. (1986)
ILs in nucleophilic aromatic substitution reaction	Phosphonium halide ions	Fry and Pienta (1985)
Formation of layer of ions at charged surface	Ethylammonium nitrate	Horn et al. (1988)
Solvents for homogeneous transition metal catalysts (Ni catalysts for dimerization of propene)	Weakly acidic chloroaluminate ions	Chauvin et al. (1990)

(continued)

Table 7.1 (Continued)

Milestone	Ionic liquid	References
Ethylene polymerization with Ziegler–Natta catalysts	Weakly acidic chloroaluminate ions	Carlin and Osteryoung (1990)
First RTIL stable in water and air	Tetrafluoroborate ions	Wilkes and Zaworotko (1992)
Introduction of the $[NTf_2]^-$ ion	1-Ethyl-3-methylimidazolium bis((trifluoromethyl)sulfonyl)amide	Bonhote et al. (1996)
Ionic liquid for catalysis	Polymer gel composed of 1-n-butyl-3-methylimidazolium hexafluorophosphate and poly(vinylidene fluoride)–hexafluoropropylene copolymer	Carlin and Fuller (1997)
Task specific ionic liquids		Wierzbicki and Davis (2000)
Dissolution of cellulose in ILs	1-Butyl-3-methylimidazolium chloride	Swatloski (2002)

in the field of ionic liquids. Dispersion forces also vary with the identity of the constituent ions and contribute to the formation of IL structures.

It is evident from the wider applicability of the "designer solvents" and increasing number of publications in the field of these "neoteric solvents" that ILs have been one of the major research areas in modern chemistry in the past few decades; however, the first IL was reported more than a century ago when Gabriel and Weiner (1888) observed the potential benefits of a solvent ethyl ammonium nitrate that has a melting point of 52 °C. The work was first noticed when Paul Walden introduced the same solvent as an RTIL with melting point of 12 °C. It would be worthwhile to know how far these research efforts have brought ILs to illustrate these unique solvents as "solvents for sustainable development" in a variety of applications. Table 7.1 lists the milestones in the development of ionic liquids as new green media.

7.3 Characteristic Properties of Ionic Liquids

Owing to the wide range of ILs and their applications, it is not possible to make generalization about the properties of ILs. The physical and chemical properties of ILs can be dramatically affected by the presence of any impurities such as water and organic solvents during the synthesis or purification. The low melting point and >99% ionicity are widely generic properties, whereas some of the previously identified properties such as thermal stability, negligible volatility, "green solvent," non-coordinating, large electrochemical window, and nontoxicity are now subject to controversy due to advancements in experimental, characterization, and quantification techniques. This section illustrates various generic (though not universal) physical and chemical properties of ionic liquids including the recent literature available on IL chemistry where the "well-known" IL properties sometimes have been deemed "not accurate."

7.3.1 Melting Point

Melting point is considered as the key feature of an ionic liquid to differentiate it from molten salts. While a molten salt is generally referred to as a high melting point, highly viscous, and very corrosive medium, ILs are liquids at low temperature and have relatively low viscosities. The melting point of ILs is greatly influenced by the presence of cations and anions and therefore this characteristic property can be directly correlated with the structure and composition of ILs. Ionic liquids with highly symmetric cations exhibit high melting points than the one with low symmetric cationic structure. Low symmetry (Seddon 1997), weak intermolecular interactions (Elaiwai et al. 1995), and a good distribution of charge (Stegemann et al. 1992) in the cation are the key features of low melting point cations. It was illustrated in the literature (CRC 1992) that the chloride salts with alkali metal cations showed very high melting points (NaCl: 803 °C; KCl: 772 °C) whereas chlorides with organic cations showed significant difference in the melting point (<150 °C) of the salt. Figure 7.1. demonstrates the widely studied cations for the development of novel ILs.

Anions can also significantly affect the melting point of ionic liquids. The presence of different anions on imidazolium salts suggested that an increasing size of the anion leads to a further decrease in the melting point (Wasserscheid and Keim 2000). Some of the anions dominantly cover the ILs range: Cl^-, Br^-, I^-, $Al_2Cl_7^-$, $Al_3Cl_{10}^-$, $Sb_2F_{11}^-$, Fe_2Cl^-, $Zn_2Cl_5^-$, $Zn_3Cl_7^-$, $CuCl_2^-$, $SnCl_2^-$, NO_3^-, PO_4^{3-}, HSO_4^-, SO_4^{2-}. Table 7.2 demonstrates the difference in melting point by varying the cations and anions in the salt. Many ILs can undergo supercooling due to the potential presence of impurities, which eventually may affect the melting point of the salt. Also, molar ratio of two reactants may influence the melting points of a quasi-binary system; thus it is possible to design a specific IL with low melting point for a particular application by combining two species in a certain fraction.

7.3.2 Vapor Pressure and Nonflammability

Since ILs are completely made up of ions, the negligible vapor pressure in the liquid state was considered their characteristic feature, which makes them distinguishable from molecular solvents. Based on this consideration, ILs are believed to be nonvolatile and thus cannot be distilled at all. In addition, the vanishingly small vapor pressure has been proved an incentive to the process engineers since the problem of azeotrope formation at the time of distillation does not arise with ILs and thus, product separation by distillation of a reaction mixture becomes very effective. Considering ILs as nonvolatile and nonflammable at ambient or higher temperatures garnered interest for ILs on their possible use as "green"

Table 7.2 Influence of different anions and cations on the melting point of salts.

Salt	Melting point (°C)	References	Salt	Melting point (°C)	References
NaCl	803	CRC (1992)	[EMIM]NO_2	55	Wilkes and Zaworotko (1992)
KCl	772	CRC (1992)	[EMIM]NO_3	38	Wilkes and Zaworotko (1992)
*[MMIM]Cl	125		[EMIM]$AlCl_4$	7	Fannin et al. (1984)
*[EMIM]Cl	87	Wilkes et al. (1982)	[EMIM]BF_4	6	Holbrey and Seddon (1999)
*[BMIM]Cl	65	—	[EMIM]CF_3SO_3	−9	Bonhote et al. (1996)
			[EMIM]CF_3CO_2	−14	Bonhote et al. (1996)

Source: Wasserscheid and Keim (2000). Reproduced with permission of John Wiley & Sons.

alternatives to volatile organic solvents. Instead of being inflammable, ILs are combustible, and therefore can be fine-tuned for energetic content (Schneider et al. 2008) and may facilitate multiphasic applications (Feng et al. 2010).

However, this widely accepted property of ILs has now been nullified in recent research articles. Recent studies (Earle et al. 2006) suggest that ILs can be distilled at 200–300 °C at low pressure, even at very low distillation rate (0.01 g/h). MacFarlane et al. (2006) demonstrated the distillation of protic ionic liquids of trifluoroacetate anion under normal pressure and temperature using the standard method. Some of the researchers even developed a correlation between distillation temperature and volatility for imidazolium cations based ILs series (Paulechka et al. 2005; Zaitsau et al. 2006). It is evident from the recent studies that vapor pressure of ILs is measurable and thus the characteristic property of "non-volatility" or "negligible vapor pressure" should not be considered a valid feature for ILs.

7.3.3 Thermal Stability

Thermal stability of ILs depends on the strength of their heteroatom-carbon and heteroatom-hydrogen bonds (Wasserscheid and Keim 2000). Fast thermogravimetric analysis (TGA) suggests high thermal stability for many ILs, generally >350 °C. The stable generation ILs represented by 1,3-dialkyl imidazolium salts with [BF_4], [PF_6] anions are thermally stable compounds that will not decompose below 300 °C. [EMIM]tetrafluoroborate has been reported to be stable up to about 300 °C (Mutch and Wilkes 1998) and [EMIM](CF_3SO_2)$_2$N is stable up to even more than 400 °C (Bonhote et al. 1996). ILs based upon tetraarylphosphonium cations exhibit superior long-term thermal stability compared to familiar salts based upon imidazolium, quaternary ammonium, or tetraalkylphosphonium cations (Cassity et al. 2013). Villanueva et al. (2013) demonstrated a slight decrease in thermal stability with an increase in the length of cations. Also, ILs based on nitrogen cations cannot be decomposed completely due to the presence of cyano group and generate char residue (Bhatt et al. 2006).

However, several studies have been published that question the thermal stability of ILs. Imidazolium-based ILs are not stable under basic conditions. Phosphonium ILs are more inert against strong bases, so that reactions using strong bases, such as Grignard reagents, can be carried out among phosphonium ILs. The results, however, are not better than molecular solvents such as tetrahydrofuran (THF) (Ramnial et al. 2005). Increase in hydrophilicity of anions leads to a decrease in thermal decomposition of the melts. Phosphonium ILs with $[NTf_2]^-$ or $[N(CN)_2]^-$ anions decompose completely to volatile products in a single step by dealkylation reactions (Olivier-Bourbigou et al. 2010). Many melts with trialkylammonium ions already decompose at a temperature below 80 °C in vacuum.

7.3.4 Density

ILs, in general, possess higher density than water and organic solvents. The magnitude of density depends upon the structural features of ILs (cations and anion). The typical density range for ILs lies between 1 and 1.6 g/cm³. Although slight structural changes do not affect the density significantly, yet the choice of anions and cations significantly influences the density range of comparable ILs. With the increase in the bulkiness of the organic cation, density of the ILs decreases linearly as shown in Figure 7.2. Here, a linear decrease in the density was observed with increase in the length of N-alkyl chain on the imidazolium cation in chloroaluminate melts (Fannin et al. 1984). Similarly, the dependence of the density of an ionic liquid on the constituent anions can be illustrated by chloroaluminate and bromoaluminate salts as shown in Figure 7.2. A significant difference in the density was observed between both melts when the concentration of aluminum trihalide was increased between mole fractions of 0.3 and 0.7 (Sanders et al. 1986).

7.3.5 Viscosity

Viscosity of ILs can affect transport properties such as diffusion and plays a major role in stirring, mixing, and pumping operations. The viscosity of ILs can be determined based on hydrogen bonding and van der Waals interactions (Bonhote et al. 1996). The effect of hydrogen bonding of ILs on their viscosity is widely illustrated in the literature by X-ray spectroscopy and NMR studies, which suggest that the increase in viscosity of more than a factor of 10 with less than 0.5 of the mole fraction of aluminum trichloride is due to the formation of hydrogen bonds between the hydrogen atoms of imidazolium cation and the chloride ion.

The effect of van der Waals interactions on viscosity is illustrated by comparing the viscosity of different hydrophobic ILs with BMIM ions as shown in Table 7.3. It is clear from Table 7.3 that stronger van der Waals interactions (for $C_4F_9SO_3^-$ and $C_3F_7COO^-$) result in a higher viscosity of the ionic liquid. Lower viscosity

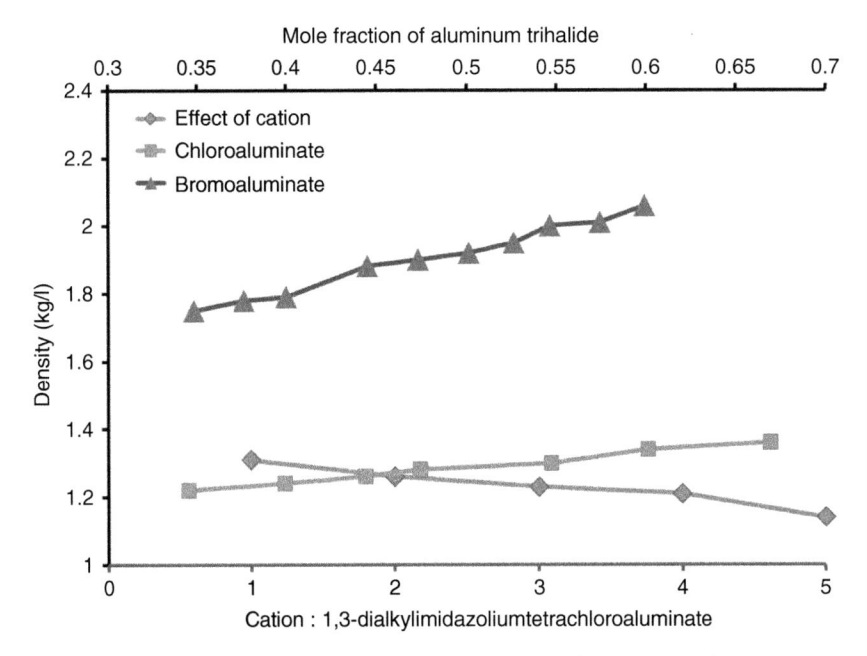

Figure 7.2 Effect of various cations and anions on density of ILs. Horizontal axis – cations group (1: R = Me, R′ = Me, 2: R = Me, R′ = Et, 3: R = Me, R′ = Pr, 4: R = Me, R′ = Bu, 5: R = Bu, R′ = Bu). Source: Adapted from Fannin et al. (1984) and Sanders et al. (1986).

Table 7.3 Dynamic viscosities of 1-n-butyl-3-methyl imidazolium (BMIM) salts at 20 °C.

Cation	Anion	Viscosity (cP)
[BMIM]⁺ [A]⁻	CF_3SO_3	90
	$n\text{-}C_4F_9SO_3$	373
	CF_3COO	73
	$n\text{-}C_3F_7COO$	182
	$(CF_3SO_2)_2N$	52

Source: Wasserscheid and Keim (2000). Reproduced with permission of John Wiley & Sons.

for $CF_3SO_3^-$ and $(CF_3SO_2)_2N^-$ despite strong van der Waals interaction was also observed due to the complete suppression of hydrogen bonding. The replacement of alkyl groups by oligoether groups has been shown to decrease the ionic liquid's viscosity significantly. This effect has been demonstrated for both substituents at the anion (such as sulfate) (Himmler et al. 2006). Novel ILs with Si substituted cations were also demonstrated with a reduction in viscosity due to more flexible side chains than an ether (Chung et al. 2007). The viscosity of many ILs is relatively high compared to conventional solvents, 1–3 orders of magnitude higher. For a variety of ILs it has been reported to range from 66 to 1110 cP at 20–25 °C, which may certainly affect transport characteristics such as diffusion and mixing.

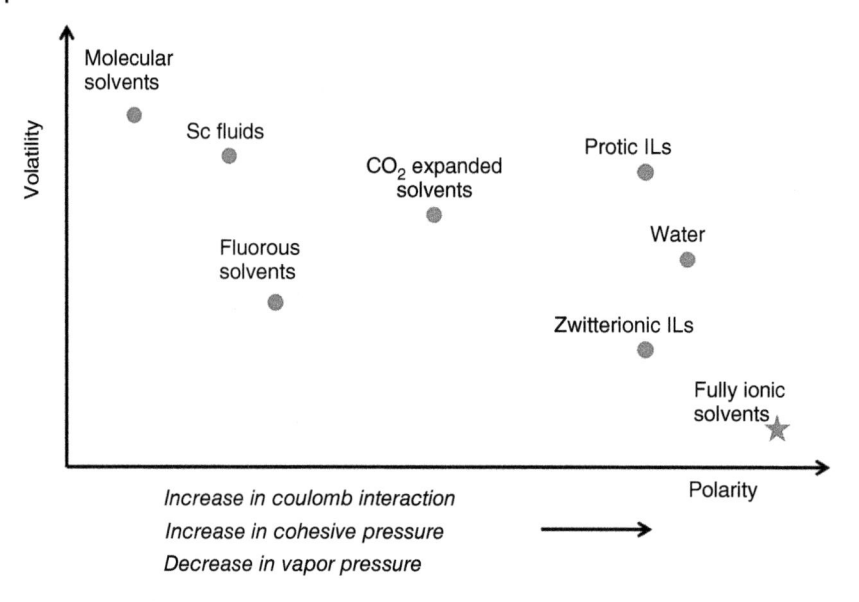

Figure 7.3 Polarity and volatility characteristics of alternative solvents.

7.3.6 Polarity

Ionic liquids are considered highly polar due to their >99% ionicity. It is a physical property of any compound that is related to other properties such as melting point, intermolecular interactions, vapor pressure, and solubility characteristics. Figure 7.3 illustrates the polarity and volatility characteristics of alternative solvents. Solvatochromic dye method was first used in the year 2000 to investigate the polarity of ILs assuming that imidazolium based ILs possess a polarity similar to lower alcohols (Carmichael and Seddon 2000). Various methods such as fluorescent dye method (Mandal and Samnta 2005), electro-paramagnetic resonance (EPR) spectroscopy (Kawai et al. 2004), 2-nitrocyclohexanone tautomerism method (Angelini et al. 2005), microwave spectroscopy (Wakai et al. 2005), and FT-IR spectroscopy (Köddermann et al. 2006) were subsequently developed to determine the polarity of ILs. Most of the research studies suggested a low polarity of ILs than expected before, yet a definitive standard has not been established for ILs. Very few studies have been reported concerning the polarity of ILs; therefore, more research work is needed in this direction.

7.3.7 Coordination Ability

Coordinating ability is a tendency of donating electron(s) in order to form a chemical bond. The cations of ILs, in general, exhibit minor coordinating ability; thus coordination ability of ILs is strongly affected by the nature of their anions. ILs have been perceived as "non-coordinating" liquids for a long time since the negative charge of the anion is believed to be highly delocalized over electron-negative atoms. Delocalization of the negative charge within the anionic core structure accounts for the weak columbic attraction between the anion and cations (Feng et al. 2010). Many intermediate levels between "strongly basic/strongly coordinating" and "strongly acidic/practically non-coordinating" are based on the choice and ratio of an anion in an IL. Acidity of chloroaluminate melts, for example, are designated as basic when the molar ratio of $AlCl_3$ is smaller than 0.5, whereas when $AlCl_3$ ratio is exactly 0.5 (i.e. only $AlCl_4^-$ is present), it is considered a neutral melt. When $AlCl_3$ ratio is large than 0.5 such as the presence of $Al_2Cl_7^-$ and $Al_3Cl_{10}^-$, these chloroaluminate melts behave like very strong Lewis acid.

Some research studies suggest ILs as "weakly coordinating" liquids (Wasserscheid et al. 2001); nevertheless, some of the studies clearly demonstrated the coordinating ability of ILs (Babai and Mudering 2006; Williams et al. 2005). The halide anions (Cl^-/Br^-) display strong coordinating abilities (Elaiwai et al. 1995); however, little information is available for other anions, such as $[BF_4]^-$, $[SO_3CF_3]^-$. From the fact that these ILs can donate electrons to H atoms, it is reasonable to imply that they can likewise coordinate to Lewis acidic metals and exhibit a coordinating property. In addition, ILs with various coordinating functionalities can be designed to coordinate by incorporating coordination groups.

7.3.8 Conductivity

Conductivity of ILs is an interesting property to consider as ILs can play the role of both solvents and electrolytes in electrochemical reactions. ILs exhibit a broad range of conductivities spanning from 0.1 to 20 mS/cm. In general, higher conductivities are found for imidazolium-based ILs in comparison with the ammonium ones. Many factors can affect their conductivity, such as viscosity, density, ion size, anionic charge delocalization, aggregations, and ionic motions. Concerning their electrochemical window, it is typically found in the range 4.5–5 V, which is similar to or slightly larger than that found in conventional organic solvents, but larger than that of aqueous electrolytes (water: 1.23 V) (Olivier-Bourbigou et al. 2010). The challenge here is still to design ILs with a wide electrochemical window along with good electrical conductivity.

7.3.9 Solubility

Solubility of ILs has been considered an appealing research aspect in order to develop extractive separation processes. ILs that have miscibility gap with water are preferentially investigated for liquid–liquid extraction process. Solubility of different acids and bases in water and $[BMIM]PF_6$ at different pH values of the aqueous phase indicates that neutral substrates tend to dissolve in ILs whereas ionic species showed higher solubility in aqueous layer. Some ionic liquids with specific cation/anion combinations demonstrate high miscibility with organic solvents on the basis of their dielectric constants. Still, systematic investigation is needed to understand the solubility characteristics of ILs. Table 7.4 lists some of the physical properties of various ILs.

Table 7.4 Physical properties of various ILs.

Salt	Cation	Molecular Weight	Melting point	Density	Refractive index	Viscosity
Bis(trifluoromethylsulfonyl)imide						
MMIM NTfO$_2$	1,3-Dimethyl imidazolium	377	22	1.559	1.422	44
EMIM NTfO$_2$	1-Ethyl-3-methyl imidazolium	391	−17	1.52	1.4231	18
EEIM NTfO$_2$	1,3-Diethyl imidazolium	406	14	1.452	1.426	35
EM2MIM NTfO$_2$	1-Ethyl-2,3-dimethyl imidazolium	406	20	1.495	1.4306	88
EMM5IM NTfO$_2$	1-Ethyl-3,5-dimethyl imidazolium	406	−39	1.47	1.4275	37
BMIM NTfO$_2$	1-Butyl-3-methyl imidazolium	419	−4	1.429	1.4271	80
BEIM NTfO$_2$	1-Butyl-3-ethyl imidazolium	433	−8	1.404	1.4285	48
Dicyanamide						
EMIM DCA	1-Ethyl-3-methyl imidazolium	177	−21	1.06		21
BMIM DCA	1-Butyl-3-methyl imidazolium	206	−6	1.06		37
BMPyrrol DCA	*N*-Butyl-*N*-methyl pyrrolidinium	208	−55	0.93		50
Tetrafluoroborate						
EMIM BF$_4$	1-Ethyl-3-methyl imidazolium	197.8	6	1.248		66
BMIM BF$_4$	1-Butyl-3-methyl imidazolium	225.8	−82	1.208	1.429	233
BMMIM BF$_4$	1-Butyl-2,3-dimethyl imidazolium	239.8	37	1.2		780
DMIM BF$_4$	1-Decyl-3-methyl imidazolium	309.8	−25	1.072		930
Hexafluorophosphate						
BMIM PF$_6$	1-Butyl-3-methyl imidazolium	284	10	1.373	1.411	400

Source: Berthod et al. (2008). Reproduced with permission of Elsevier.

7.4 Classification of Ionic Liquids

ILs, in general, are classified into protic and aprotic ionic liquids. Protic ionic liquids (PILs) are formed by direct proton transfer from a Brønsted acid (AH) to a Brønsted base (B) to yield a $[BH^+][A^-]$ species as shown in Eq. (7.1):

$$[AH] + [B] \rightarrow [BH^+][A^-] \quad (7.1)$$

Here, in Eq. (7.1), if the final product $[BH^+][A^-]$ is considered a pure liquid, the amounts of the neutral [AH] and [B] precursors should be negligible. In other words, it can be stated that a pure PIL should not contain more than 1% of the neutral species in its composition. Proton transfer is essential in order to consider the derived ionic liquid as an ionic product; otherwise, the final product of the given reaction would be described as an adduct sustained by a AH···B hydrogen bond. Many of the PILs involve very strong acids and hence the equilibrium is heavily shifted to the right and results in the formation of an ionic salt. In addition, the higher boiling point of PILs than the average boiling point of acid and base precursors (1 : 1 mixtures) suggests significant and fast proton transfer between acid and base molecules (Greaves and Drummond 2008). Nevertheless, recent studies suggest that the formation reaction of PILs can be more complex due to the possible existence of multiple equilibria that may lead to the formation of electrically charged aggregates (Canongia Lopes and Rebelo 2010). Research is being carried out to quantify the extent of proton transfer and the ionic characteristic of PILs by using differences in the aqueous pK_a of [AH] and $[BH^+]$ (MacFarlane et al. 2006; Yoshizawa et al. 2003), Walden plots (Yoshizawa et al. 2003), 1H-NMR, vibrational spectroscopy (Nuthakki et al. 2007), and cyclic voltammetry experiments (Bautista-Martinez et al. 2009).

Aprotic ILs contain substituents other than a proton (typically an alkyl group) at the site occupied by the labile proton in an analogous protic ionic liquid. These ILs require synthetic strategies that are different from the simple acid–base reactions used to obtain most protic ionic liquids. The applicability of ILs in diverse areas extended research toward the design of new IL families. The tuneabiliy of combinations of cations and anions and the possibility to achieve modification offer access to ILs with targeted properties. Figure 7.4 demonstrates the new ILs family, which have been categorized on the basis of their characteristics.

Task specific ionic liquids (TSILs) are those class of ILs in which the functional group is covalently tethered to the cation or anion (or both) to impart to them specific properties or reactivities. These can be considered as the liquid version of solid-supported catalyst with the added advantage of kinetic mobility and large operational surface area (Sawant et al. 2011). Davis (2004) illustrated the concept of TSILs to interact with a solute in a specific manner while investigating the applicability of thiazolium IL as solvent-catalyst for the condensation of benzoine (Davis and Forrester 1999). Olivier-Bourbigou et al. (2010) suggested that Brønsted acidity can also be introduced by the addition of Brønsted acids (HF or HCl) into halide based ILs in order to reduce the volatility of the acid due to the formation of $X(HX)_n$ anion. Alkane sulfonic or carboxylate acid groups have been covalently tethered to different cations such as imidazolium, benzimidazolium (Wang et al. 2008), pyridinium (Bittner et al. 2012), ammonium, or phosphonium.

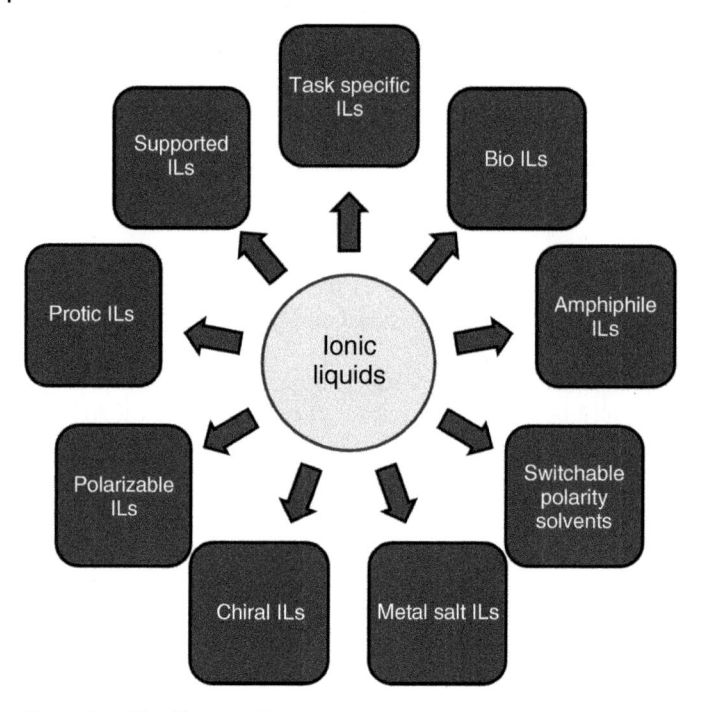

Figure 7.4 Classification of ILs.

Mono-charged diamine based ILs with both thermal stability and low melting point can be obtained by associating with $[NTf_2]^-$ anion (Yoshizawa-Fujita et al. 2006).

Switchable polarity solvents (SPS) are those neutral liquids that are capable of reversing their properties between a nonionic liquid and a highly polar and viscous ionic liquid. Nonionic to ionic transformation generally happens in the presence of CO_2 whereas the reverse action takes place in the presence of N_2/argon gas or heat. These solvents are able to dissolve both the organic and inorganic components of the reaction while highly polar and then change properties for easier separation and effective product isolation once the reaction is complete. Some deep eutectic mixtures with properties similar to those of ILs are also represented as SPS. These mixtures can simply be obtained by mechanically mixing two different components with less toxic emission and high mass efficiency.

Chiral ILs have been developed to improve the recovery of the catalyst, which is often used in substantial quantities. A large range of chiral ILs have been synthesized based on chiral amino acid anions and ammonium, imidazolium, and phosphonium cations (Suzuki et al. 2013; Pastre et al. 2010) or by asymmetric synthesis; they can bear central, axial, or planar chirality. Chiral ILs can be used as chiral solvents and as sole inducer of chirality due to their polymer-like behavior and potential high degree of organization.

Biological ILs, entirely composed of biomaterials, have shown acceptability to fine-tuning for better energy content and physical properties due to their biological properties. Hypergolic fuels in which hydrazine is replaced with ILs based on dicyanamide anions have been proposed as bio ILs. ILs such as dialkylimidazolium formate were produced as liquids having strong hydrogen bond accessibility. They are good solvents for polysaccharides dissolution under mild conditions and high concentrations.

Deep eutectic solvents (DESs) are widely acknowledged as a new class of ionic liquid (IL) analogs due to a range of similar characteristics with ILs. In fact, DES and ILs are two different types of solvent, and thus cannot be used interchangeably. As mentioned before, ILs are essentially composed of only ions, with no molecular species present in them. DES, the nearly ionic liquids, actually do have molecular constituents that make a difference between the DESs and ILs. Smith et al. (2014) explained DESs as those systems that are formed from a eutectic mixture of Lewis or Brønsted acids and bases. They can contain a variety of anionic and/or cationic species in contrast to ILs that are composed primarily of one type of discrete anion and cation. It is illustrated here that although the physical properties of DESs are similar to those of other ILs, their chemical properties suggest application areas that are significantly different. The term DES was first proposed by Abbott et al. (2003) while mixing choline chloride and urea in a 1 : 2 molar ratio. The hydrogen bond between the donor molecule and the chloride ion was observed to be so strong that it generates an "ion" that is much larger, breaking down the cation–anion interactions and lowering the mixture's melting point, thus generating an ionic liquid-like material. The solvent properties of DESs can be adjusted by changing the hydrogen-bond donor, giving 10^5 possible different liquids and allowing the possibility of tailoring the properties of the liquid to a specific process (Abbott et al. 2004).

7.5 Environmental Scrutiny of Ionic Liquids

"Green solvents" is a widely used term in the literature to address ionic liquids due to their negligible vapor pressure, nonflammability, and versatile solubility (Armand et al. 2009; Lewandowski and Swiderska-Mocek 2009). However, questions have been raised on the "greenness" of these solvents by various researchers (Cevasco and Chiappe 2014; Jessop 2011). A precursor to ionic liquids, these were labeled as toxic and environmentally hazardous in recent investigations. Jastorff et al. (2003) employed structure–activity relationships to theoretically discuss the toxicity associated with ILs for the first time in literature. Later, several other investigations were carried out on microorganisms such as *Lactobacillus rhamnosus, Escherichia coli, acetylcholinesterase,* and plants/animals such as algae, rabbits, snail, and many more (Feng et al. 2010). An increase in toxicity of aquatic organisms due to increase in the number of nitrogen atoms in the cation was reported on the commonly used ionic liquids (Couling et al. 2006).

Biodegradability of ILs is another subject of research interest in the field of ILs. Several standard biodegradability tests are available as mentioned in Table 6.6 to investigate the ecotoxicity of chemical reagents. Modified Sturm, closed bottle tests (OECD 301B and D, respectively), the DOC Die-Away Test (OECD 301A), and the CO_2 headspace test (ISO 14593) are the reference methods for laboratory

Table 7.5 Applicability of OECD test methods.

Test	Analytical method	Suitability for compounds		
		Poorly soluble	Volatile	Adsorbing
DOC Die Away (301A)	Dissolved organic carbon	−	−	±
CO2 Evolution (301B)	Respirometry: CO_2 evolution	+	−	+
MITI (I) (301C)	Respirometry: oxygen consumption	−	±	+
Closed bottle (301D)	Respirometry: dissolved oxygen	±	+	+
Modified OECD screening (301E)	Dissolved organic carbon	−	−	±
Manometric respirometry (301F)	Oxygen consumption	+	±	+
CO2 headspace test (ISO 14593)	CO_2 evolution	+	+	+
OECD 309	14C labeling	±	+	+
ASTM 5988	CO_2 production/BOD	−	−	±

Suitable method to screen compound: +; unsuitable method to screen compound: −.
Source: Coleman and Gathergood (2010). Reproduced with permission of Royal Society of Chemistry.

testing of ultimate biodegradability of the ionic liquids. Table 7.5 lists the widely adapted OECD test methods to investigate the biodegradability of ionic liquids.

Gathergood's research group investigated the degradation potential of imidazolium cations combined with a variety of anions $[Br]^-$, $[BF_4]^-$, $[PF_6]^-$, $[N(CN)_2]^-$, $[(CF_3SO_2)_2N]^-$, and octylsulfate (Gathergood et al. 2004, 2006) using Sturm and closed bottle test. None of the compounds examined showed significant degree of biodegradation except octylsulfate-containing IL. The biodegradation efficiency was observed to decrease in the order $[PF_6]^- > [BF_4]^- > [Br]^-$ with 60%, 59%, and 48% of CO_2 evolution values, respectively. Kumar et al. (2006) suggested in their studies that it is not possible to degrade $[C_4mim][BF_4]$ abiotically.

New ILs containing biodegradable side chains were also designed (Gathergood and Scammells 2002) by the inclusion of potential sites of enzymatic hydrolysis and oxygen in the form of hydroxyl, aldehyde, or carboxylic acid groups, unsubstituted linear alkyl chains, and phenyl rings to improve the degradability of ILs. However, addition of oxygen containing functional groups such as alcohols, aldehydes, and carboxylic acids can restrict the ILs performance as reaction media whereas the incorporation of phenyl rings may increase the melting points of IL solvents. Wells and Coombe (2006) explored the microbial degradation with ammonium, imidazolium, phosphonium, and pyridinium compounds by measuring the biological oxygen demand. The authors observed no biodegradability of cations incorporated short chains (C ≤ 4) within this test series, which was in agreement with Docherty et al. (2007) and Stolte et al. (2008). For longer alkyl chains (C_{12}, C_{16} and C_{18}) containing ILs, a strong inhibitory effect of these compounds on the inoculum used was found, indicating that the active microbial consortium was significantly impacted by ILs toxicity.

Although these research studies indicate the toxicity and nonbiodegradability concerns with this leading area of research, it should still be kept in consideration that ILs are designer solvents and thus can be manipulated with infinite possibilities for developing an eco-friendly reagent. Since ionic liquids are at a preliminary stage of research investigation and more research is needed to address these questions, the concept of greenness of ionic liquids can be rephrased "if ionic liquids can contribute to more sustainable production and use of chemicals, which they clearly can do" (Welton 2015).

7.6 Applications of Ionic Liquids

The journey of ILs took over a century to become the dominant research stream in the field of chemical engineering, and now, the neoteric solvents have extended scope of their applications to various domains. Figure 7.5 demonstrates the wide applicability of ILs in different research areas.

ILs emerged as novel liquids in catalysis, in refinery and chemical industry, recovery of catalytic systems, and in developing more energy-efficient separation with less environmental impact. This solvent has also been employed in hydrogenation, hydroformylation, oligomerization, Friedel–Craft alkylation and many more industrial applications based on the transition metal catalysis considering the new biphasic reactions with an ionic catalyst. Ionic liquids are also providing unexpected opportunities in life sciences (as solvents in enzymatic and whole-cell biocatalysis) and as active pharmaceutical ingredients. Table 7.6 provides a list of different applications of ILs in the field of catalysis, electrochemistry, energy, and other selected examples of industrial scale applications.

In this chapter, we are essentially focusing on the applications of ILs as a metal extractant from aqueous media, industrial waste, and waste electrical and electronic equipment (WEEE).

7.6.1 Extraction of Metals from Aqueous Media

ILs are known to have hydrophobic properties that define their potential to extract hydrophobic compounds in biphasic separations. As mentioned in Chapter 6, the metals ions tend to get hydrated in the aqueous medium by making a complex with water molecules; therefore, hydrophobic extractants are required to increase the metal's hydrophobicity in order to remove metal ions from the aqueous phase. Selectivity of the extraction process as well as the mechanism of metal removal significantly depends upon the nature of ILs, the alkyl chain associated with, and the kind of lipophilic derivatives. Concentrations, pH

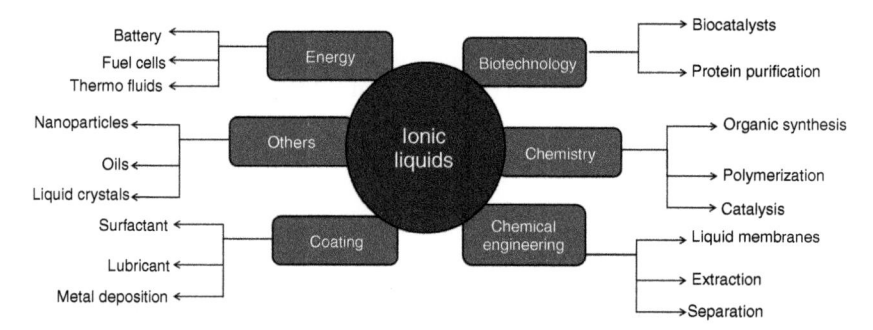

Figure 7.5 Applications of ILs in different areas.

Table 7.6 Applicability of ILs in different research areas.

Process	Role of ionic liquid	References
EtOH/H$_2$O or EtOH/THF separation	Entrainer for breaking azeotropes	Andreatta et al. (2015) and Pereiro et al. (2011)
Storage of gases	Liquid support	Watanabe et al. (2017) and Tempel et al. (2008)
Energy devices	Electrolyte	Moreno et al. (2017) and Wu et al. (2016)
Dye sensitized solar cells	Electrolyte	Yusof and Yahya (2016) and Vicent-Luna et al. (2014)
GC stationary phase	GC stationary phase	Nan and Anderson (2018) and Fanali et al. (2017)
Oligomerization	Solvent and co-catalyst	Wang et al. (2018) and Matsumoto et al. (2017)
Lubrication	Additive	Zhou and Qu (2017)
Friedel–Craft alkylation	Solvent and catalyst	Liu et al. (2009) and Xiao and Malhotra (2005)
Fluorination	Organo-catalyst	Khan et al. (2015) and Shinde et al. (2015)
Acid scavenging process	Acid scavenger	Maase and Massonne (2005)
Hydrosilylation	Catalyst	Li et al. (2017) and Kukawka et al. (2016)

of the solution, presence of interfering metal ions, and various cationic species existing in the solution also affect the extraction efficiency of ILs.

In the preliminary investigations, efforts were made to extract strontium nitrate from water using a series of 1,3-dialkylimidazolium ILs (Dai et al. 1999). Visser's research group also investigated the performance of ILs over well-known extractants 1-(2-pyridylazo)-2-naphtol and 1-(2-thiazolyl)-2-naphthol to extract metal ions Fe^{3+}, Co^{2+}, and Cd^{2+} from the aqueous phase (Visser et al. 2001a,b). These studies showed the successful extraction of metals and thus the scope of ILs started flourishing in the field of extraction of metals from contaminated aqueous stream.

Since the comparison between the performance of ILs and molecular solvents has always been a subject of research, while some of the research studies suggest that IL–extractant combinations are better than molecular organic solvents to yield metal ion extraction, other works suggest that the utility of ILs may be limited by solubilization losses and the difficulty in recovering the extracted metal ions (Domańska and Rękawek 2009). In addition, the extraction mechanism of metal ions into ionic liquids is believed to take place via ion-exchange mechanism (Jensen et al. 2002; Dietz and Dzielawa 2001), which is different from what is observed for extraction into organic solvents. Ionic liquid cations are lost with a neutral extractant during extraction of a metal ion, whereas anions can be lost during extraction of anionic metal complexes. These solubilization losses of ILs in the organic solution limit the use of ILs in liquid–liquid extraction processes. Design of TSILs (Quijada-Maldonado et al. 2018; Zhou et al. 2015; Nockemann et al. 2010) or the addition of organic ligands in biphasic solutions (Reddy et al. 2015; Vendilo et al. 2012) may help reduce the solubilization losses and can greatly increase the metal affinity for the IL phase providing extraction yields much greater than those found with conventional molecular solvents. However, some studies suggest that structural variation of the ionic liquid, e.g. by increasing the alkyl chain length or fluorination of the alkyl chain, may pose a negative effect on the distribution ratios and the extraction efficiency of the ILs. The ionic nature of these solvents can result in a variety of extraction mechanisms including solvent ion pair (IP) extraction and cationic exchange (CE), and simultaneously both of them may lead to salting out of the organic cation in the aqueous phase (Messadi et al. 2013), thus nullifying the expected benefit with these compounds. Water-immiscible ionic liquids (Smirnova et al. 2015; Vergara et al. 2014) is another option for efficient extraction of metals without losing ILs in organic phase. Rickert et al. (2007) suggested the increase in the solubility of hydrophobic ILs (Water-immiscible ILs) in the presence of neutral extractants, which affects the "greenness" of the process.

Reyna-González et al. (2010) reported an environment friendly process to extract Cu^{2+} metal ions from aqueous media using a novel RTIL [mimSBu][NTf_2] since this IL does not lose its components during extraction. Also, it is capable of extracting metal at room temperature without adding any complexing agent or adjusting the pH of the solution. The work was further extended (Reyna-González et al. 2012) to develop a new room temperature PIL [3-Bu-Pyr][NTf_2] with asymmetric properties and the capacity of pyridine derivatives to form complex with Cu^{2+}. This IL was relatively easy to synthesize and purify than the previous one ([mimSBu][NTf_2]). Strong interaction between Cu^{2+} and the N atom of the IL cation was observed in the complex. Recovery of the used ILs from the Cu^{2+}–IL–H_2O system and recycling after acid washing were illustrated.

Supported ionic liquid membranes (SILMs) are also being used in recent literature for the recovery of metals from contaminated sites, which stabilize ionic liquids by using porous supporters. Hoshino (2013) suggested the application of SILMs for the recovery of lithium from seawater. The SILMs can recover valuable metals from solutions where target metals are dissolved by selecting an adequate ionic liquid. Table 7.7 summarizes the literature available for metal extraction from various aqueous media using ionic liquids.

Table 7.7 Metal extraction from various aqueous media using ionic liquids.

Ionic liquids	(%) Metal extracted	Solution	References
Cyphos® IL 104, 102, 101	73% Pd^{3+} using Cyphos® IL 102	Aqueous HCl solution	Regel-Rosocka et al. (2015)
Betainium bis-(trifluoromethyl-sulfonyl)imide, [Hbet][Tf2N]	99% Nd^{3+}	Aqueous solution	Vander Hoogerstraete et al. (2013)
3-Butylpyridinium bis(trifluoromethanesulfonyl)imide [3-BuPyr][NTf$_2$]	83% Cu^{2+}	Aqueous solution	Reyna-González et al. (2012)
1-Methyl-1-[4,5-bis(methylsulfide)] pentylpiperidinium bis(trifluoro-methylsulfonyl)imide	99.87% Au	Aqueous solution	Lee (2012)
Trioctylmethylammonium salicylate (TOMAS)	99% Fe^{3+}, 89% Cu^{2+}	Aqueous solutions	Egorov et al. (2010)
Cyphos® IL 101, 104, 109, 111	100% Zn^{2+}, Fe^{2+} using Cyphos® IL 104.	Acid pickling solution	Marszałkowska et al. (2010)
1-(n-Alkyl)-3-methylimidazolium and tetraalkylammonium cations and hexafluorophosphate, bis[(trifluoromethyl)sulfonyl]imide, tetrafluoroborate, and chloride anion	Zn^{2+}, Cd^{2+}, Fe^{3+}	Aqueous HCl solution	de los Ríos et al. (2010)
Cyphos® IL 101 (Cytec) and Aliquat® 336 (Henkel)	Pt	Aquoeus solution	Stojanovic et al. (2010)

7.6.2 Extraction of Metals from Industrial Solid Waste/Ores

Recently, ionic liquids (ILs) have been studied as novel solvents for the sustainable dissolution, extraction, and separation of metal ions (Han and Row 2010; McCluskey et al. 2002; Kılıçarslan and Sarıdede 2015) from solid industrial waste and metal ores. Kilicarslan et al. (2014) employed [BMIM]HSO$_4$ ionic liquid for the recovery of copper and zinc from the brass ash obtained from brass manufacturing unit. Leaching experiments were performed using ([BMIM]HSO$_4$) at ambient pressure in the presence of H_2O_2 and oxone as the oxidants. More than 99% Zn was recovered using 50% IL concentration in the aqueous phase. Recovery of copper was observed to be highly affected by the presence of oxidant. Nearly 24% and 82% copper were recovered in the absence and presence of hydrogen peroxide as oxidant respectively. Oxone could not show any remarkable effect on metal recovery during the IL leaching process.

Jenkin et al. (2016) used a DES ethaline (mixture of choline chloride and ethylene glycol [mole fraction 1 : 2]) in combination with I_2 as an oxidizing agent to extract gold from a quartz vein-hosted orogenic-type gold ore and tellurium bearing gold ore. In general, these type of ores demand pyrometallurgy (roasting)

Table 7.8 Ionic liquid assisted extraction of metals from industrial waste and WEEE.

Ionic liquid	Industrial waste/Ore/WEEE	Targeted metal extraction	References
[BMIM]BF$_4$	Chalcopyrite	90% of copper	McCluskey et al. (2002)
[BMIM]HSO$_4$	Ore concentrate	87% gold and ≥60% silver	Whitehead et al. (2004)
[Cmim][CF$_3$SO$_3$]	Ore concentrate	Gold, silver, and base metals (Cu, Zn, Pb, Fe)	Whitehead et al. (2007)
[BMIM]HSO$_4$	Chalcopyrite	87.8% Cu	Dong et al. (2009)
[EMIM]HSO$_4$	WPCB	69.43% Cu	Atalay et al. (2015)
[HC1M]HSO$_4$	Brash ash	95% Cu, 100% Zn	Kılıçarslan and Sarıdede (2015)
[BMIM]HSO$_4$	WPCB	99%	Huang et al. (2014)

processing followed by hydrometallurgical treatment (preferable cyanidation) to extract gold. In addition, the rare earth metal tellurium is released to the environment during these conventional processing and cannot be recovered. Successful recovery of Au was reported by selective dissolution in DES by electrolysis process, which indicates the applicability of DES for the recovery of metals from ores.

Ionic liquid has also been used in combination with conventional extractants to improve the extraction efficiency. A novel tridentate neutral extractant TODGA was developed earlier in the year 2001 for the extraction of lanthanides and actinides in spent nuclear fuel reprocessing (Sasaki et al. 2001). Shimojo et al. (2008) employed the same extractant in combination with [C$_2$mim][Tf$_2$N] for the IL-based extraction of rare earth metals and observed a significant improvement in the extraction of metals compared to that in a conventional solvent extraction process. The extraction mechanism was demonstrated as shown in Eq. (7.2):

$$Ln^+_{aq} + 3TODGA_{IL} + 3C_n mim^+_{IL} \leftrightarrow Ln(TODGA)^{3+}_{3,IL} + 3C_n mim^+_{aq} \quad (7.2)$$

However, stripping of metal loaded IL phase was difficult because the anion [NTf$_2$]$^-$ acts as counterion to neutralize the metal complex. Table 7.8 lists some of the studies carried out to recover metals from industrial waste using a variety of ILs.

7.6.3 Extraction of Metals from WEEE

Ionic liquids have been observed as a good extractant for the recovery of rare earth metals from WEEEs. Sun et al. (2012) studied the separation of rare earth ions using the functionalized ILs [TBA][DEHP], [A336][DEHP], and [THTP][DEHP], diluted in [C$_6$mim][NTf$_2$]. The extraction efficiency was observed to be higher for those dissolved in [C$_6$mim][NTf$_2$] with relative distribution coefficients 100 times higher.

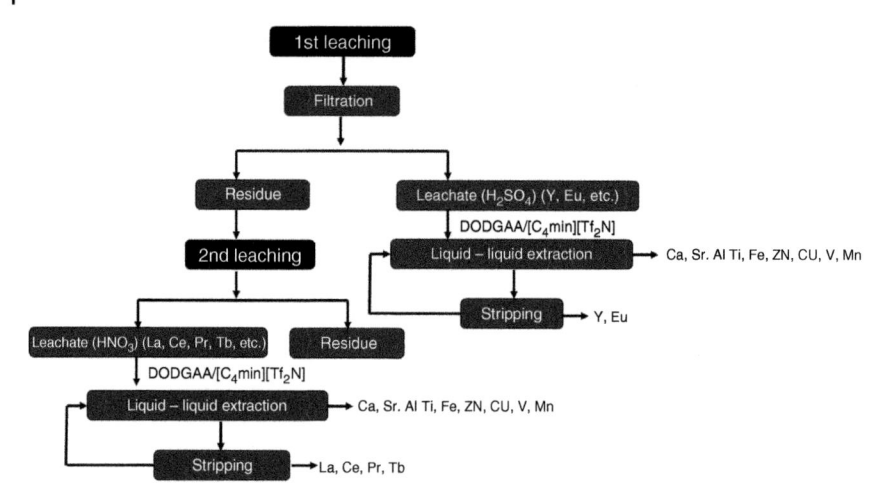

Figure 7.6 Process flow sheet for the recovery of metals from WEEE using ILs based extraction system.

Sahin et al. (2017) employed 1-methylimidazolium hydrogen sulfate (HmimHSO$_4$) for the extraction of rare earth metals from recycling slag powders of nickel metal hydride (NiMH) batteries. The slag samples were prepared by pyrometallurgical processing of spent NiMH batteries at 1350 °C for seven hours. Leaching experiments were performed with 30% [HmimHSO$_4$] concentration in the aqueous phase. 100% Fe, Mg, and yttrium were recovered at 80 °C reaction temperature and two hours leaching time using 63–90 μm particle size slag powder. Leaching efficiency for lanthanides and cerium was about 15% whereas 20% of neodymium was recovered under the same leaching conditions.

Yang et al. (2013) employed an IL based extraction system by preparing a mixture of N,N-dioctyldiglycol amic acid (DODGAA) in [C$_4$mim][Tf$_2$N] and the acid leachate obtained from the acid leaching of waste fluorescent lamps for the recovery of rare earth metals. The process flow sheet is shown in Figure 7.6. More than 90% recovery of La, Y, Tb, and Eu was obtained. Successful extraction along with an efficient stripping of metal loaded IL phase indicates a practical extraction system. In addition, the reusability of the extraction [C$_4$mim][Tf$_2$N] phase of DODGAA was examined for the five sets of extraction and stripping cycles and no change in the extraction efficiency was observed.

Li et al. (2019) synthesized [P$_{666,14}$][Cl$_3$] ionic liquids and used them as a reactive solvent for the recycling of samarium-cobalt magnets. The proposed IL has hydrophobic characteristics and therefore the aqueous solution can be used for the post-stripping processes. Oxidative dissolution followed by multiple stripping cycles resulted in the significant recovery of Sm^{3+} and Co^{2+} along with Cu(NH$_3$)$^{2+}$ and Fe(OH)$_3$. The recovered IL could be used further in subsequent cycles. Table 7.8 lists some of the studies carried out to recover metals from WEEE using a variety of ILs.

7.7 Summary and Outlook

Despite offering unique characteristics and significant benefits, ILs based extraction processes are still limited to laboratory scale studies. There are several issues such as scaling up of ILs synthesis unit, purification of ILs, sensitivity of ILs to any impurities, and stability and recycling of ILs, which need further research to make them suitable for process commercialization. Future research should concern procedures for stripping and recovery of metals and metal compounds from ILs as well as the recycling of ILs for potential use on a larger scale. In addition, the increasing concerns of ILs degradability and toxicity may constitute further barriers to scale up the production and wide application of ILs.

Although research in the field of ILs is at its infancy and more research work is needed to establish the process as a future technology, yet the current knowledge gives an impression that the ILs can be tailored to the requirement and therefore the properties as well as degradability can be adjusted to some extent. Also, some important fundamental viewpoints are now different from the original concepts, as insights into the nature of ILs become deeper that can be attributed to infinite combinations of cations and anions leading to a diverse suite of behaviors. Regardless, ILs remain more desirable than conventional volatile solvents and/or catalysts in many physical and chemical processes, often exhibiting "green" and "designer" properties to a useful degree (Olivier-Bourbigou et al. 2010). It is important to understand that process evaluation should satisfy all the three pillars of sustainability in terms of raw materials, synthesis and purification procedures, applications, and recycling possibilities.

Questions

1. How can ionic liquids be differentiated from molecular solvents?

2. Discuss the generic properties of ionic liquids.

3. "Viscosity of ILs is relatively high compared to conventional solvents." Please explain why.

4. Discuss the difference between protic ionic liquids and switchable polarizable ionic liquids. Give two examples of each.

5. "Ionic liquids are 'non–intrinsic green' solvents." Discuss your opinion.

6. Discuss the advantages and limitations of the applicability of ionic liquid in metal extraction from aqueous solutions.

7. Discuss the possible reasons that limit the industrial implementation of ionic liquid assisted extraction of metals from industrial waste.

8

Scale-up Process for Metal Extraction from Solid Waste

"Engineering refers to the practice of organizing the design and construction [and, I would add operation] of any artifice which transforms the physical world around us to meet some recognized need."

G.F.C. Rogers

Nomenclature

F	feed size (F80) (6000 µm)
W_i	standard work index (16.1)
K	constant (200 for ball mills and 300 for rod mills)
C_s	% of critical speed (70%)
S	specific gravity of feed (S = 1.4)
D_1	diameter of ball mill (1 m)
M_i	top ball size in the ball mill (in.)
g	gravitations acceleration (m/s²)
R	radius of the ball mill (m)
r	radius of the top ball size (m)
N_c	critical speed of the ball mill (rpm)
G	net mass of undersize material produced per revolution (g)
T	total amount of undersize material obtained per revolution (g)
U	(%) of undersize particles in the feed
W	energy required to grind 1 tonne of feed material to the product (kWh)
B	equivalent energy consumption per revolution (kWh)
a	correction factor for diameter inside shell lining
b	correction factor to include effect of 0% loading and mill type
c	correction factor for speed of mill
L	length in feet of grinding chamber (m)
C_1	correction for cyclone feed concentration ($C_1 = (53-V/53)^{-1.43}$)
C_2	correction for pressure drop ($C_2 = (3.28 * ((\Delta P)^{-0.28}))$
C_3	correction for Specific Gravity ($C_3 = (1.65/(G_s - G_1))^{0.5}$)
D_2	cyclone diameter
T_1	hot fluid inlet temperature (140 °C)
T_2	hot fluid outlet temperature (50 °C)
t_1	cold fluid inlet temperature (35 °C)

Sustainable Metal Extraction from Waste Streams, First Edition.
Garima Chauhan, Perminder Jit Kaur, K.K. Pant, and K.D.P. Nigam.
© 2020 Wiley-VCH Verlag GmbH & Co. KGaA. Published 2020 by Wiley-VCH Verlag GmbH & Co. KGaA.

t_2 cold fluid outlet temperature (50 °C)
V_P volume of precipitator unit when 80% tank is filled with slurry (m³)
H_1 height of the cylindrical section of dechelator unit (m)
R_P radius of the dechelator (precipitator) tank (m)
H_2 height of the conical section of the dechelator unit (m)

8.1 Introduction

Industrial extraction of metals from solid/electronic waste has been notably few in numbers. Often, traditionally tested pyrometallurgical methods are used for large scale operations in combination with some of the hydrometallurgical processes such as chemical leaching. Although a large number of research articles have been published in recent years on the significance of green engineering in solid waste management, industrial setup has not been explored so far for any green process. Improved control of reaction, high efficiency, and low cost with highest safety are the demands of any process industry. Still, the key barriers for the adoption of newer green technologies has been the lack of knowledge about these green technologies along with few pilot scale studies, scanty information on continuous operational strategies, and the risk-averse attitude of industries. Conversely, strict environmental regulations and global awareness about the hazards associated with the conventional waste treatment and metal extraction methods have made it quite difficult for the industries to continue the same practices in the coming years.

There are various kinds of waste containing metals around us as discussed in Chapters 1 and 2. Several green approaches to extract metals from solid waste have also been discussed in the Chapters 4–7. However, the bench-top research of the same needs to be scaled up to industrial level. Some pilot scale investigations on these technologies have been performed, while during implementation stages, the practical hurdles lead to reduced efficiency. Therefore, it is important to understand the concept of "process intensification" (PI) of these green engineering approaches. The key objective of a green PI is to design a metal extraction plant such that the process is not only clean, safe, and energy efficient but also has a lower cost.

The chemical industry is witnessing a major transformation using PI for various unit operations such as heat exchanger and distillation unit and is resulting in more compact intensified process plants. Various biochemical industries, especially food production and effluent treatment plants, have also been intensified by the contribution of PI technologies. In fact, the reactors based on intensified processes are designed by our research group at IIT Delhi as well, resulting in the development of a pilot plant for metal extraction from e-waste.

Although simultaneous optimization of process efficiency, product quality, and production flexibility can be achieved through intensification of individual unit operations, economic optimization can be achieved only after switching from batch to continuous process. The development of continuous operation also leads to fewer batch to batch product quality variation and enables tight control over its quality. The challenge lies in the development of an appropriate pilot plant, which can lead to the development of cost, energy, and environment efficient industrial plants for large scale operation.

In the present chapter, a theoretical scale-up design for metal extraction from the spent catalyst is discussed in detail. Various ways in which batch process can be converted into continuous process are also illustrated. One of the recent examples of PI for metal extraction is CFIs. The principle and applications of CFI for metal extraction process have been described in detail in Chapter 9.

8.2 Process Intensification

The most common way to achieve high efficiency at reduced cost is the use of PI techniques, which is defined as any chemical engineering development that leads to a substantially smaller, cleaner, safer, and more energy efficient process. PI at each unit operation involved in the process results in reduced size of the plant, cleanliness, and energy efficiency. Hazardous chemicals coming out of the plant undergo this process and result in added safety to not only human beings but also the environment. Thus, PI is considered as Greener technology, which meets the vision of sustainable development, i.e. meeting the needs of the present generation without compromising on the ability of future generations to meet their own needs (WCED 1987).

Leading scientists and industrialists agree on seven key benefits of PI:

1. *Green processes:* The most unique feature of PI is the development of efficient reactor design with maximum yield and minimum by-products. Intensively designed reactors have higher conversion efficiency and higher selectivity too. This permits the production of the green process without much of the purification sequences.
2. *Greater safety:* PI results in lower consumption of raw materials, processes are optimized, and there is obviously reduction in the amount of toxic chemical inventories. Unit operations are optimized, the size of the plant is reduced, and thus the process is simplified and the risk of explosion reduced. Batch processes are converted into continuous operations, which are known to have better heat handling capacity and are safer.
3. *Energy efficient process:* Reactions that occur at the micro level have generally resulted in better energy efficiency, mixing, and product yield. Design using alternative energy sources also has energy benefits. Using the tools of PI, advanced compact heat exchangers, bio-reactors, and distillation units can be designed, aiding more energy efficient processes.
4. *Reduced cost:* Industries utilizing PI report significant reduction in the overall cost of the process. Combining multiple units into single unit has added benefits. Compact equipment require relatively less land requirement and have lower initial investment cost. The equipment are also less expensive. Less inventory, feedstock, and utility materials are used for the same throughput.

5. *Reduced raw material cost:* Processes designed using PI have high yield and selectivity. Using the same amount of raw material, higher amount of product can be obtained.
6. *Higher process flexibility:* Processes are more flexible and achievable.
7. *Quality improvement:* With the production of minimum by-products, quality of the main product improves. The need to purify the product also reduces.

There are four principles of an efficient process:

1. Process kinetics is a key to obtaining higher process performance. The limiting factor responsible for less conversion can be optimized and high selectivity can be achieved.
2. Similar processing time for each individual molecule may give products with uniform properties. Waste may be reduced, eliminating the need for product separation.
3. Efficient processes can be an outcome of the optimization of the driving force at every step. Less enabling materials can be utilized and thus reduction in equipment sizes can be obtained.
4. Synergy between partial processes leads to multitasking. Several processing tasks can be grouped together. As compared to individual counterparts, higher process efficiencies can be achieved (Van Gerven and Stankiewicz 2009).

An effectively intensified process can achieve all the above four principles, using any one or a combination of two or more PI fundamental approaches, clustered in four domains of structure (spatial), energy (thermodynamic), functional (synergy), and temporal (time) as shown in Figure 8.1.

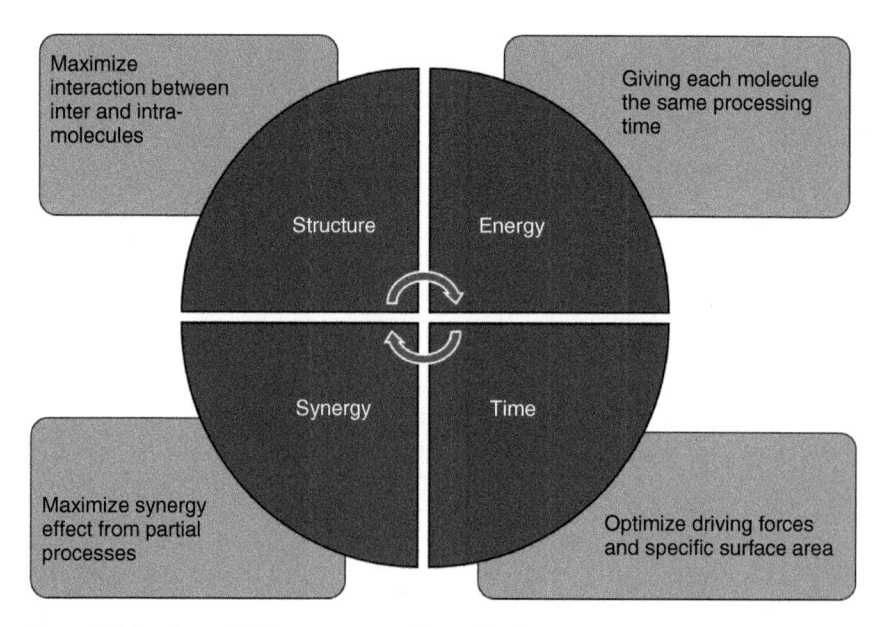

Figure 8.1 Fundamental views on process intensification.

8.3 Intensification of Metal Extraction Processes

Enhancing the performance of a process can be done in two ways. One is the integration of various operations, functions, and phenomena involved. Secondly, the targeted intensification of operations using batch to continuous process can be done.

There are a number of separation processes involved that can be intensified. The methods used to intensify a process are specific to the application that must be considered. A range of technological innovations is required for process-specific intensification of unit operations involved in metal extraction processes, including mixing, filtration, leaching, liquid–liquid extraction, evaporation, drying, concentration of liquid, absorption, adsorption, centrifugation, etc.

8.3.1 Centrifugation

Centrifugation is a process in which particles are separated on the basis of their size, shape, density, and viscosity of the solution using centrifugal forces. While working with fine drops and particles, they are found to be highly effective. The principle involved is the simple conversion of the inertia force of gas particle flows to a centrifugal force by means of a vortex generated in the body. They are available in many shapes and sizes. The conventional type of centrifuges includes the disk and bowl centrifuge. The intensified centrifugation units are reported to use 20% less surface area with 12% residence time than a falling film unit. The range of fluids that can be handled are also high with viscosity up to as high as 20 000 cP. The cleaning in place (CIP) is an additional feature. The high concentration ratio of 25 : 1 can be achieved in a single pass (Anlauf 2007). Use of ultracentrifuge is recommended for isotope separations. Diffusion, as an alternative technique, is known to be much less effective. To achieve greater separation efficiency, a cascade of centrifuges is recommended than a single unit.

The key to successful conversion of batch to continuous process lies in

1. phase development
2. consistent manufacture to high tolerance
3. conservative design based on experience.

8.3.2 Liquid–Liquid Extraction

Liquid–liquid extraction consists typically of two process steps, namely, mixing and phase separation. Effective phase separation can be achieved, for example, with polytetrafluoroethylene (PTFE) membrane or wettability based separator units. Mixing and contacting of the two liquid phases is typically achieved in a straight capillary tube (or slightly bended for reasons of compactness). Using the latter deliberately and tightly, however, enhanced mixing performance can be achieved with curved channels that enable the generation of vortices. Because of the action of the centrifugal force that acts perpendicular to the flow direction of the fluid, a secondary flow develops in curved channels, which is called Dean vortices. Dean vortices enhance the radial mixing in the case of single-phase flow, and thus narrower residence time distribution compared to straight capillaries can be achieved. Mixing can be further enhanced by complete flow inversions by inducing 90° bends at equal intervals along the length of the tube.

CFI is based on this principle (details have been presented in Chapter 9). The direction of the centrifugal forces inducing the formation of secondary flow profile in a helically coiled tube is changed with the 90° bends. Therefore, the secondary flow profile is developed after the 90° bend in a different plane that is perpendicular to the previous plane. As a result, the radial mixing can be further enhanced in CFI. Hence, a narrow residence time distribution close to an ideal plug flow reactor can be achieved even at laminar flow regimes. Recently, a milliscale CFI was employed by Gursel et al. (2016) for continuous flow liquid–liquid extraction. This device has plug flow characteristics, is simple in fabrication, and also has small space requirements, and is therefore a potential candidate for exploring the metal extraction from waste electrical and electronic equipment (WEEE). The research work proposed a continuous, uninterrupted metal scavenging unit where copper-catalyzed azide–alkyne cycloaddition click reaction was performed in a CFI to eliminate the need to isolate and handle potentially explosive azide. This flow reactor was coupled with a downstream unit where the copper scavenging process was optimized by investigating the performance of various chelating agents (ethylenediaminetetraacetic acid [EDTA], diethyltriaminepentaacetic acid [DTPA], ethylenediamine-N,N'-disuccinic acid [EDDS]) for the formation of Cu–chelate complex.

8.3.3 Mixing

Mixing is an important unit operation in metal extraction processes, and helps to obtain a uniform and efficient reaction. Conventional mixing techniques involve a mechanical agitator and coalescence by gravity. However, these conventional mixing methods need to be intensified because they involve long mixing time, and mechanical entertainment. The disadvantage due to large plant footprint for the case of coalescence makes the process less interesting. In order to reduce floor area and provide more compactness to mixing devices, use of helical coil is recommended in place of straight coils. Improvement in mixing pattern results in superior performance in various other unit operations as well.

There are various designs of coils proposed by researchers that can provide enhanced mixing using less space and time. A standard helical coil includes a square duct, with helical coiled arms as its edges, with sufficient empty space in between. The use of curved channels helps in the generation of vortices. Narrower residential time distribution curves can be achieved. CFIs are based on observations by various researchers that have proved that mixing can be further enhanced by using multiple vertical bend of 90 °C, and complete flow inversion and better radial mixing can be obtained.

The direction of the centrifugal forces inducing the formation of secondary flow profile in helically coiled tube is changed with the 90° bends. Therefore, the secondary flow profile is developed after the 90° bend in a different plane that is perpendicular to the previous plane. As a result, the radial mixing can be further

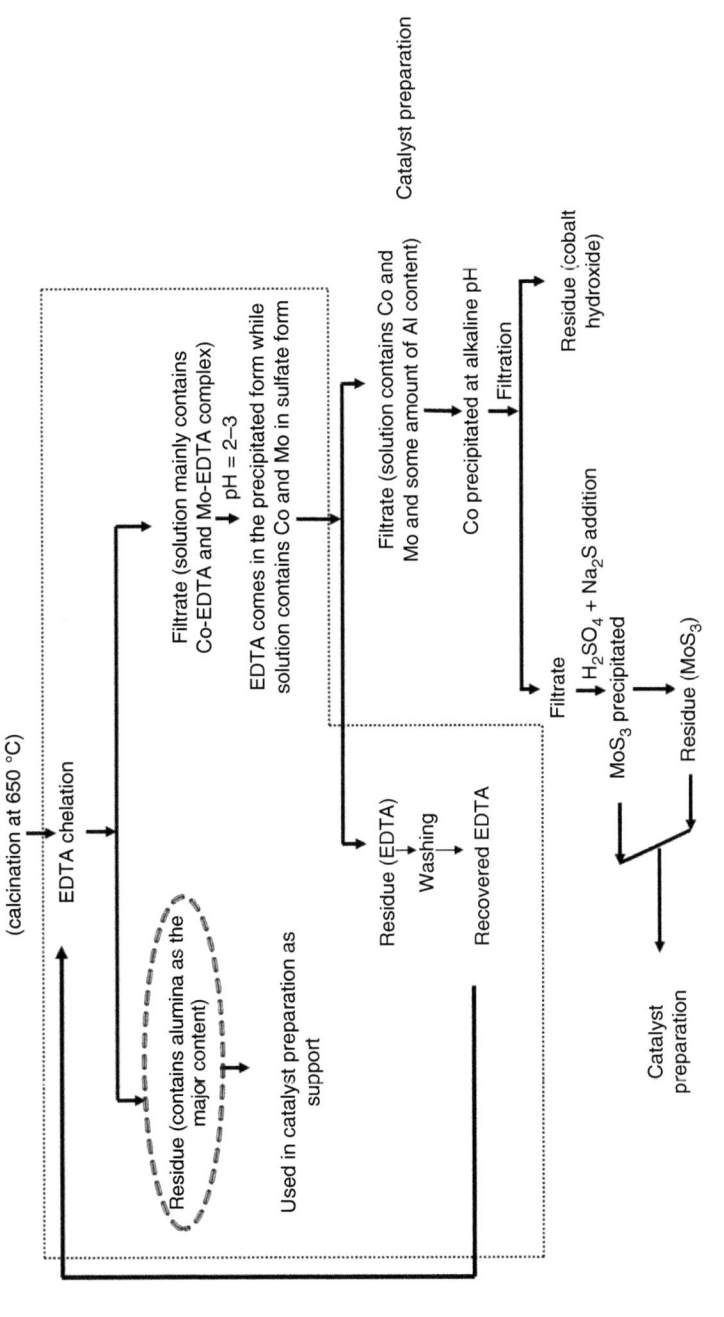

Figure 8.2 Process flow chart to extract heavy metals using chelation technology.

enhanced in CFI. Hence, a narrow residence time distribution close to an ideal plug flow reactor can be achieved even at laminar flow regimes.

8.3.4 Reactors

Reactors are known to be the "heart" of a chemical engineering process operation, the efficiency of which depends on the mixing and heat/mass transfer rate. In case of slow rate of mixing, the interaction of reactants is poor, and the proportion of available volume is quite less. There will be poor control over the local stoichiometry ratio and the rate of reaction will vary across the reactant. While selectivity and product yield are compromised, by-product yield is uncontrolled. It is easily believed that the high rate of mixing will lead to greater yield, which is not true. In case of very high rate of mixing, the reactant can mix immediately on entering the reaction, but significant reaction cannot be achieved. Therefore, the desired control over the reaction will not be obtained. In the case of ideal radial and axial mixing, each element of the fluid spends equal time in the reactor, and the yield of the desired product is maximum, along with minimum formation of the by-product. In a novel intensified reactor, the volume of the reactor is minimized and elements do not spend any time longer than required for optimum conversion.

Micro-reactors are the reactors optimized for greater yield and high selectivity using a reactor of very small size. The fluid flows through it in multiple sub-millimeter channels, allowing the reaction to take place. High rates of reactions can be achieved in these small volume reactors, which also have high heat and mass transfer rates. The basic principles of chemical reaction engineering apply to micro-reactors; however, the use of channels for allowing the fluid to flow provides high surface to volume ratio and thus makes the process more efficient.

There are various micro-reactors designed for chemical process operation using stainless steel, glass, silicon, and polymer. A micro-reactor was designed by Kestenbaum et al. (2002) using a stainless steel housing to be operated at temperatures up to 300 °C and pressure 25 bar. The reactor was divided into three sections, namely, a mixing unit at the top, a catalytic area at the bottom, and the diffusion path of 1 mm length in between. The mixing unit structure was made of polymethyl methacrylate (PMMA). Inlet for the reaction gases was provided from the opposite sides and the flow was toward the diffusion path using special channel geometry. The researchers found the selectivity of this process using micro-reactor reaction promoters to be up to 70%, which was comparable to that with promoter based industrial reactor (80%) while the space–time yield of the micro-reactor (0.78 tonne/h) was reported to be much higher than with the industrial reactor (0.26 tonnes/h). The major advantages of micro-reactor technology are the use of less utilities and waste, accompanied by high throughput and high safety. The studies available in the literature have highlighted the need for directing more efforts toward the development of industrial scale reactors.

8.3.5 Comminution

Comminution is the process of breaking down the raw catalyst particles into intermediate or finer sized particles to minimize the effect of mass transfer. It is a combination of grinding and particle classification to obtain an optimum mesh of grind. Comminution circuits carry responsibilities that can have an impact on all phases of operation, including efficient mill operation and successful production with the main purpose of exercising close control on product size. A good control over the grinding circuit is essential to facilitate easy downstream processing and to minimize the energy consumption in the unit process. It is considered that grinding circuits are the largest power consumers and most expensive phase of many industrial operations. Therefore, responsible and efficient operation of grinding circuits is an important consideration for designing of the pilot plant to perform the chelation process.

8.3.6 Drying

Drying is a very important unit operation to remove water or other solvents from solids. Sometimes, gases also need to be dried of liquids or dehumidified. Drying of gases is done using adsorption or vapor compression. Sometimes, membranes are used for drying of gases. For the drying of solids, mechanical means such as pressure can be applied. However, most of the dryers are based on thermal means such as hot air to dry solids. The intensification of the process of drying can be done using optimization of parameters such as size and the amount of particles to be dried, drying time, and energy required, without compromising on product quality. It is very important to understand the nature of the material to be dried as many metals change their nature of application of heat. Dryers such as freeze dryers and fluidized bed dryers are known to effectively dry a product using less space and time. Microwave drying in combination with hot air dryers can bring additional savings in the energy used.

A few of the successfully installed units have reported membranes to be effective in intensifying the drying operation of gases. During pyrometallurgical treatment of metals, outlet gases should be dried using intensified drying operations. Shell Global Solutions company has used hollow fiber based membranes for intensified drying of natural gases. The company has reported the removal of water at as low a concentration as less than 1%. As only 0.5% of methane can permeate through it, selectivity of the membrane is reported to be very high (Rijkens 2000).

8.4 Scaling Up from Batch to Continuous Process

Based on the unit operations discussed in the Section 8.3, a theoretical design of the pilot plant to process a batch size of 1 tonne of spent catalyst using chelation technology is discussed. All the required reaction parameters (molar concentration of chelating agent, solid to liquid ratio, pH, stirring speed, reaction temperature, reaction time, and particle size) were determined on the basis of our previous laboratory scale studies (Chauhan et al. 2012, 2013a). Process flow diagram (PFD) is prepared to propose the most economical path for the process. Equipment selection and their design parameters are estimated to perform various unit operations involved in pilot plant designing such as comminution, chelation, filtration, and precipitation. The economics of the metal extraction process was evaluated by estimating the operational cost to process 10 g of spent catalyst, which suggests lower process cost than with other available technologies.

8.4.1 Process Design Fabrication

Experimental studies on metal extraction from the spent catalyst have been conducted at laboratory scale and have been reported in our research publications (Chauhan et al. 2012, 2013a,b). Spent catalyst, obtained from the refinery or fertilizer industry, is pretreated first, which includes grinding to a size of 300 μm followed by calcination at 550 °C to drive off all the coke deposited over it. Chelation experiments were performed at 130 °C under autogenous reaction conditions for four hours at 500 rpm to form a metal–ligand complex. Vacuum filtration was performed for solid–liquid separation and alumina was recovered as residue. Dechelation was performed to recover the chelating agent from the solution, which can be recycled in the process. Metals present in the filtrate were precipitated by selective precipitation method (Chauhan et al. 2013a,b) as mentioned in Chapter 6. The complete experimental procedure is depicted in Figure 8.2 for the reader's ready reference.

Building upon the experimental investigation, the basic scheme of the process was fabricated and is demonstrated in Figure 8.3. The process design comprises comminution scheme for the grinding of spent catalyst for which wet ball mill is the preferred choice to obtain the slurry for further downstream processes. Classification of particles for proper size control using hydrocyclones was also incorporated in the design. Continuous stirring was observed as an important requirement for effective mass transfer during the process; therefore, stirred tank reactor with sufficient agitation facility was designed for the pilot plant. Vacuum rotary drum filter was considered the best choice for the filtration. A shell and tube heat exchanger was designed in order to obtain efficient heating and cooling utilities. Dechelation was performed in a cylindrical tank with conical bottom to facilitate solids handling.

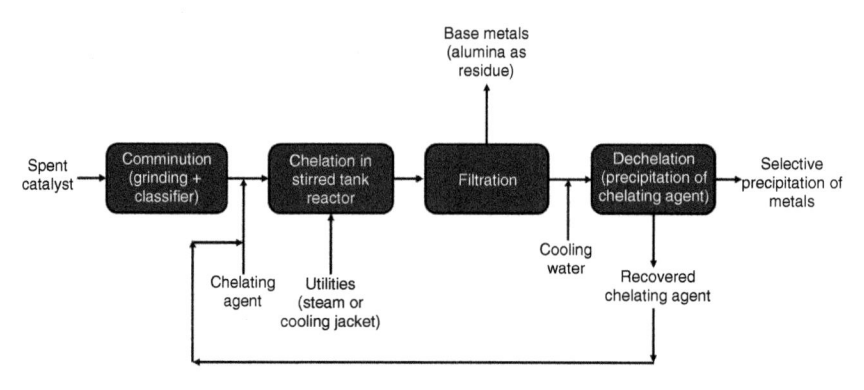

Figure 8.3 Block flow diagram of process design.

8.4.2 Designing of Pilot Plant

Basic engineering is performed to develop process deliverables such as equipment selection, size, and specification of each equipment, material balance in the process, and utilities requirement. A PFD was prepared to propose the most economical route for the chelation process as shown in Figure 8.4, which demonstrates the sequencing operation of the comminution circuit, chelation process in the stirred tank reactor, filtration, dechelation with interstage cooling, and separation of metal in solution by filtration. The complete designing of the pilot plant covers the following sections:

a. Material balance
b. Development of comminution circuit (ball mill + hydrocyclones)
c. Reactor sizing and agitator selection
d. Filtration selection and sizing
e. Design of heat exchanger
f. Design of dechelator unit
g. Batch scheduling.

8.4.2.1 Material Balance

Certain assumptions have been made for the theoretical consideration of material balance up to reactor circuit on the basis of laboratory observations:

i. The feed rate of (~6 mm) spent catalyst is 1 tonne/h.
ii. Specific gravity (S.G.) of the spent catalyst powder and liquid (distilled water + EDTA solution) is 1.8 and 1.0, respectively.
iii. Spent catalyst is mixed with the chelating agent solution in the ball mill to obtain the slurry for effective wet grinding operation at solid to liquid ratio (S/L) = 1 : 1.5.
iv. Input feed (fresh feed) contains 40% of the solids to ball mill.
v. A pressure drop of 50 kPa is assumed in the cyclone for efficient particle separation.
vi. The feed, overflow, and underflow from cyclones are at 60%, 40%, and 75% solid consistency.
vii. The recirculation load is 225% of the feed rate for the cyclones.
viii. The overflow is at 60% of 300 µm particle diameter.
ix. (S/L) ratio for the feed into the reactor is 1 : 15 at 0.4 M concentration of chelating agent.
x. The specific gravity of the slurry is calculated by the following method:

$$S.G.(slurry) = 1/(\% \text{ solids}/S.G.(solids) + \% \text{ liquid}/S.G.(liquid))$$

Based on the above assumptions, calculations were made for the material balance of the chelation process, which is shown in Figure 8.5.

Figure 8.5 indicates that an S/L ratio of 1 : 1.5 was taken to prepare the slurry in ball mill. This slurry is mixed with chelating agent solution at alkaline pH (pH = 9) before entering into the reactor to make the total volume of the reactor at the S/L of 1 : 15. Particle size was maintained in hydrocyclones at 300 µm and the particles above 300 µm size were sent back to the ball mill for regrinding.

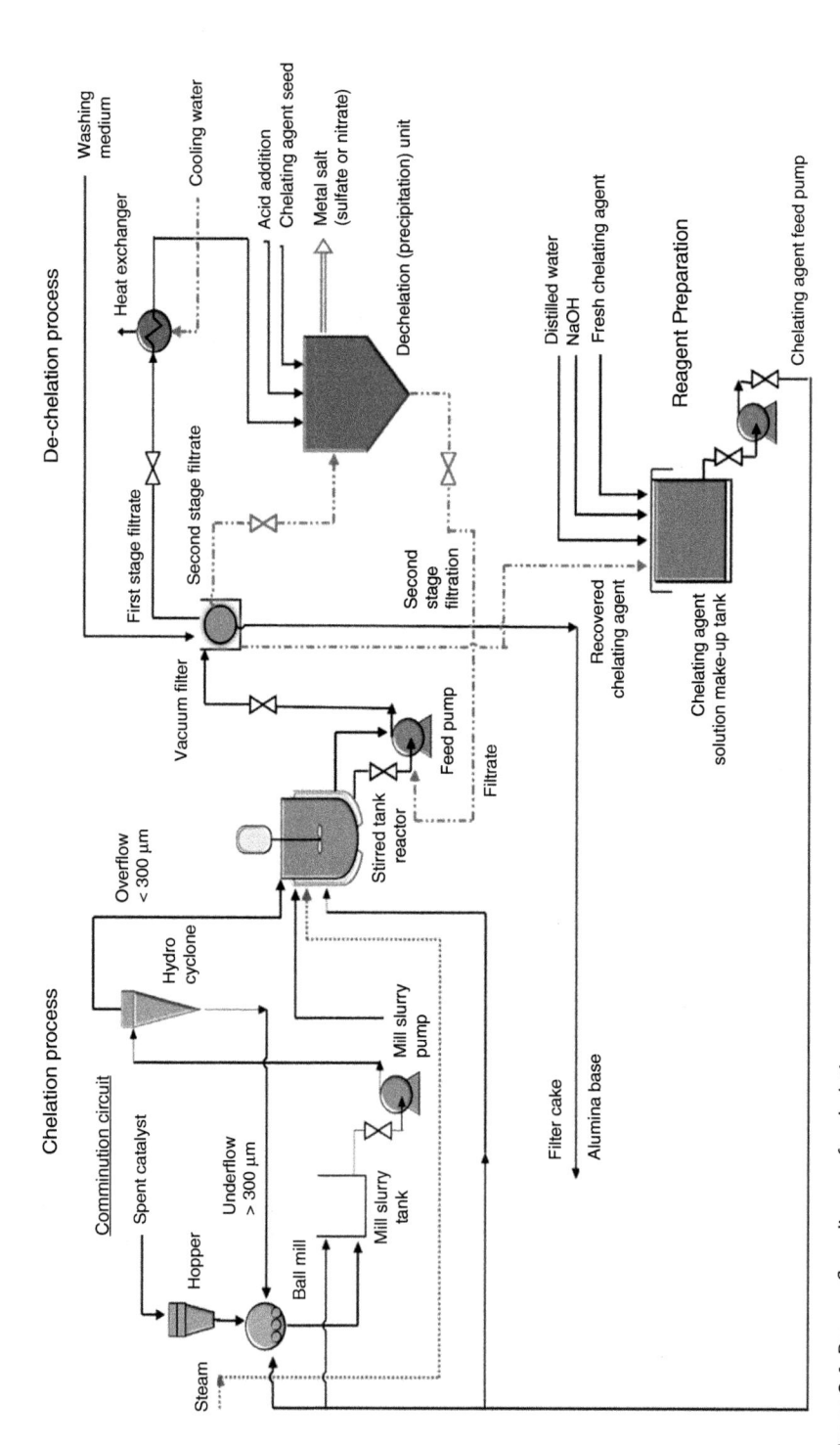

Figure 8.4 Process flow diagram for chelation process.

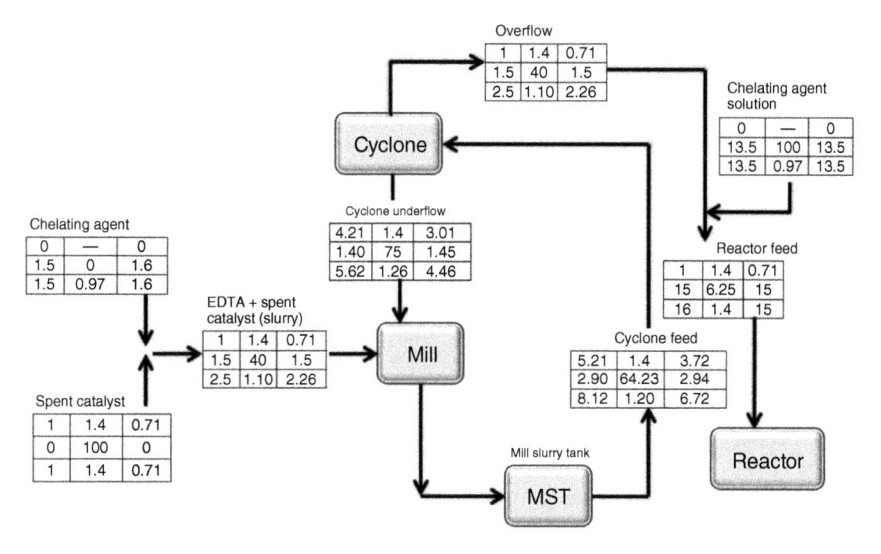

Figure 8.5 Material balance for the chelation process.

8.4.2.2 Development of Comminution Circuit

In the present study, ball mill is considered for the grinding because the desired product size (300 μm) falls in the size range for the ball mill, i.e. 420 to 20 μm. Wet grinding with overflow discharge arrangement is advantageous in this process due to the wet slurry requirement in the downstream process. Secondly, wet grinding is less energy intensive than the grate discharge arrangement. The ball mill was operated in the closed circuit mode with the particle classifier. A closed circuit with overflow discharge arrangement is designed for the comminution scheme as shown in Figure 8.6. It is evident from Figure 8.6 that comminution circuit is basically made of two important sections: (i) ball mill and (ii) hydrocyclone before transferring the material into the reactor. Design of both of these equipment is discussed in the following subsections.

Designing of the Ball Mill An efficient design consideration is essential to estimate the energy consumption and efficiency of the grinding mill. Numerous theories and experimental methods have been reported in the literature for sizing calculation of the grinding mill; however, the Bond method (Levin 1989) has enjoyed wide acceptance in the industrial applications due to its simplicity and wide practicability. Bond method comprises the standardized grindability test, Bond work index (W_i), and Bond comminution law. The Bond work index of a material is defined as the energy needed to reduce one short tonne of that material from a notional infinite size to a d_{80} size of 100 μm. It is determined by the Bond grindability test, and is expressed in kilowatt-hours per short tonne. In the present study, value of the W_i is taken from the literature, which is 16.1 kWh/tonne (Levin 1989).

During grinding, most of the reduction is done by the repetitive hitting impact of the balls to the bottom due to rotation of the mill; therefore, calculations are needed to estimate the desired ball size, ball charge distribution, and critical

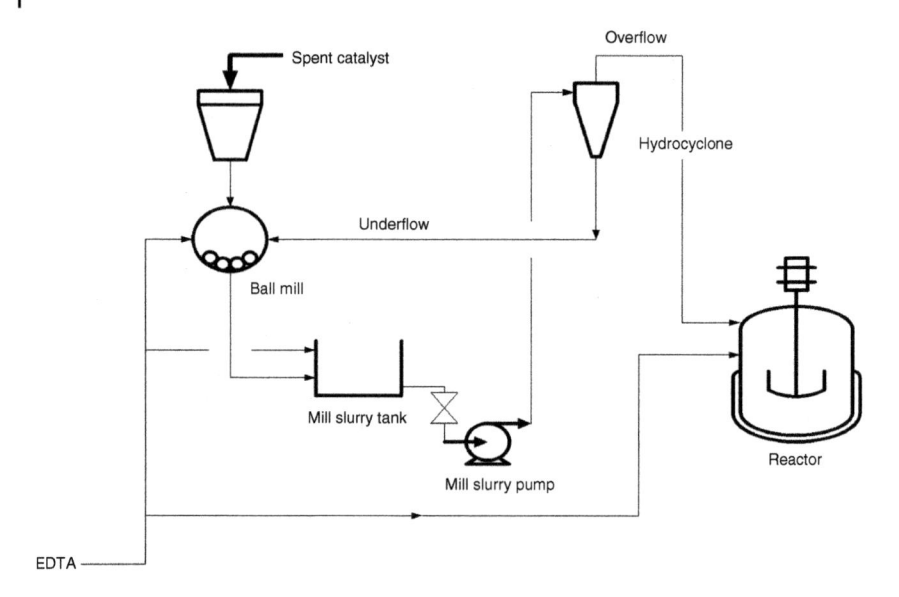

Figure 8.6 Closed circuit with overflow discharge arrangement.

speed of the mill. The top ball size in the ball mill can be calculated using Eq. (8.1):

$$M_i = (\sqrt{(FW_i/KC_s)} - \sqrt{(S/\sqrt{(D_1)})}) \tag{8.1}$$

Here the value of K is taken from the literature (Levin 1989), which is 200 for ball mills. Values of the other factors are given in the nomenclature. Since, no specific value of the diameter is mentioned in the literature, a range of diameter values was investigated to find the most suitable diameter for the minimum energy consumption. The reader is referred to the graphs reported in the research article by Levin (1989) for the diameter values and correction factors. The diameter of the ball mill was considered 1 m by extrapolation of the available data given by the supplier.

Thus, the top ball size in the ball mill is calculated as $M_i \sim 2.50$ in. (6.35 cm) and the ball charge distribution in the ball mill can be given as follows:

Balls of 2.5 in. = 32% weight

Balls of 2 in. = 39% weight

Balls of 1.5 in. = 22.6% weight

Balls of 1 in. = 2.9% weight

Balls of 0.5 in. = 1.4% weight

Ball of < 0.5 in. = 2.1% weight

Critical speed of the ball mill (N_c) was calculated using Eq. (8.2):

$$N_c (\text{rps}) = 1/(2^*\pi) \text{ sqrt } (g/(R-r))$$
$$N_c = 0.728\ 846\ 323 \text{ rps}$$
$$= 43.730\ 779\ 36 \text{ rpm} \qquad (8.2)$$

Thus, approximately 44 revolutions are required per minute as the critical speed for the ball mill with the given specification.

Efficiency of the ball mill can be related with the net mass (g) of the undersized material produced per revolution (G) of the Bond grindability test mill. The total amount of undersized material (T) obtained per revolution consists of G and a proportional amount of undersized material present in the feed (-6 mesh) to the mill.

If U is the (%) undersize in the feed, the total amount of undersize obtained per revolution (T) is given by Eq. (8.3):

$$T = G \times (100/(100 - U)) \qquad (8.3)$$

Then, the energy (W) required in an operating plant to reduce 10^6 g of material from a feed size F to a product size P can be given as Eq. (8.4):

$$(10^6 \times (100 - U))/(G * 100) \text{ revolutions} = W_i (1/\text{sqrt}(P) - 1/\text{sqrt}(F))\text{kWh} \qquad (8.4)$$

The reader is referred to the curve reported in the research article by Levin (1989) for the apertures of various limiting screens. The value of G was calculated from Eq. (8.4) and substituted in Eq. (8.3) to calculate the total amount of undersized particles per revolution, which predicts the efficiency of the grinding mill.

The energy required to grind 1 tonne of feed material to the product was calculated using Eq. (8.5):

$$W \text{ (kWh)} = 10 * W_i * (1/\sqrt{(P)} - 1/\sqrt{(F)}) \qquad (8.5)$$
$$W \text{ (kWh)} = 8.314\ 004\ 25 \text{ kWh}$$

The equivalent energy consumption per revolution (B) was evaluated using Eq. (8.6):

$$B = (4.9 \times 10^{-3} * G^{0.18})/(P_1)^{0.23} * (100 - U) \text{ kWh} \qquad (8.6)$$
$$B = 1.643\text{E}(-5) \text{ kWh}$$

Mill sizing depends on certain factors such as the size distribution of feed and products, Bond's work index, type of milling and discharge, and capacity of the plant. A number of correction factors are applied to the power required to grind 1 tonne of catalyst sample. Based on these correction factors and the energy consumption, the length of the ball mill can be calculated as shown in Eq. (8.7):

$$L = W(HP)/(a * b * c) \qquad (8.7)$$

$a = 3$ (by extrapolating the graph for Factor A)
$b = 40\%$ media charge $= 5.02$ (wet overflow grinding).

Most overflow discharge mills operate with 35–45% charge, the values of a and b were taken from the Levin 1989.

$c = 70\%$ of the critical speed $= 0.1657$.

Speed of the operation should not be very low or high to optimize the capital cost and to avoid the wear on media. Therefore, 70% was assumed as the critical speed of the mill.

Literature suggests that length to diameter ratio for the ball mill should be in between 1 : 1 to 2 : 1 for efficient grinding (Levin 1989). The calculated values for the length of the ball mill are 4.46 ft or 1.5 m, which is in concordance with the range suggested in the literature, between 1 : 1 and 2 : 1 for efficient grinding.

Design of the Particle Classifier (Hydrocyclones) Ball mills are operated in combination with the particle classifier (cyclones, screen, spiral classifier) in close circuit to provide homogeneity in the particle size. In the present study, hydrocyclones are considered the best choice due to the practical range of classification, which lies between 40 and 400 μm. The theoretical calculations performed in this section are intended to provide an estimation for selecting the proper number and size of cyclones and to determine the proper level of operating variables. A "standard cyclone" should have a geometrical relationship between the cyclone diameter, inlet area, vortex finder, apex orifice, and sufficient length providing retention time to properly classify the particles. Designing of hydrocyclones involves the following calculations to compute all the desired parameters.

D50c point defines the particle classification where 50% particle size refers to the overflow and 50% to the underflow. In the present study, overflow is specified at 60% of 300 μm particles diameter; so here, a multiplier of 60% passing was calculated from the information provided in Arterburn 1976.

Multiplier of 60% passing $= 2.08$

Specified micron size $= 100 - 300$ μm

D50c(application) $=$ Particle size * multiplier of 60%passing

$= 300 * 2.08 = 624$ μm

The cyclone diameter depends on the D50c(base), i.e. the micron size that a "standard cyclone" can achieve operating under the base conditions, which was calculated from Eq. (8.8).

$$\text{D50c(application)} = \text{D50c(base)} * C_1 * C_2 * C_3 \tag{8.8}$$

$C_1 =$ Correction for cyclone feed concentration ($C_1 = (53 - V/53)^{-1.43}$)

$C_2 =$ Correction for pressure drop ($C_2 = (3.28 * ((\Delta P)^{-0.28}))$)

$C_3 =$ Correction for specific gravity ($C_3 = (1.65/(G_s - G_1))^{0.5}$)

Thus D50c(base) $=$ D50c(application)/($C_1 * C_2 * C_3$)

D50c(base) $= 11.588$

The values of the correction factor can be calculated from the formulae given in the Nomenclature. The relation between cyclone diameter and D50c base value can be given by Eq. (8.9):

$$D50c(base) = 2.84 * (D_2)^{0.66} \tag{8.9}$$

$$D_2 = 8.42 \text{ cm or } 3.35 - 4 \text{ in.}$$

Thus, the diameter of the cyclone is recommended to be 4 in. for pilot plant studies.

The flow rate of 6.12 m³/h for 4 in. diameter cyclone with 50 kPa pressure drop is recommended (Arterburn 1976). The total feed flow (6.72 m³/h) was calculated from the material balance and has been shown in Figure 8.5. Therefore, two cyclone units are recommended to design a pilot plant of 1 tonne processing capacity. The apex size depends on the flow rate of the underflow. Material balance calculations (Figure 8.5) suggest that the flow rate for the underflow is 4.46 m³/h. Therefore, underflow per unit can be given as (4.46/2) = 2.23 m³/h. The apex diameter was calculated from Figure 8.5, at the given underflow rate and it is recommended that the apex diameter should be 1 in. for the specified parameters. Area of the inlet nozzle was approximated to be 0.05 times the cyclone diameter squared whereas the size of the vortex finder equals 0.35 times the cyclone diameter. Thus, the overall specification for the hydrocyclone is listed in Table 8.1.

8.4.3 Reactor Sizing and Agitator Selection

A stirred tank reactor is selected to operate in the batch mode with a holding time of four hours to achieve 90% extraction of metals. The chelation process is designed to be accomplished at 130–140 °C under autogenous reaction condition

Table 8.1 Hydrocyclone specification for processing of 1 tonne of spent catalyst.

Design parameter	Values (units)
Inside diameter of cyclone feed chamber	4 in. (0.1 m)
Area of inlet nozzle	0.8 in.² (0.000 51 m²)
Size of the vortex finder	1.4 in. (0.035 m)
Diameter of cylindrical section	4 in. (0.1 m)
Length of the cylindrical section	4 in. (0.1 m)
Angle of conical section	10–20°
Apex diameter	1 in. (0.025 m)
Number of units required	2 (+1 as spare capacity) = 3
Feed flow rate	6.723 813 m³/h
Overflow rate	2.260 677 m³/h
Underflow rate	4.463 135 m³/h
Material of construction	Ceramic liners inside apex orifice, vortex finder, and lower cone liner

where the pressure is expected to reach up to 4 atm. These reaction conditions have been marked on the basis of the optimum reaction parameters obtained during laboratory investigations (Chauhan et al. 2013a). Volume of the liquid content of a vessel and the dimensions and arrangement of impellers, baffles, and other internals were determined in order to minimize the amount of energy required and to achieve an efficient mixing. The internal arrangements may also depend on the homogeneity/heterogeneity of a reaction mixture to enhance heat or mass transfer.

In the present study, the solid (spent catalyst) to liquid (EDTA solution) ratio (S:L) is optimized at 1 : 15 by performing laboratory experiments. On the basis of this S:L optimization, material balance suggests a volume of 15.3 m³ as shown in Figure 8.5, which is closer to a volume of 19 m³ (5000 gallons) available off the shelf. The reactor is sized for 90% extraction of metals in four hours based on laboratory studies performed under autogenous reaction conditions (Vuyyuru et al. 2010). Stirred tank with a dished bottom is recommended to minimize the power requirement. Single impeller arrangement at the center of the vessel and 1/3 times the tank diameter off the bottom were employed whereas the liquid level in the vessel is maintained equal to the tank diameter. Pitch blade turbine was considered a most suitable impeller for the chelation process in this example, which can give both axial and radial flows. Four baffles 90° apart should also be installed with the impeller to avoid vortex formation of the slurry. The speed of the agitator is recommended to be kept slightly higher than the calculated value to negate the effect of nonideal behavior in the system. Figure 8.7 demonstrates the dimensions of the reactor and its internals to provide effective heat and mass transfer. An impeller in a reactor provides a certain volumetric rate at each rotational speed and the corresponding power input, which is influenced also by the geometry of the equipment and the properties of the fluid. Therefore, flow pattern and the degree of turbulence can be considered the key aspects of the quality of mixing.

The design of the impeller, thus depends on the kind of impeller and operating conditions described by the various dimensionless numbers (Reynolds, Flow, Froude, Power numbers) as well as the individual characteristics whose effects have been correlated. The calculations for these dimensionless numbers were performed using Eqs. (8.10)–(8.14).

$$\text{Reynolds number } N_{Re} = 10.75 N d^2 S/\mu \tag{8.10}$$

$$\text{Power number } N_p = 1.523(10^{13}) P/N^3 d^5 S \tag{8.11}$$

$$\text{Flow number } N_Q = 1.037(10^5) Q/N d^3 \tag{8.12}$$

$$\text{Dimensionless blend time } t_b N(d/D)^{2.3} = N_{Re} \tag{8.13}$$

$$\text{Froude number } N_{Fr} = 7.454(10^{-4}) N^2 d \tag{8.14}$$

The Froude number is pertinent when gravitational effects are significant, as in vortex formation; in baffled tanks its influence is hardly detectable. The power, flow, and blend time numbers change with Reynolds numbers in the low range, but tend to level off above $N_{Re} = 10\,000$ or so at values characteristic of the kind of

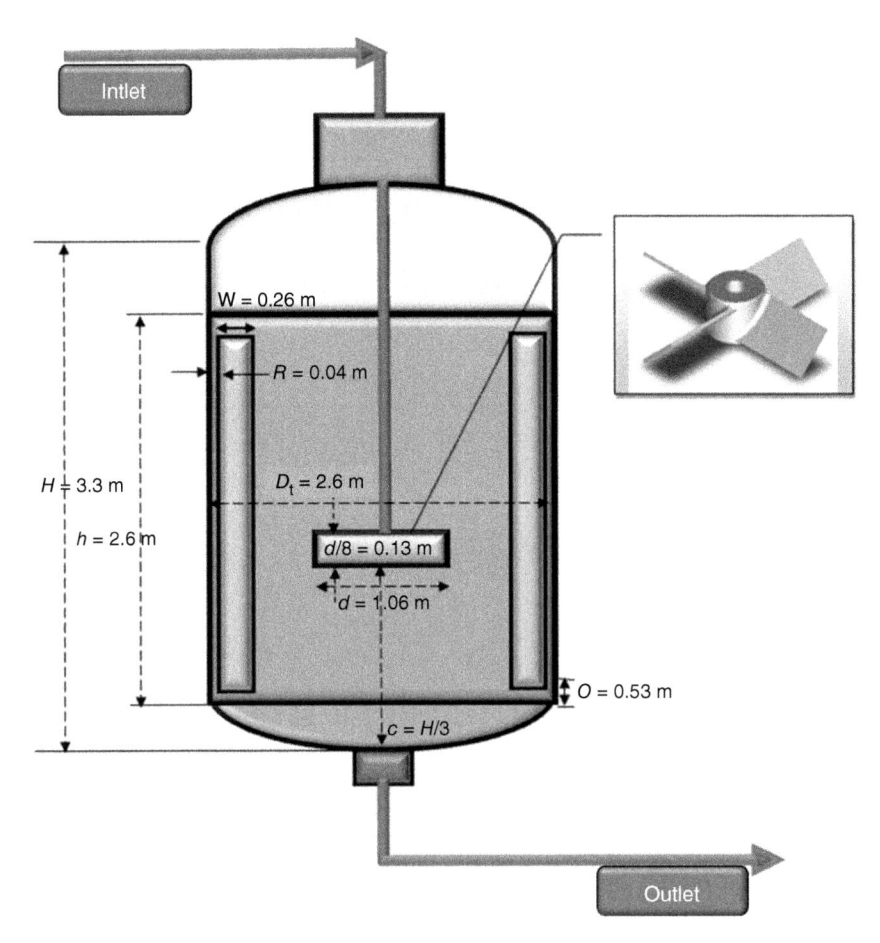

Figure 8.7 Reactor dimensions.

impeller. The recommended specifications of the stirred tank reactor and agitator selection obtained on the basis of the calculation are given in Table 8.2.

8.4.4 Design of Filtration System

Solid–liquid separation in the present study is proposed to be accomplished by vacuum filtration in a continuous mode. The filtrate is selected on the basis of quality of filtrate and filter cake, space requirement, and ease of operation and maintenance, and most importantly, the filtration operation should be cost effective. Therefore, based on the above considerations, rotary vacuum filter was recommended for the chelation process. This kind of filter offers simultaneous washing of the cake along with the filtration operation, which reduces the cost of the downstream washing equipment and washing medium. The cake produced using this filter contains nearly 10–15% moisture; therefore, drying is not necessary at this stage. The filtrate solution does not contain more than 0.5% solid content. In the proposed design, the same filtration unit is being used twice

Table 8.2 Proposed design of the reactor for the chelation process.

Parameter	Proposal
Type of vessel	Dished bottom
Liquid level	Equal to the diameter of the vessel for single impeller arrangement
Volume capacity (m^3)	19
Size ratio (h/D)	1
Height of the reactor (m)	3.3 m
Diameter of the reactor (m)	2.6 m
Settling velocity (m/min)	3.048
Solid contents	6.25% (obtained from material balance)
Baffles	Four baffles 90° apart (width (W) = 0.20 m), offset from the wall = 0.033 m
Impeller type	Pitch blade turbine with axial flow
Impeller location	Centrally located
Impeller speed (rpm)	125 (calculated)
Impeller diameter (m)	1.04
Impeller thickness (m)	0.13
Height of the impeller (m)	1.10
Power input (hp)	30

to separate alumina from the metal–chelate complex (first) and to separate the recovered chelating agent from the filtrate solution. This kind of arrangement reduces the installation and maintenance cost of one extra filtration unit for the dechelation operation.

The basic principle of filtration can be explained using Eq. (8.15):

$$Q = dV/dt = A\Delta P/\mu R = A\Delta P/\mu(R_f + \alpha c V/A) \tag{8.15}$$

$$\text{Or in other form:} \quad V = A/(\mu\alpha c) * [(A\Delta P/Q) - \mu R_f] \tag{8.16}$$

The time required to filter a specified amount of sample is given by Eq. (8.17):

$$t = \int dV/Q \tag{8.17}$$

Filtration calculations are always performed based on the Bench scale filter test; therefore, the following assumptions have been made on the basis of literature:

i. Viscosity of the slurry is 5 cP.
ii. Filter cloth resistance is 10^{-10} m^{-1}.
iii. Specific cake resistance is 1.6×10^{-12} m/kg.
iv. Cake thickness = 1 cm.
v. Peripheral drum speed = 0.5 m/min.
vi. The amount of wash equals the pore space of the cake.

The proposed specifications for the filtration unit are listed below and all the calculations are given in the Supporting Information.

- Rate of filtration = 40–60 kg/(m² h)
- Filtration time = 1 h
- Peripheral speed = 30 m/h
- Vacuum pump capacity = 0.6 m³/m²/min
- Vacuum applied = 300 mmHg
- Pressure difference = 0.8 bar
- Diameter = 1.85 m
- Length = 1.79 m
- Washing media requirement = 0.31 m³

8.4.5 Design of Heat Exchanger

Shell and tube heat exchanger is proposed for the present study in order to cool down the slurry so that the required temperature can be achieved for the dechelation operation. A fixed tube sheet with one shell and two tube pass arrangement was taken as an initial assumption to design the heat exchanger. Specific heat capacity of the slurry was assumed to be the same as that of water. Cold water was taken as the cooling media and placed in the tube side. Dimensions for the shell and tube heat exchanger were assumed as 1 in. outer diameter (OD) tubes (d_o) on 1.25 in.² pitch (P_T), 8 ft tube length (L), 0.875 in. inner tube inner diameter (ID) (d_i), and 25% cut-segmental baffles. Log mean temperature difference (LMTD) and its correction factor (F_T) were calculated using Eqs. (8.18) and (8.19):

$$\text{LMTD} = (T_1 - t_2) - (T_2 - t_1)/\ln(T_1 - t_2)/(T_2 - t_1) \tag{8.18}$$

$$F_T = [\sqrt{(R^2 + 1)} \ln[(1 - S)/(1 - RS)]]/ \left[\begin{array}{l} (R - 1) \ln \left\{ (2 - S(R + 1 - \sqrt{(R^2 + 1)}))\, / \right. \\ \left. (2 - S(R + 1 + \sqrt{(R^2 + 1)})) \right\} \end{array} \right] \tag{8.19}$$

where

$R = (T_1 - T_2)/(t_2 - t_1)$
$S = (t_2 - t_1)/(T_1 - t_1)$

The overall heat transfer coefficient (U_{asm}) is assumed to be 130 Btu/(hr ft² °F). The total surface area and the required number of tubes were calculated on the basis of U_{asm}, which suggests nearly 350 ft² surface area (A_1) and 200 tubes (n_t) for the proposed heat exchanger. It is worth mentioning that all the calculations are done with 10% standard tolerance on each value. All the calculated parameters are shown in Table 8.3. It can be deduced from Table 8.3 that all the values lie in the permissible range and therefore, the proposed design parameters can be considered feasible for the design of the shell and tube heat exchanger in the present study.

Table 8.3 Design consideration for shell and tube heat exchanger.

Design parameter	Calculated values
LMTD (°C)	38.82
Heat transfer area (ft²)	350
Number of tubes	200 tubes
Tube side heat transfer coefficient (h_i) (Btu/(h ft °F))	300.99
Shell side heat transfer coefficient (h_o) (Btu/(h ft °F))	361.64
Equivalent diameter for the shell side for square pitch (ft)	0.0825
Tube clearance ($C = P_T - d_o$)	0.25
Shell side cross flow area (ft²)	0.667
Overall heat transfer coefficient $U_{O,cal}$ (Btu/(h ft² °F))	138.85
Error estimation for $U_{O,cal}$ (%)	6.81
Tube side pressure drop (psi)	0.053
Shell side pressure drop (psi)	0.078 789 442
Number of baffles	7
Clean overall heat transfer coefficient U_C (Btu/(h ft² °F))	152.390 225
Over surface check = $U_C - U_{O,cal}/U_C$	8.8% error
Over design check	4% error

8.4.6 Design of Precipitator Unit

Dechelation was performed in the laboratory by the addition of sulfuric acid to adjust the pH of the solution. The chelating agent can be precipitated at acidic pH; therefore, pH 2–3 is maintained for the dechelation process and then nearly six hours residence time is provided. The precipitation process is generally performed in the conical tanks for easy maintenance as the chelating agent is precipitated in semisolid phase. Precipitator design is based on the following parameters:

Amount of slurry being handled = 15 m³/h

Residence time = 6 h

Total slurry in the tank at a time = 90 m³

Considering 80% tank filled with slurry(V_P) = 112.5 – 115 m³

The volume of the tank can be divided into two sections: one is the cylindrical section and the other is the conical section in the bottom. Thus, the total volume is given by Eq. (8.20):

Volume of tank = Volume of cylinder + Volume of the conical bottom

$$V_P = (\pi * R_P^2 * H_1) + (\pi/3 * R_P^2 * H_2) \tag{8.20}$$

Here, it is assumed that the height of the cylindrical section (H_1) is four times the radius of the tank (R), whereas the height of the conical section (H_2) is considered

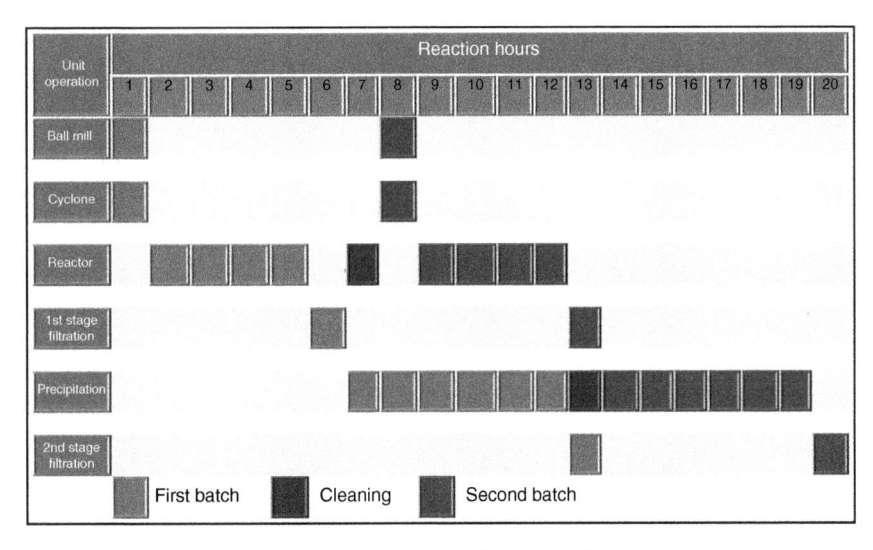

Figure 8.8 Batch scheduling to process 1 tonne of spent catalyst.

as 20% of the H_1. Design parameters for the precipitator unit were calculated using Eq. (8.20) and are given below:

$R = 2.04 - 2.5$ m

$H_1 = 10$ m

$H_2 = 2$ m

8.4.7 Batch Scheduling

Batch processing time to process 1 tonne of spent catalyst using chelation technology on pilot scale was estimated as shown in Figure 8.8 on the basis of PFD, which covers the following activities.

At time $t = 0$:
 i. Feeding of spent catalyst into the ball mill hopper is started.
 ii. EDTA in 1.5 and 13.5 m³ at required pH is filled in the ball mill and the reactor respectively.

The agitator of the reactor should be started at 500 rpm after 15 minutes, i.e. 0.25 hour to provide efficient mixing.

At time $t = 1$ hour:
 i. Grinding and classification are completed and particle size of 300 μm is obtained.
 ii. Ball mill and slurry pump are stopped.
 iii. Steam is provided to the reactor.

At time $t = 5$ hours:
 i. Reaction is completed, achieving nearly 90% extraction of metals.
 ii. Vacuum pump is started to allow the filtration process.

iii. Cooling water is sent to the heat exchanger unit for cooling down the filtrate temperature.
iv. Steam addition and agitator should be stopped.
v. Product in the form of filtrate starts to come out for first stage vacuum filtration.

At time $t = 6$ hours:
i. First stage filtration and interstage cooling operation are completed.
ii. Alumina along with wash filtrate contains nearly 15% of metals, so is recycled back to the reactor.
iii. Dechelation is started in the precipitator tank by the addition of sufficient amount of H_2SO_4 and EDTA seeds.

At time $t = 12$ hours:
i. Precipitation is completed.
ii. Metal sulfates are obtained in the filtrate, which can be sent for selective precipitation according to further application requirements.
iii. Vacuum pump is started again for second stage filtration unit.

At time $t = 13$ hours:
i. Second stage filtration is completed with the EDTA cake sent to EDTA collection tank from where it can be recycled to the reactor for the second batch.
ii. The wash filtrate is recycled back to the dechelator unit for further precipitation.
iii. First batch cycle is completed.

Thus, from Figure 8.8 and all the required steps of the chelation–dechelation process, it can be deduced clearly that a complete batch requires 13 hours to process the given amount of catalyst sample.

8.5 Summary and Outlook

PI presents a highly dynamic field of chemical engineering, which is based on the approach of optimization of the process using optimization of all unit operations involved in the process. However, it is the conversion of batch to continuous process where successful implementation of PI lies. The chapter described the conversion process using the example of metal chelation technique. This conversion is a tool box for the adoption of technology at industrial scale operation. Nevertheless, studies on PI for green metal extraction processes lack detailed investigations. Case studies where batch processes of metal extraction are optimized and industrialized are scanty. This generates quite a challenge with regard to the readiness of the process to be industrialized. More pilot scale studies on the green processes mentioned in this book can open the door to these green technologies as a standard choice for leaching of metals from various kinds of waste.

Preferably, using process integration and green technologies, metals can be extracted from waste in an eco-friendly manner using less time and energy instead of being compelled to use traditional toxic chemicals and processes for lack of alternatives.

Questions

1 Why is scale-up of technology essential, especially in the field of metal recovery from waste?

2 What are the various unit operations that need to be intensified for the recovery of metals from waste?

3 Explain the significance of CFI for waste treatment.

4 Elaborate on the process of pilot plant design using the example of chelation technology for metal recovery from waste.

5 Processes should be intensified at the batch scale. Comment on the above statement.

9

Process Intensification for Micro-flow Extraction: Batch to Continuous Process

Jogender Singh[1,2,3], Loveleen Sharma[1,4], and Jamal Chaouki[4]

[1] Indian Institute of Technology Delhi, Department of Chemical Engineering, Hauz Khas, New Delhi, 110016, India
[2] Tecnologico de Monterrey, Escuela de Ingeniería y Ciencias, Ave. Eugenio Garza Sada 2501, Monterrey, Nuevo León, 64849, México
[3] Sardar Vallabhbhai National Institute of Technology Surat, Department of Chemical Engineering, Keval Chowk, Surat, Gujarat, 395007, India
[4] Process Engineering Advanced Research Lab (PEARL), Department of Chemical Engineering, Polytechnique Montreal, P.O. Box 6079, Station Centre-Ville, Montreal, Quebec, H3C 3A7, Canada

"Science is where knowledge extends, not where it begins."
David L. Katz

"Science And Sense In A Post-Truth World: How Do We Know?"
Huffington Post

Abbreviations

CFI	coiled flow inverter
De	dean number
DHA-Et	docosahexaenoic acid
DMSO	dimethyl sulfoxide
DTPA	di ethyl tri amine penta acetic acid
ETFE	ethylene tetrafluoroethylene
FEP	fluorinated ethylene propylene
HDEHP	bis-2-ethylhexyl phosphoric acid
HEDTA	hydroxyethylethylenediaminetriacetic acid
HEH [HEP]	(2-ethylhexyl) phosphonic acid mono(2-ethylhexyl) ester
ID	tube inner diameter
OD	tube outer diameter
PDMS	polydimethylsiloxane
PFA	perfluoroalkoxy alkanes
PTFE	polytetrafluoroethylene
PUFA-Et	polyunsaturated fatty acid
Re, N_{Re}	Reynolds number

Sustainable Metal Extraction from Waste Streams, First Edition.
Garima Chauhan, Perminder Jit Kaur, K.K. Pant, and K.D.P. Nigam.
© 2020 Wiley-VCH Verlag GmbH & Co. KGaA. Published 2020 by Wiley-VCH Verlag GmbH & Co. KGaA.

9.1 Introduction

Process industry deals with a wide range of multiphase systems including liquid–liquid/gas–liquid operations, viz boiling, evaporation, condensation, distillation, absorption, and electrolysis; and solid–fluid operations, viz fluidization, dust disengagement, separation, filtration, and extraction. Extraction of valuable metals/minerals from process waste (pharmaceutical waste, chemical waste, and metallurgical waste, etc.) not only minimizes the environmental hazards but also results in the commercial use of various extracted metals/minerals along with their different by-products including uranium from gold tailings, tin recovered from zinc tailings, pyrrhotite from nickel tailings, vanadium from tar sand tailings, magnesium from asbestos tailings, and phosphorus from moon rock (Chalkley et al. 1989; Ritcey 1989). As far as separation is concerned, the miniaturized devices (extraction, distillation, adsorption, and absorption equipment) offer higher separation efficiency caused by enhanced heat and mass transfer rates of multiphase micro-flow systems (Gunther and Jensen 2006; Ciceri et al. 2014). The miniaturized separation and purification equipment varies from micro channels, micro tubes, and micro mixers to micro-packed beds and phase separators with high efficiency (Zhao and Middelberg 2011; Langsch et al. 2014). Thus, the working principles of the equipment are significantly different from those of conventional separation equipment such as vessels and columns.

To understand the working mechanism of these devices, several studies have been carried out investigating the flow pattern evolution, liquid–liquid extraction, and solid–liquid extraction (Gürsel et al. 2016; Zhang et al. 2017). Therefore, this chapter highlights the intensified micro-flow extraction devices and their applications and working mechanism in terms of flow patterns, residence time distributions (RTDs), overall mass transfer coefficients, and extraction efficiencies. The industrial viability of these continuous micro-flow extraction devices is also analyzed.

9.2 Miniaturized Extraction Devices

In the last few decades, multiphase extraction process has received a great amount of attention due to its importance in miniaturization of the micro-flow devices for process industry. Miniaturization of any flow device must accomplish the following:

a. Reduced use of energy
b. Smaller footprint of the equipment and reduced plant size
c. Reduced capital expenditures (build lower cost plants)
d. Improved process safety (due to the reduced amount of toxic chemicals in the process)
e. Enhanced environmental benefits.

Thus, the Sections 9.2.1 and 9.2.2 are focused on the process intensification of micro-flow extraction devices, their enhancement techniques, and the various applications of micro-flow extraction.

9.2.1 Intensification in Miniaturized Extraction Devices

The intensification of any process is based on an enhancement mechanism that causes improvement in the heat and/or mass transfer efficiencies. As illustrated in Figure 9.1, the enhancement mechanism for a micro-flow system can be categorized into active and passive techniques. Active techniques include mechanical utilities, fluid vibration (including ultrasonic), surface vibration, jet impingement, suction or injection, electrostatic fields, grooves and rivulets, rotation, induced flow instabilities (e.g. pulses), and other electrical methods. On the other hand, the extended surfaces that do not require any direct application of external forces fall into the "passive" category. The different mixing elements and process conditions of a device can cause different flow patterns and these flow patterns play a significant role in improving the mixing or separation efficiency of the device. Passive techniques including extended surfaces, rough surfaces, swirl flow, surface tension, porous structures, additives (for liquids and gases), and coiled tubes have opened new opportunities for process intensification of the micro-flow extraction.

A recent study has summarized the different miniaturized micro-flow extractions devices (Wang and Luo 2017), as shown in Figure 9.2. The different micro-flow extraction devices were classified as (a) microchannel chips, (b) tubular extractor, and (c) high capacity mini extractors. Microchips were mainly studied at the laboratory scale due to the ability of a microscope to detect the integrated flow paths (Chen et al. 2012). As shown in Figure 9.2a-1, microchips were prepared from polydimethylsiloxane (PDMS) using a soft lithography method (Huh et al. 2010). However, due to their rapid swelling in most of

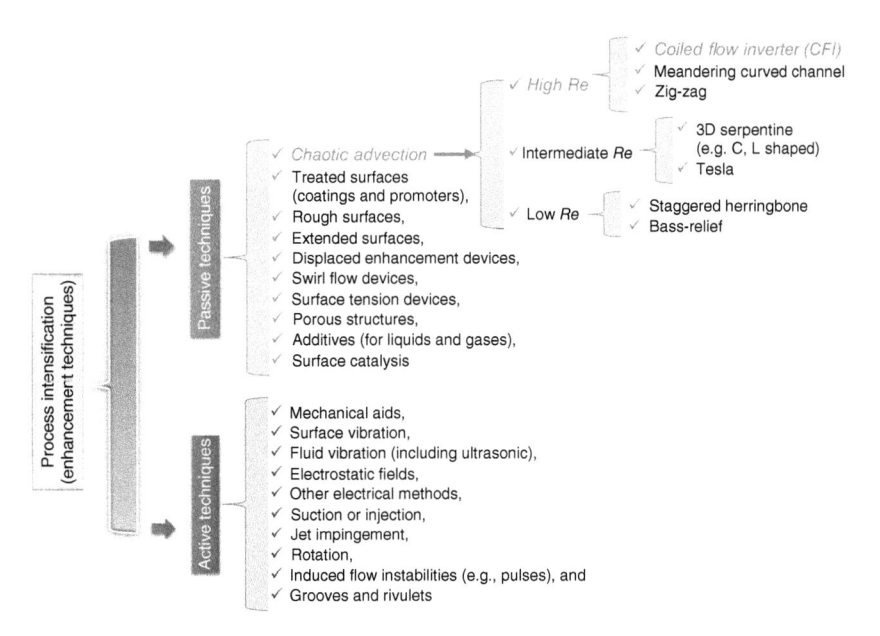

Figure 9.1 Illustration of the different enhancement techniques and the position of coiled flow inverter as a micro-mixer.

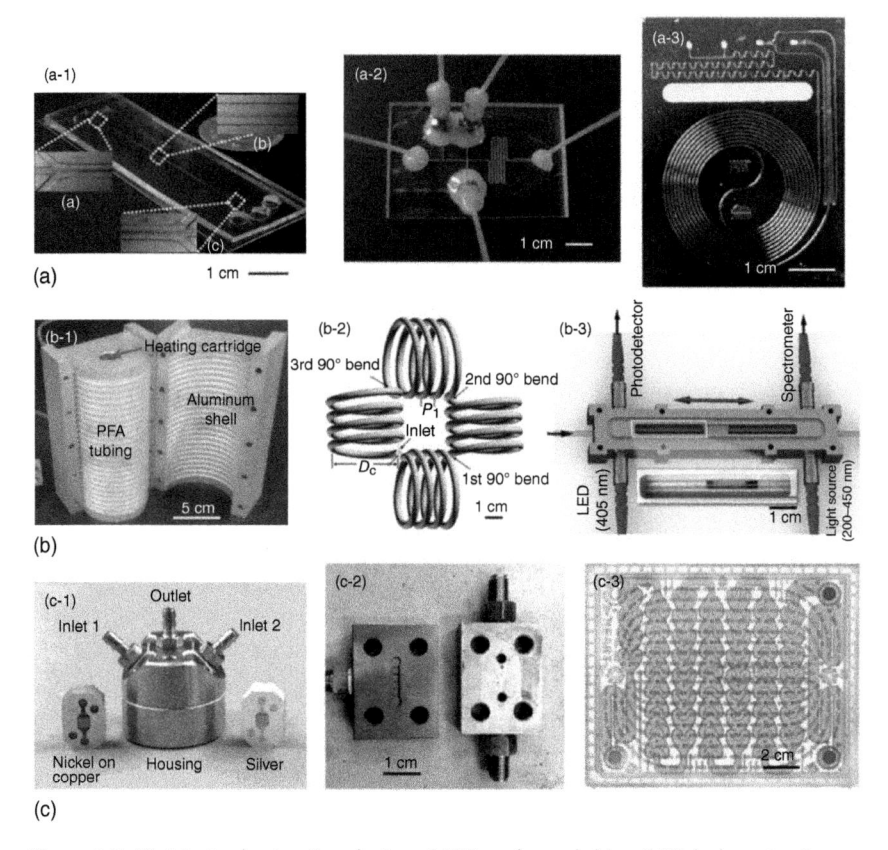

Figure 9.2 Miniaturized extraction devices. (a) Microchannel chips. (b) Tubular extractors. (c) High capacity mini extractors Source: Wang and Luo (2017). Reprinted with permission of Elsevier.

the solvents, these PDMS-based microchips are preferred for the processes comprising aqueous solutions (Novak et al. 2012). Therefore, the previous studies have proposed polymethylmethacrylate (PMMA), polycarbonate (PC), and polyvinylidene fluoride (PVDF) as the potential plastic materials for solvent extractions (Li et al. 2009; Ogonczyk et al. 2010; Liao et al. 2015). As outlined in Figure 9.2a-2, mechanical milling or thermoforming is another method to prepare microchips (Miserere et al. 2012; Sackmann et al. 2015). However, micro-extractors made from silicon wafer/Pyrex glass via wet/dry etching followed by sealing via anodic bond are still preferred over the plastic chips, which are weak for extreme operating conditions at high temperature and high pressure (Marre et al. 2010; Assmann et al. 2012).

The second category of micro-flow extraction devices are the micro tubes and capillaries. These devices can be made with relatively cheaper and solvent resistant fluorine plastics including ethylene tetrafluoroethylene (ETFE), fluorinated ethylene propylene (FEP), and perfluoroalkoxy alkanes (PFA). The micro-mixers and flow control systems can be connected to these tubes with the help of commercially available fittings (Jensen et al. 2014). As shown in Figure 9.2b-1, the temperature control of these devices can be achieved by applying a heating shell around the tubes (Adamo et al. 2016). For further intensification of heat and mass transfer in tubular devices, a multifunctional device known as coiled flow inverter (CFI) was initially proposed by Saxena and Nigam (1984). In CFI, the direction of centrifugal force can be changed periodically by 90° due to the equidistant 90° bends, which generates chaotic advection and further enhances the mixing between two phases. The working principle and the mechanism caused enhancement of heat transfer, mass transfer, mixing, and separation in CFI as demonstrated in Section 9.3. Furthermore, the periodic movement of a droplet in the forward and backward directions inside a short straight tube also augments the mass transfer rate (Abolhasani et al. 2015). An oscillatory flow extractor with periodic movement system for a single droplet is shown in Figure 9.2b-3. In this method, the partition coefficient can be measured rapidly with a lower substance cost due to the lower volume of a single droplet in microliters.

The scale-up of a miniaturized device or process can be achieved based on the intensification techniques. To scale up the micro-mixers for the formation of the swarms of droplets, two different multiphase micro-mixers, namely slit interdigital micro-mixer (Benz et al. 2001) and microsieve dispersion mixer (Wang et al. 2011; Shao et al. 2012), have been studied as shown in Figure 9.2c-1,c-2, respectively. The maximum treatment capacity of a microsieve dispersion mixer was reported to be 100 ml/min. The miniaturized extractors with high capacity have higher velocities as compared to that of micro tubes or microchips caused by a lower resistance time. Also, the size of the droplets can be reduced to micrometers with volumes to the picoliters with the application of stronger shear and inertial forces. Generally, micro-mixers are preferred for mass transfer operations due to their larger surface area for better contact between the phases (Xu et al. 2005). However, secondary dispersion devices can offer a constancy of the dispersed size of droplets in the systems where a longer extraction time is desired. Figure 9.2c-3 shows the corning advanced flow reactor with a treatment capacity up to 200 ml/min (Woitalka et al. 2014).

9.2.2 Application of Miniaturized Extraction Devices

The miniaturized flow extraction devices have wide applications ranging from environmental science to microelectronics and computer science due to their advantages including enhanced flow pattern control, mass transfer, and phase separation. Figure 9.3 presents a SCOPUS analysis of the applications of the micro-flow extraction devices. In addition, a comparative analysis of these miniaturized micro-extraction devices is presented in Table 9.1. Comparative analysis of the previous studies shows potential application of these devices for small scale purification processes such as in the preparation process of fine chemicals (Fries et al. 2008). Another application was demonstrated for the purification of the plant alkaloid strychnine by a Ψ–Ψ microchannel employing the parallel flows of aqueous–organic–aqueous solutions (Tetala et al. 2009). In Ψ–Ψ micro-flow system, the combination of extraction and back extraction in the same microchip offered a 79.5% yield of the strychnine with a residence time of 25 seconds. Another study on the purification of tanshinone IIA, which

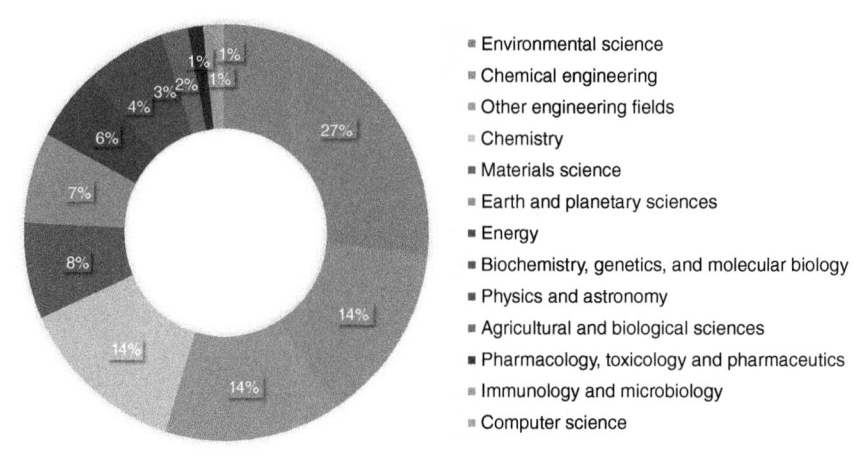

Figure 9.3 Application of the micro-flow extraction devices (SCOPUS analysis).

- Environmental science 27%
- Chemical engineering 14%
- Other engineering fields 14%
- Chemistry 14%
- Materials science 8%
- Earth and planetary sciences 7%
- Energy 6%
- Biochemistry, genetics, and molecular biology 4%
- Physics and astronomy 3%
- Agricultural and biological sciences 2%
- Pharmacology, toxicology and pharmaceutics 1%
- Immunology and microbiology 1%
- Computer science 1%

is used for the treatment of cardiovascular and cerebrovascular diseases, was purified with a 92% yield by employing a similar microchip (Mu et al. 2010). The continuous micro-flow extractors were found to be less labor intensive with enhanced yield over conventional methods of mechanical shaking or separating funnels. In solvent extraction processes, the temperature control is as important as that of mass transfer enhancement. For example, the extraction of ethyl ester from docosahexaenoic acid (DHA-Et, a polyunsaturated fatty acid, PUFA-Et) using silver ions solution is exothermic in nature and should be carried out at low temperatures (268–298 K) to prevent the oxidation of DHA-Et. The oxidation of DHA-Et may result in toxic products, which can induce hazardous diseases such as diabetes, rheumatoid arthritis, and cancer. Thus, an extraction system was developed with low temperature conditions using a water bath to employ in the systems where plug flow patterns are desired (Kamio et al. 2011). In another study, a continuous online phase separator was employed to separate vanillin from water using supercritical carbon dioxide (Assmann et al. 2012). These micro-extractors can also be employed in the high pressure processes as they are characterized by their small volumes (Marre et al. 2010; Ohashi et al. 2011). Thus, miniaturized extraction devices offer potential future applications for phase separation and purification processes.

9.3 CFI for Continuous Micro-flow Extraction

Several purification and separation devices have been developed that are dedicated for different chemical operations including absorption (Niu et al. 2009; Ye et al. 2012; Gao et al. 2011), chromatography (Tran et al. 2010; Culbertson et al.; 2000; Jemere et al. 2009), distillation (Sundberg et al. 2009; Ziogas et al. 2012; Hartman et al. 2009; Lam et al. 2011), and liquid–liquid extraction (Kenig et al. 2013; Assmann et al. 2013). Until recently, liquid–liquid extraction and multiphase metal extraction have gained great interest due to their wide

Table 9.1 Applications of miniaturized extraction devices for separation and purification processes.

Serial numbers	Authors	Working system	Device	Purpose
1.	Sahoo et al. (2007)	Toluene/organic azide/NaCl aqueous solution	Microchip with a membrane phase separator	To generate an intermediate in situ
2.	Tetala et al. (2009)	Chloroform/strychnine/strychnos seed extract	Ψ–Ψ channel microchips	To purify the small scale natural product
3.	Kamio et al. (2011)	Heptane/DHA-Et/AgNO$_3$, NaNO$_3$ solution	Tubular extractor with T-connector for phase separation	Separation of different products at low temperature
4.	Noël et al. (2011)	HCl aqueous solution/Et$_3$N/toluene	Tubular extractor with membrane phase separator	Quenching a reaction via extraction process
5.	Assmann et al. (2012)	ScCO$_2$/vanillin/aqueous solution	Microchip with microchannel array for phase separation	Fast and safe extraction using microchip at high pressure
6.	O'Brien et al. (2012)	H$_3$PO$_4$ aqueous solution/phenylhydrazine/hydrazone dichloromethane (DCM) solution	Miniaturized stirred mixer with gravity settler	Separation of the reactant in excess
7.	Snead and Jamison (2013)	Hexane/diphenhydramine NaCl aqueous solution	Tubular extractor and membrane phase separator	Separation of the product followed by the reaction
8.	Launiere and Gelis (2016)	DTPA or hydroxyethylethylenediaminetriacetic acid (HEDTA) aqueous solutions/Am or lanthanide ions/HEH [EHP] or phosphinic acid (bis(2-ethylhexyl) phosphinic acid (BEPA) or bis-2-ethylhexyl phosphoric acid (HDEHP) dodecane solutions	Capillary extractor with a membrane phase separator	To study the extraction kinetics at various pH values
9.	Kolar et al. (2016)	Cyanexs® 572/REEs/aqueous solutions	Ψ–Ψ channel microchip	To study the extraction rates and selectivity
10.	Zhao et al. (2016)	trioctylamine (TOA) octanol solution/HCl, H$_3$PO$_4$/H$_3$PO$_4$, KCl aqueous solution	Tubular extractor with a gravity settler	Kinetic study of selective extraction

(continued)

Table 9.1 (Continued)

Serial numbers	Authors	Working system	Device	Purpose
11.	Gürsel et al. (2016)	Toluene/water + acetone n-Butyl acetate/water + acetone	PTFE membrane separator and slit shaped separator	To achieve a multistep liquid–liquid extraction at pilot scale
12.	Kurt et al. (2016a,b)	n-Butyl acetate/acetone/water	T-junction (T-mixer) with helically coiled tubular device (HCTDs) as residence time units (RTUs),	Enhancing liquid–liquid extraction via chaotic advection
13.	Zhang et al. (2017)	Cyanex 272 (Diisooctylphosphinic acid, $C_{16}H_{35}O_2P$, 90%)	CFI and a T-type phase separator	Segmented micro-flow extraction and separation
14.	Zhang et al. (2019)	Toluene–sulfuric acid, toluene–water, and ethyl acetate–water	Circular PTFE capillary	To study the flow patterns for optimized reactor design

application in process industry, especially to produce the pharmaceutical products (Cervera-Padrell et al. 2012). Furthermore, for enhanced liquid–liquid contact, microchannel devices with higher surface density are advantageous (Assmann et al. 2013). Liquid–liquid extraction is carried out in two different parts, i.e. mixing and phase separation. In the micro- and milliscale devices, surface tension governs the separation and dominates over the inertial forces and gravitational forces (Cervera-Padrell et al. 2012). Thus, surface tension is employed as a driving force for separation in micro-flow devices (Kralj et al. 2007). The slug flow-based liquid–liquid extraction in T-mixer and straight tube was performed using a polytetrafluoroethylene (PTFE) membrane phase separator for copper-catalyzed azide–alkyne cycloaddition reaction system at a maximum flow rate of 1 ml/min (Gürsel et al. 2015). The extraction system containing T-mixer offered higher extraction efficiency than a straight tube because of secondary flow.

Recent studies have demonstrated that a novel design CFI can significantly enhance the extraction efficiency in the liquid–liquid and metal extraction operations (Gürsel et al. 2016; Kurt et al. 2016a,b; Zhang et al. 2017). CFI was design and invented by Saxena and Nigam (1984), Indian patent 159/DEL/2005. The working principle of CFI is based on flow inversion, which combines the two effects, i.e. coiling and 90° bends, as shown in Figure 9.4. Previously, CFI was studied for various applications in process industry as highlighted in recent studies (Soni et al. 2019; Singh et al. 2019a,b). Owing to its superior performance, the CFI was employed in various unit operations, as shown in Figure 9.5 (Klutz et al. 2015a,b, 2016; Lobedann et al. (2015); Rathore and Nigam 2015; Kateja et al. 2016;

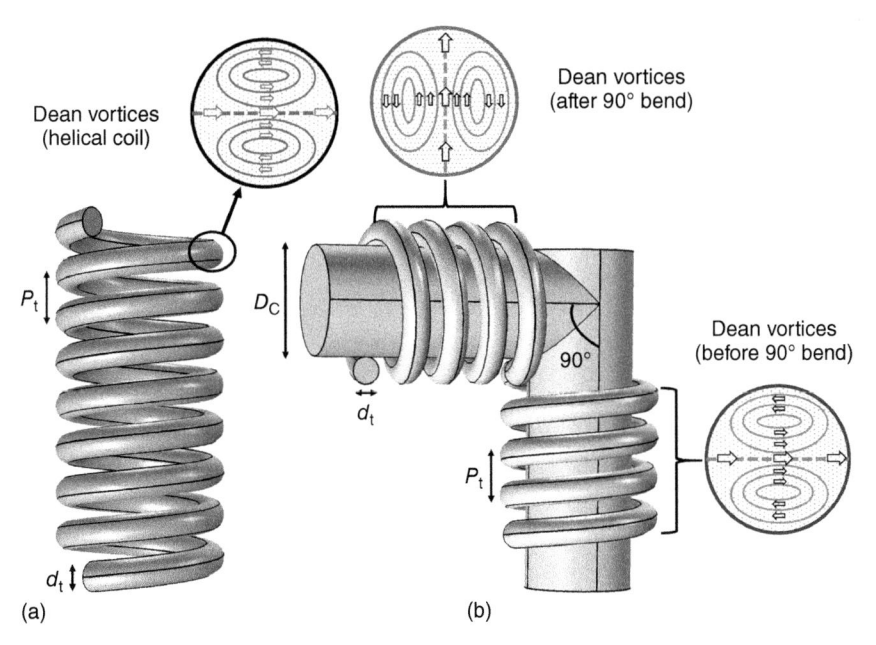

Figure 9.4 Working principle of coiled flow inverter (CFI). (a) Helical coil tube. (b) Coiled flow inverter.

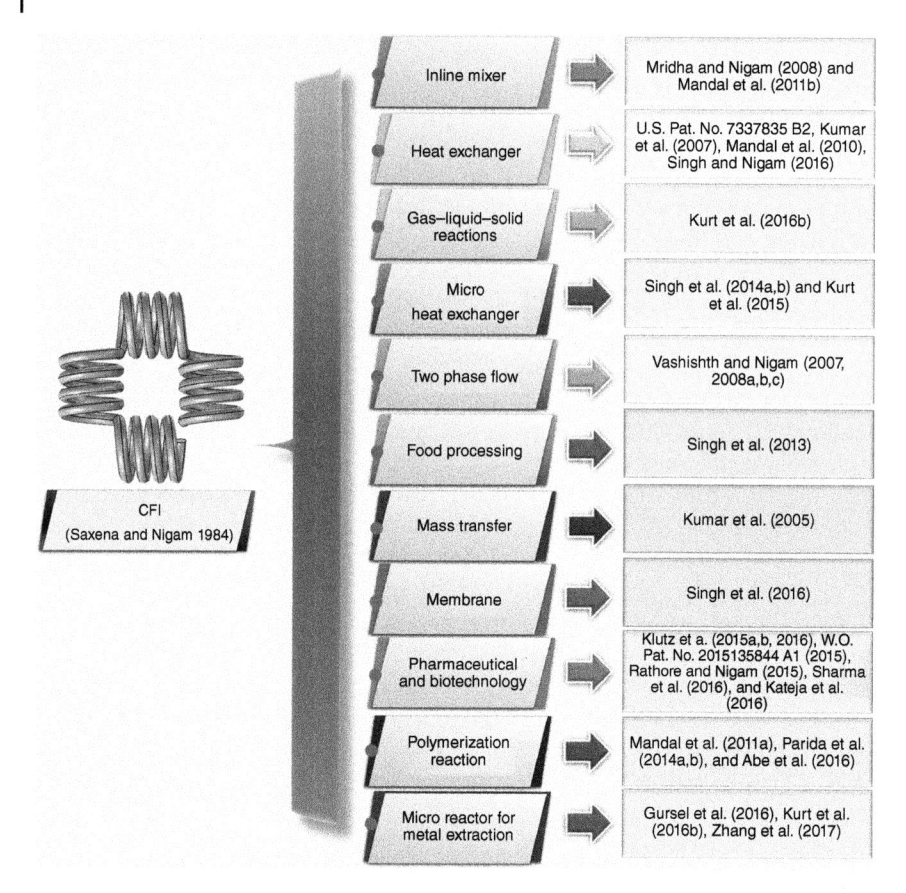

Figure 9.5 Illustration of the various applications of coiled flow inverter (CFI).

Abe et al. 2016; Sharma et al. 2016; Kumar et al. 2007; Kurt et al. 2015, 2016a,b; Gürsel et al. 2016; Mandal et al. 2010, 2011a,b; Mridha and Nigam 2008; Parida et al. 2014; Singh et al. 2012, 2014a,b, 2016; Singh and Nigam 2016; US Patent No. 7337835 B2 2008; Vashisth and Nigam 2007, 2008a,b; Vashisth et al. 2008; Kumar and Nigam 2005).

Thus, this section of the chapter discusses the application of CFI as a continuous micro-flow extraction device in liquid–liquid and metal extraction processes. As reported, CFI can offer maximum mixing efficiency per unit area at the expense of small increase in the pressure drop (Soni et al. 2019; Sharma et al. 2017). The extraction process was accomplished in a multistep synthesis for continuous flow at a pilot scale. The scale-up process of flow separation unit is also discussed in terms of micro-effects and higher throughput in the Section 9.3.5.

9.3.1 Designing CFI as an Extractor

As mixing is an inherent step in liquid–liquid extraction (as discussed in Section 9.3), the mixing characteristics of CFI should be discussed before employing it as an extractor. The mixing characteristics of CFI are explained through RTD

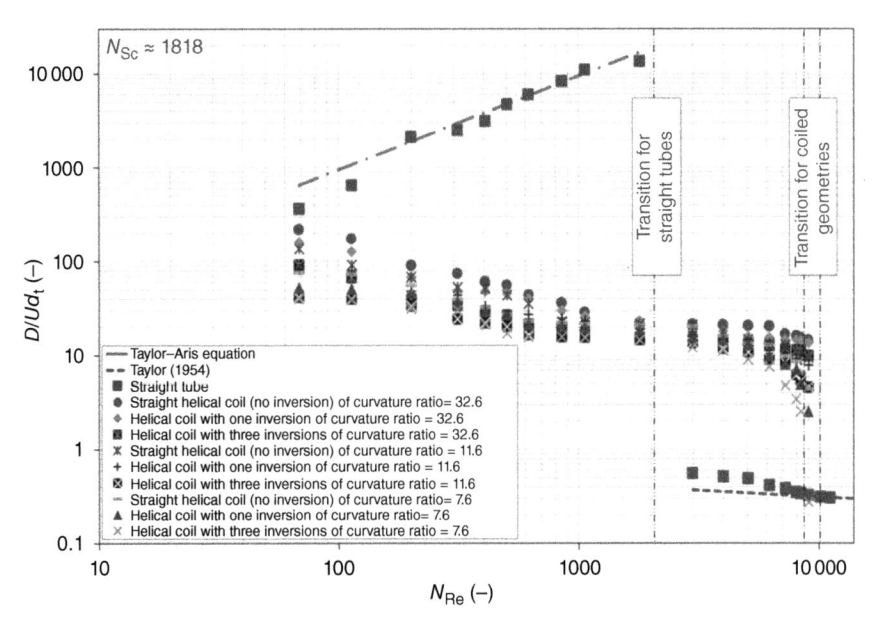

Figure 9.6 Comparison of the dispersion of fluids in straight tubes and different coiled geometries. Source: Sharma et al. (2017a). Reprinted with permission of Elsevier.

study by Sharma et al. (2017a) and depicted in Figure 9.6. In well-defined CFIs, the Dean vortices are well developed (as shown schematically in Figure 9.4), so that Figure 9.6 leads to clear well-developed trends. For the pre-transition region (before transition), the gap between laminar coiled tube and straight tube is very significant and there is an increasing gap with Reynolds number (N_{Re}). Thus, one can gain by effecting better radial mixing by involving a certain degree of curvature in otherwise straight tubes. Their study has established the changing nature of Dean vortices (secondary flow), which can be used in deciding the zone of efficient operation of CFI by considering geometric factors and flow factors. Another study by Sharma et al. (2017b) established the effect of viscous forces on the mixing through CFI, and postulated that the effect of viscous forces on the extent of axial dispersion is weakened by incorporating coiling in a straight tubular flow. Moreover, a correction factor of $(\lambda)^{1/2} \cdot (Sc)^{2/3}$ is presented to deduce the intensity of axial dispersion from coiled tubes to straight tubes. Thus, even for highly viscous fluids, CFI is an appropriate choice as an extractor. While designing CFI as an extractor, some design factors should be considered. Details about all parameters for designing an effective CFI are well documented by Saxena and Nigam (1984). Here are a few recommendations:

a. Minimum number of coil turns on each arm of CFI = 4.
b. Uniform volume on each arm of CFI.
c. Curvature ratio of the CFI geometry should be kept as minimum as possible, without causing deformation of the coil tubing.
d. No gap between the coil turns.

9.3.2 Extraction Parameters

For comparison of continuous and batch micro-extraction systems, extraction efficiency (E), extraction ratio (E_r), separation factor ($\beta_{Co/Ni}$), and distribution ratio (D) are defined as follows.

The ratio of the solute transferred to the maximum transferable amount is known as extraction efficiency (E):

$$E = \frac{C_{org, out} - C_{org, in}}{C_{org, eq} - C_{org, in}} \times 100 \tag{9.1}$$

where $C_{org, out}$, $C_{org, in}$ are the outlet and inlet concentrations of acetone in the organic phase, respectively. The extraction ratio (E) is defined as ratio of the metal ion concentration in the organic phase to that in the feed aqueous phase:

$$E = \frac{C_{aq, in} - C_{aq, out}}{C_{aq, in}} \times 100 \tag{9.2}$$

The separation efficiency of Co from Ni is described by the separation factor ($\beta_{Co/Ni}$), defined as one distribution coefficient divided by another:

$$\beta_{Co/Ni} = \frac{D_{Co}}{D_{Ni}} = \frac{C_{aq, in} - C_{aq, out}}{C_{aq, out}} \tag{9.3}$$

The distribution coefficient (D), defined as the ratio of the metal ion concentration in the organic phase to that in the aqueous phase after extraction, is expressed as

$$E = \frac{C_{org, out}}{C_{aq, out}} = \frac{C_{aq, in} - C_{aq, out}}{C_{aq, out}} \tag{9.4}$$

The Dean's number (De) is calculated using the formula given in Eq. (9.5).

$$De = Re\sqrt{\frac{1}{\lambda}} \tag{9.5}$$

where Re is the Reynolds number and λ is the curvature ratio defined as the ratio of curvature diameter to the tube diameter.

9.3.3 Methodology and Setup for Micro-flow Extraction

A generalized setup has been explained based on the different experimental setup for continuous micro-flow extractions used in previous studies (Gürsel et al. 2015, 2016; Kurt et al. 2016a,b; Zhang et al. 2017). The experimental setup of continuous micro-flow system can be arranged in two separate units including an extraction or separation unit and a flow visualization unit, as shown in Figure 9.7a,b, respectively. Further, the extraction/separation unit was composed by the coupling of the mixing/reaction with a separation unit. A syringe pump was used to deliver the reagents (phenylacetylene, benzyl bromide, sodium azide (NaN$_3$) and dimethyl sulfoxide [DMSO]) to the reactor. Another syringe pump was used for delivering the aqueous phase (waste metal source and DMSO) to the reactor. For laboratory scale experiments, a tubular reactor made of PFA tube (1.59 mm

OD, 0.5 mm ID) was used and the reagents were pre-mixed using a T-mixer with 0.5 mm ID. The flat bottom flangeless fittings were used to make the connections between different fluidic streams. To control the temperature of the reaction, the reactor was submerged in an oil bath in which the temperature was maintained by a heating plate. A pressure regulator with a capacity of 40 psi was used to regulate the backpressure.

The reacting solution coming out of the reactor was quenched with ethyl acetate (EtOAc) and ethylenediaminetetraacetic acid (EDTA). Syringe pumps were used to deliver the EtOAc and EDTA, which were separately loaded into two 5 ml Becton Dickinson (BD) plastic syringes. These two streams were mixed with reaction outlet solution by using a polyether ether ketone (PEEK) cross mixer. Then, the solution flowed in the PFA tube (1000 mm length, 1.59 mm OD, 0.5 mm ID) and was finally delivered to the flow separator. The organic phase containing the solvent and the aqueous phases containing metals were separated in the flow separator. Similar experimental setups were used for continuous liquid–liquid separation (Gürsel et al. 2016) and for the extraction of Co and Ni (Zhang et al. 2017).

The flow patterns in the micro-flow extraction system were investigated employing a flow visualization unit, which comprises a high-speed imaging camera and a computer, as shown in Figure 9.7b. To investigate the flow patterns, equal volumes of organic and aqueous phase were fed for a range of total flow rates from 24 to 80 ml/h. The flow patterns were recorded with a high-speed camera after achieving a table flow regime. For clear pictures of the flow patterns, the camera was positioned at a defined distance from the inlet and outlet of the

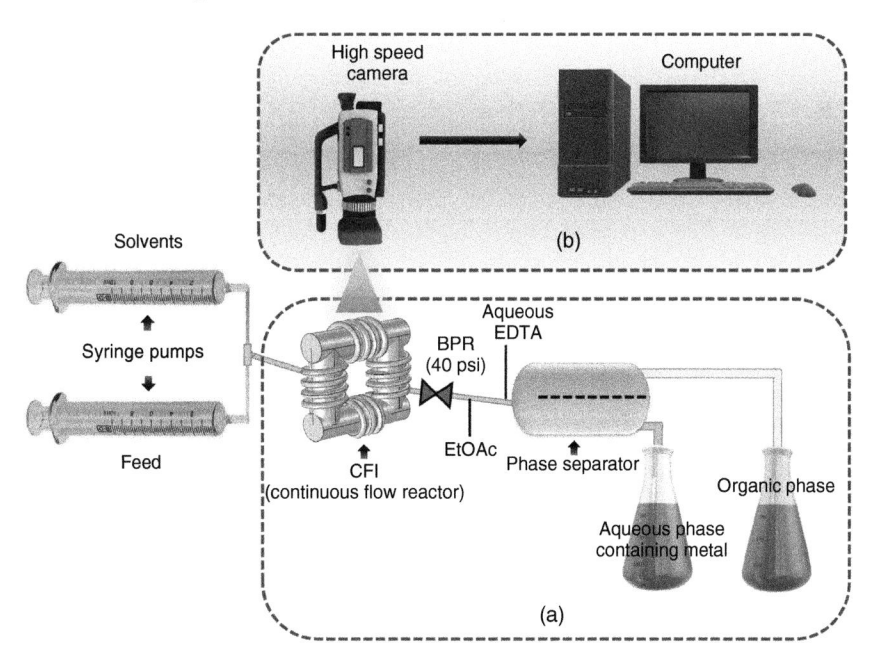

Figure 9.7 Illustration of the micro-flow extraction system: (a) extraction/separation unit and (b) flow visualization unit. BPR–Back Pressure Regulator.

CFI. Also, the image analysis can be performed using an image analysis software (MotionPro) in the computer as shown in Figure 9.7b.

9.3.4 Liquid–Liquid Micro-flow Extraction

As explained previously, liquid–liquid extraction is an important part of the process industry. Thus, this section presents a discussion on liquid–liquid extraction process in terms of the flow patterns and extraction efficiencies of batch and continuous micro-extraction in CFI.

9.3.4.1 Typical Flow Patterns

The liquid–liquid flow extraction process in a micro-extraction device is governed by its larger surface area density and the flow patterns. As illustrated in Figure 9.8, generally annular flow, parallel flow, bubbly flow, slug flow (Taylor flow), slug-dispersed flow, bubbly-dispersed, and annular-dispersed flow are the typical flow patterns observed in the microchannels for immiscible liquid–liquid two-phase flow (Cao et al. 2018; Salim et al. 2008; Zhao et al. 2006). The flow patterns in microchannels are governed by the interfacial forces and different fluid properties. For smaller differences between the interfacial tension and fluid properties, more complex flow patterns such as slug-dispersed flow were characterized by the flow of continuous phase (in small size bubble form) in the dispersed slugs at higher velocities (Zhang et al. 2019). In addition, a dispersed flow pattern with very fine bubbles of continuous phase filled into the dispersed phase was observed on further increase in the flow rates. According to the dominance of forces in the flow systems, these flow patterns can be classified into three regimes as follows:

Dispersed inertia or viscous force dominated regime: Typical flow patterns observed under this flow regime were annular flow and parallel flow, as shown in Figure 9.8a,b, respectively. The formation of these flow patterns takes place due to strong disturbance of the flow caused by increased flow rates of both the fluids, which generates dispersion of the bubbles into the annular core, which is known as annular-dispersed flow, as shown in Figure 9.8g.

Continuous inertia or viscous force dominated regime: The spherical bubbles form with a size smaller than the tube diameter for stronger inertia or viscous force over the interfacial tension. This flow pattern is known as bubbly flow, as shown in Figure 9.8c.

Interfacial tension dominated regime: The slug flow pattern is mainly observed under the regime dominated by interfacial tension at relatively low flow rates of both the liquids. As shown in Figure 9.8d, slug flow is characterized by generation of the elongated droplets with a length longer than the width of microchannel. If interfacial tension has complete control over the flow system, the sizes of the continuous and dispersed segments are uniform with a stable slug flow. Further, the inertia and viscous forces lead to the formation of droplets of irregular lengths with increase in flow rates of both the liquids. If the flow rates of both liquids are increased further, the slug-dispersed flow is observed for the water–toluene system with lower viscosity, as shown in

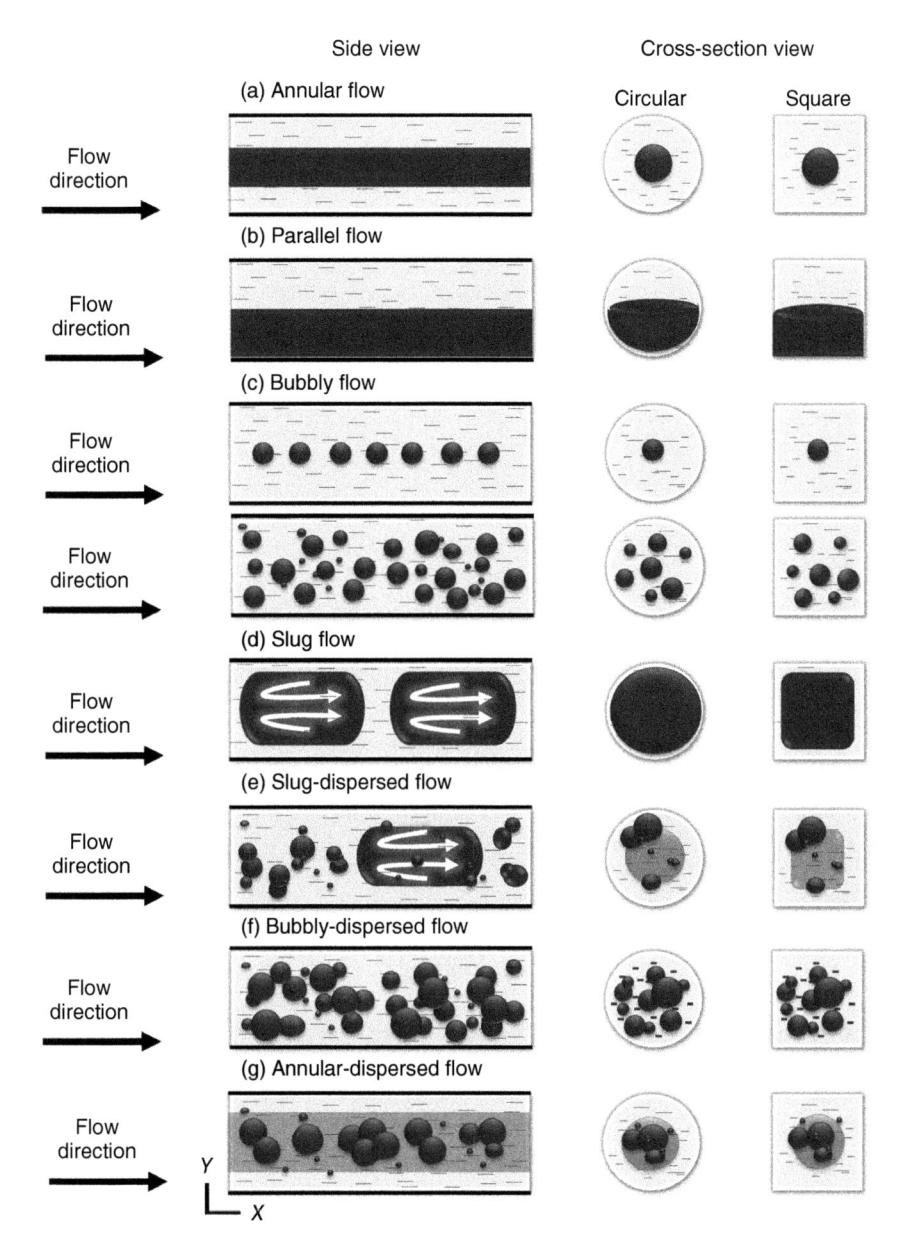

Figure 9.8 Flow pattern map for toluene–water system (a–d) and the ethyl acetate–water system (a–g).

Figure 9.8e. Generalized flow patterns maps are covered under the above classification. However, the properties of the fluid also have strong effects on the characteristics of the flow. The flow patterns for toluene–water and toluene–sulfuric acid systems are generally same, except the slug-dispersed flow. For toluene–water system, the fine toluene droplets generated by the rough interface were easily encapsulated into the water droplets. Similar flow patterns were reported for n-butyl acetate/acetone/water flow system in CFI. Furthermore, an additional flow pattern shown in Figure 9.8f was observed for the ethyl acetate–water system in the T-mixer. Therefore, the flow patterns are strongly influenced by the geometry and properties of the fluids.

9.3.4.2 Extraction Efficiency

The liquid–liquid extraction for a model system of toluene–water–acetone was carried out in CFI and straight tube in combination with the PTFE membrane flow separator or slit shaped flow separator (Gürsel et al. 2016). A milli CFI with length and tube diameter equal to 210 cm and 3.2 mm was used in the places where partitioning of the slug flow takes place. The extraction efficiency (E) was investigated as a function of the total flow rate at equal flow rates of both the aqueous and organic phases (i.e. the volume fraction of aqueous phase was kept at 0.5). A comparative analysis of the CFI with straight tube for two different systems, viz toluene–water–acetone system and butyl acetate–water–acetone are shown in Figures 9.9 and 9.10, respectively. A maximum flow rate up to 120 ml/min (7.2 l/h) was studied for the toluene–water–acetone system. The results shown in Figure 9.9 reveal that extraction efficiency of both the CFI and straight tubes increases with increase in flow rate up to 30 ml/min. The increase in efficiency was due to the enhanced mass transfer rate caused by the internal circulations within the slug. Up to a flow rate of 30 ml/min, the extraction efficiency of CFI was about 8% higher than that of the straight tube. However, the extraction efficiency of the straight tube started dropping with further increase in total flow rates due to decrease in the contact time. Other studies also observed the same behavior for the straight tube (Kashid et al. 2007, 2010). On the other hand, extraction efficiency of the CFI did not drop

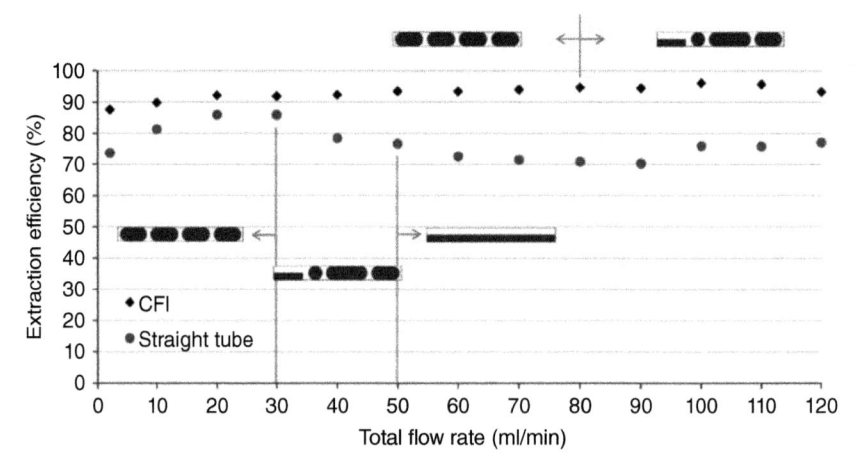

Figure 9.9 CFI vs. straight tube: extraction efficiency as a function of the total flow rate for the toluene–water–acetone system. Source: Gürsel et al. (2016). Reprinted with permission of Elsevier.

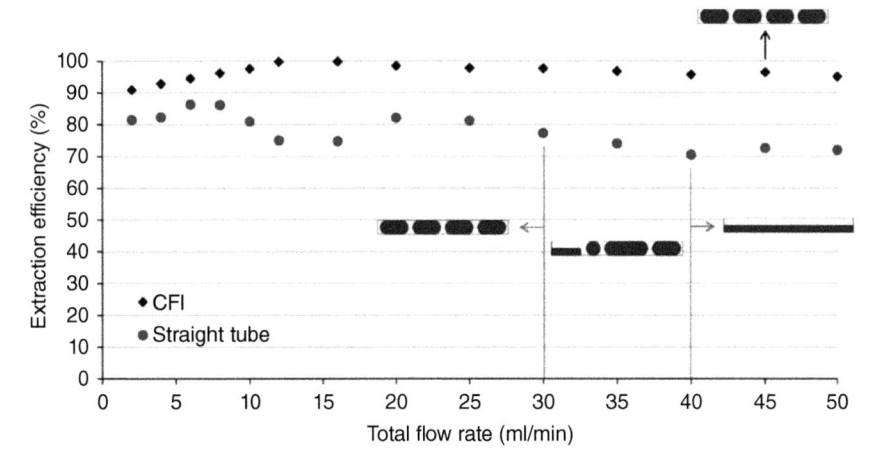

Figure 9.10 CFI vs. straight tube: extraction efficiency as a function of the total flow rate for butyl acetate–water–acetone system. Source: Gürsel et al. (2016). Reprinted with permission of Elsevier.

even at higher flow rates (>30 ml/min) and thus, CFI provides up to 20% higher extraction efficiency than the straight tube. As mentioned earlier, secondary flow is generated due to centrifugal forces inside coiled tubes. Dean number is a measure of the magnitude of the secondary flow. The reason behind these phenomena is the enhanced radial mixing caused by utilization of enhanced secondary forces and chaotic advection due to multiple flow inversion in the CFI. Similar results were observed for the butyl acetate–water–acetone system, as shown in Figure 9.10. For the butyl acetate–water–acetone system, the CFI offers 10% higher extraction efficiency up to a flow rate of 30 ml/h. Further increase of flow rate (up to 50 ml/h) results in 21% higher extraction efficiency of CFI than that of straight tube caused by chaotic advection.

9.3.4.3 Effect of Aqueous Phase Volume Fractions on Extraction Efficiency

The effects of aqueous phase volume fractions (aqueous phase flow rate divided by the total flow rate) on the yield of acetone and extraction efficiencies were investigated by Gürsel et al. (2016), as shown in Figure 9.11a,b, respectively. The aqueous phase volume fraction was varied from 0.1 to 0.9, while the total flow rates were kept constant at 20 ml/min for toluene–water–acetone system and 10 ml/min for butyl acetate–water–acetone system.

As shown in Figure 9.11a, acetone yield decreases with increase in the volume fractions of the aqueous phase. The reason for the above observation was the decreased amount of toluene that extracts the acetone from water, due to increase in the aqueous phase volume fractions. Furthermore, the acetone yield was close to the experimentally measured value in the CFI for a constant value of total flow rate (20 ml/min). As shown in Figure 9.11b, a similar trend was observed for the extraction efficiency of toluene–water–acetone system. The extraction efficiency for toluene–water–acetone system decreases with increase in aqueous phase

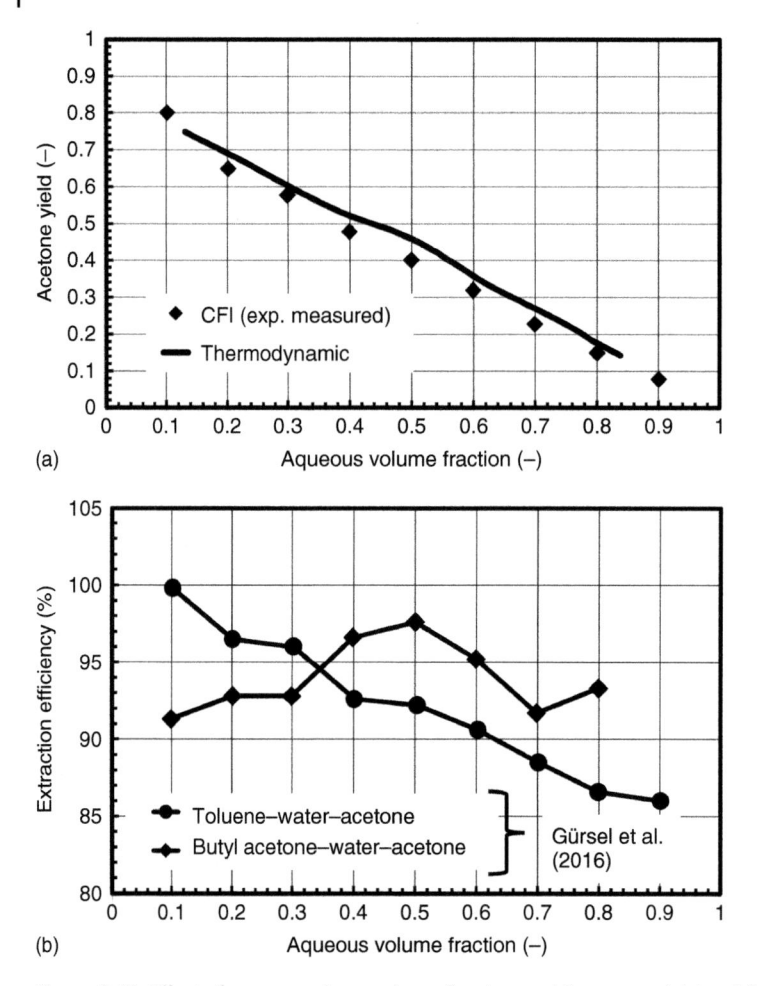

Figure 9.11 Effect of aqueous phase volume fraction on (a) acetone yield and (b) extraction efficiency at constant total flow rates.

volume fractions. As explain previously, the reason is the availability of less amount of toluene due to the increased volume fractions of the aqueous phase. However, the effect of the aqueous phase volume fractions was not that significant for butyl acetate–water–acetone system, as shown in Figure 9.11b. It may be noted that the value of the extraction efficiency was equal to 94% and varied only by ±3% with a variation of the aqueous phase volume fraction. Further, a maximum of extraction efficiency was observed due to the opposing effects of water slug and butyl acetate slug caused by increased volume fractions of aqueous phase. The mass transfer in both the water slug and the butyl acetate slug is governed by internal circulation flow. The water slug length increases with increase in aqueous phase volume fraction and thus decreases the mass transfer, while with increasing aqueous phase volume fraction the butyl acetate slug length decreases and therefore increase the mass transfer.

9.3.5 Micro-flow Extraction of Co and Ni

In this section, the segmented miniaturized extraction of Co and Ni sulfate solutions with Cyanex 272 has been discussed from a recent study (Zhang et al. 2017). The micro-extraction of Co and Ni in CFI was compared with the batch system of different batch time (30 seconds, 10 minutes), considering the effect of pH, effect of residence time, and effect of extractant concentration. The effects of these factors on extraction ratio and separation factor have been discussed in this section.

9.3.5.1 Effect of pH

The metal cations were extracted using a proton-exchange mechanism with release of H^+ during the extraction process, as Cyanex 272 extractant is an organophosphorus acid. The production of H^+ would increase with increasing rate of metal extraction, which causes a decrease in pH and therefore the amount of the metal extracted also decreases. So, an optimum value of pH_{final} must be maintained to increase the rate of the forward extraction reaction. Thus, to maintain the pH of the reaction, the acid extractant must be pre-neutralized by adding sodium hydroxide. The effect of the pH_{final} on the extraction ratio of Co and Ni is shown in Figure 9.12. It may be noted that the extraction in both the micro-flow CFI and batch system was a function of pH_{final}. In the case of micro-flow CFI, the extraction ratio of Co improved significantly from 2.13% to 97.54% with an increase of pH_{final} from 3.25 to 5.15 and remains constant thereafter. Similarly, the extraction of Ni also enhanced from 2.31% to 13.19% with increase in pH_{final}. Similar trends were observed in two different cases (30 seconds, 10 minutes) of batch system. Additionally, in all three cases, the separation factor ($\beta_{Co/Ni}$) increases to the highest value and then declines, as shown in Figure 9.12b. The drop in the separation factor at higher pH_{final} is caused by the depletion of cobalt, while the extraction of nickel continuously increases. Therefore, for the considered experiments the range of pH_{final} must be between 5.8 and 5.9. Furthermore, micro-flow CFI offers higher separation efficiency than the batch system. The enhanced performance of CFI micro-flow extraction system is much more visible in the Sections 9.3.5.2 and 9.3.5.3.

9.3.5.2 Effect of Residence Time

The extraction efficiency and separation factors were plotted against residence time, as shown in Figure 9.13a,b, respectively. It may be noted that effect of residence time on the extraction efficiency of CFI micro-flow extractor was insignificant. The extraction efficiency of CFI microextraction was 96.7% for Co and 6.3% for Ni at 15 seconds. However, in batch system, the extraction ratio of Co was influenced by an increase in residence time. Similarly, the extraction of Ni also increases with increase in the residence time. However, the results suggested that the higher extraction efficiency of one metal would eventually lead to a lower extraction of other metals. In addition, the separation factor ($\beta_{Co/Ni}$) increases up to 748 at 60 seconds in CFI micro-flow extraction, much higher than that from 426 at 600 seconds in the batch system. Furthermore, the CFI offers a 43% higher separation factor than that of the batch system with 10 times lower residence time.

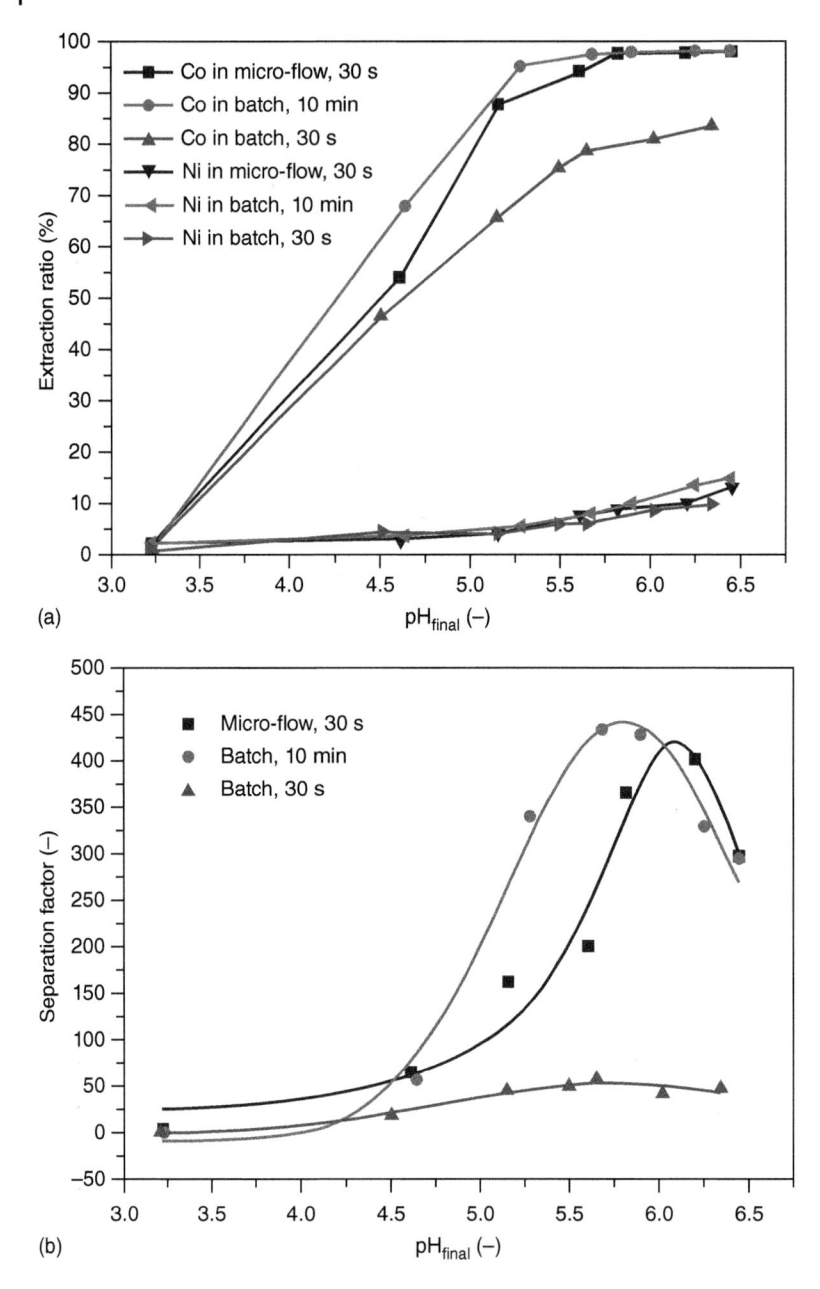

Figure 9.12 Effect of pH_{final} on (a) extraction ratio and (b) separation factor of Co and Ni, for 20% v/v Cyanex 272 by varying level of pre-neutralization as organic phases. Source: Zhang et al. (2017). Reprinted with permission from Elsevier.

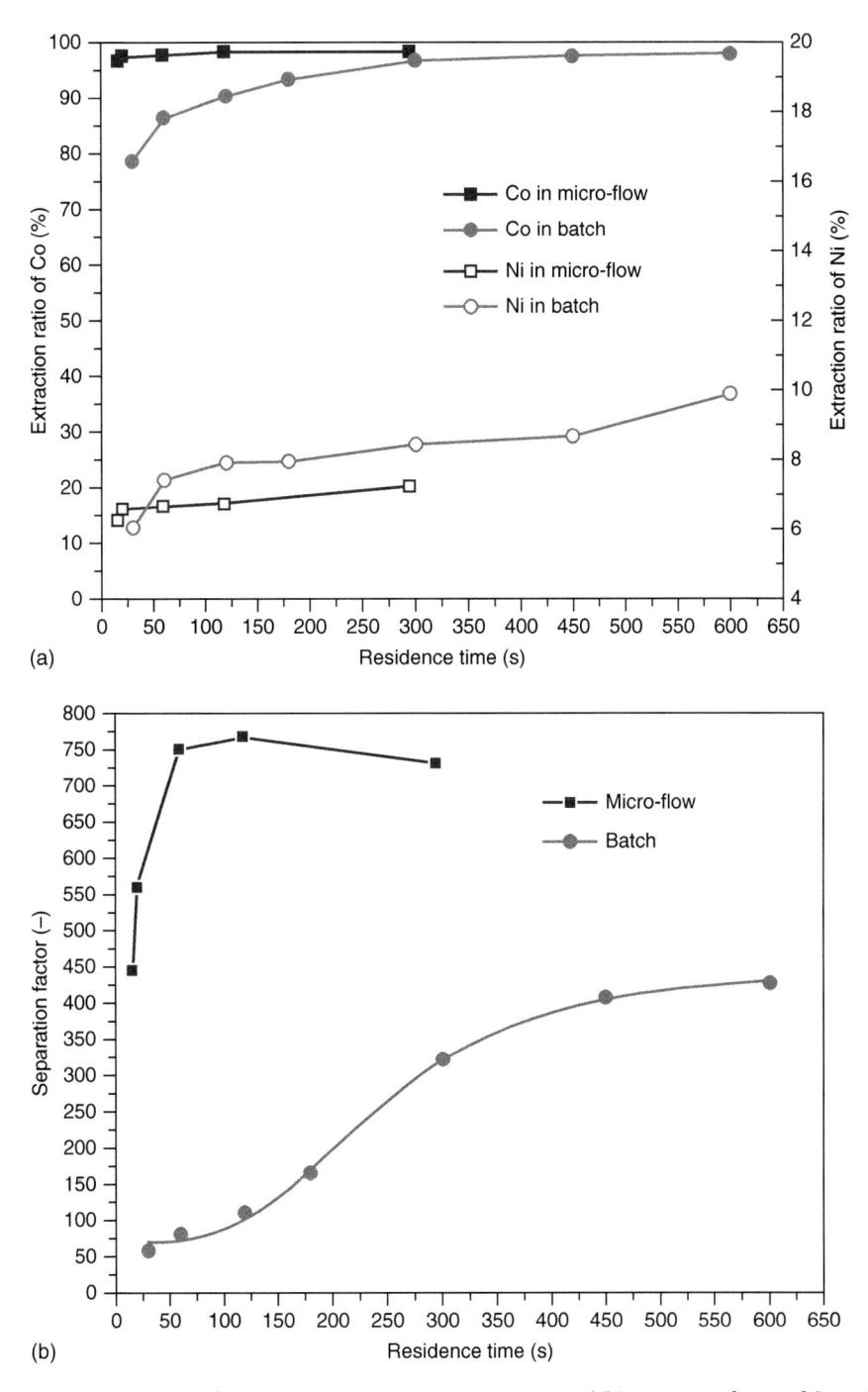

Figure 9.13 Effect of residence time on (a) extraction ratio and (b) separation factor of Co and Ni, for 20% v/v Cyanex 272 with 40% pre-neutralization. Source: Zhang et al. (2017). Reprinted with permission of Elsevier.

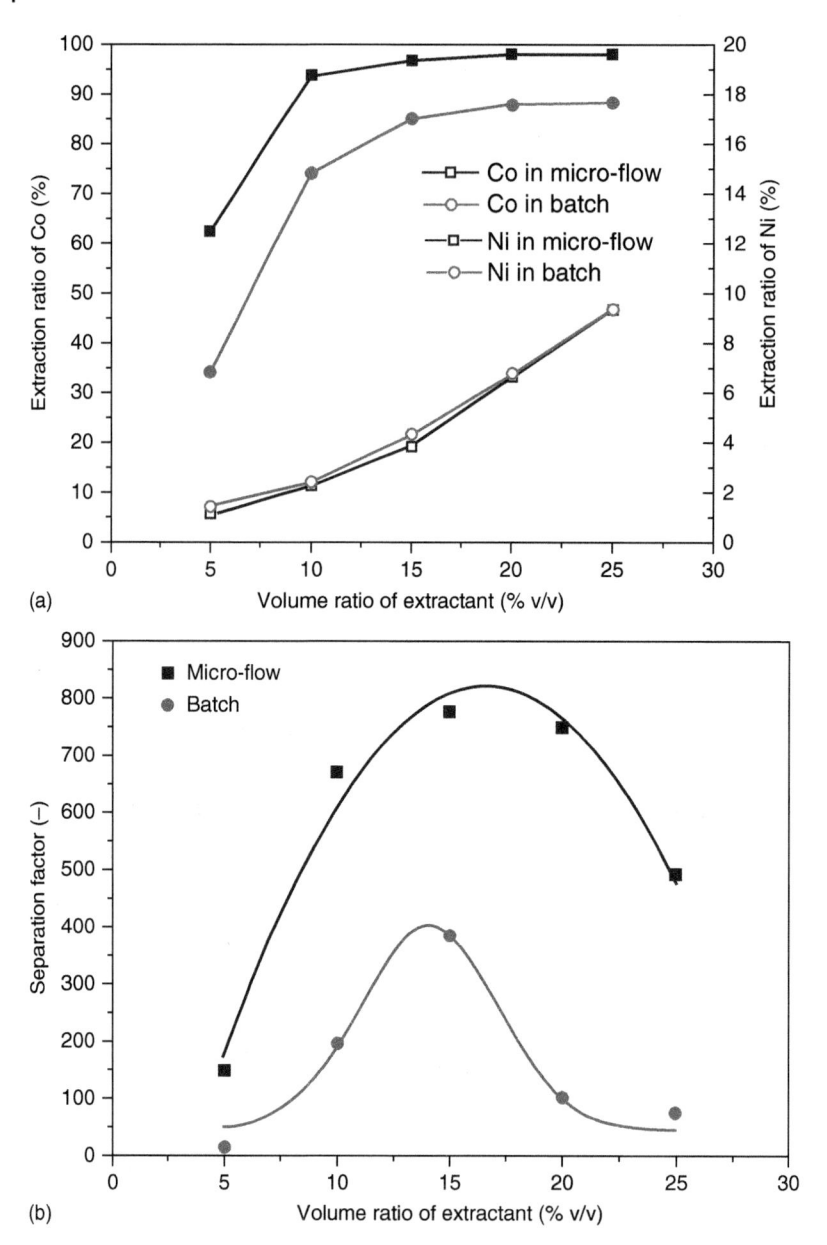

Figure 9.14 Effect of extractant concentration on (a) extraction ratio and (b) separation factor of Co and Ni, for 40% pre-neutralization of Cyanex 272 and 60 seconds residence time. Source: Zhang et al. (2017). Reprinted with permission of Elsevier.

9.3.5.3 Effect of Extractant Concentration

The effects of Cyanex 272 on the extraction ratio and separation factor was investigated by varying its concentration from 5% v/v to 25% v/v, as shown in Figure 9.14a,b. In thenCFI micro-flow extractor, the Co extraction ratio increases from 62.6% to 96.9% with increase in concentration of Cyanex 272 from 5% v/v to 15% v/v. Similar trends were observed for the batch system. Furthermore, it can be noted that CFI micro-flow extractor provides much better extraction ratio of Co than the batch system at any extractant concentration. In contrast, there was little difference between micro-flow extractor and batch system for the extraction of Ni. Thus, continuous extraction of Co and Ni can be successfully achieved in CFI continuous micro-flow extractor with enhanced extraction efficiency at reduced residence time.

9.4 Summary and Future Challenges

Micro-flow extraction for liquid–liquid and metal extraction operations can offer a significant contribution to the miniaturization of separation and purification processes. In the last two decades, scientists have acquired great knowledge on the different unit operations involving multiphase system, which has positively resulted in the understanding of multiphase micro-flow processes. Therefore, intensified miniaturized extraction systems including droplet generators and in-line phase separators have been developed with significantly improved extraction efficiency in fine chemical synthesis and small scale purification. In addition, continuous micro-flow extraction has offered a great enhancement in extraction and separation factors.

However, the scale-up and industrial application of these miniaturized separation devices is still a concern. The micro-extraction of precious metals and other products from the real process waste (metallurgical waste, chemical waste, and pharmaceutical waste) offers novel opportunities and poses challenges for industrial viability of these micro-flow extraction devices. 3D-printing may be effectively utilized to design truly miniaturized micro-flow extraction devices for the separation and purification of novel target substances in the biological, chemicals, and pharmaceuticals products.

Questions

1. Define process intensification and its importance in miniaturized extraction processes.

2. What are the advantages of the micro-flow extractors in different unit operations?

3. Discuss the methodology and experimental setup of continuous micro-flow extractions. What are the advantages of the continuous micro-flow extractors?

4 Discus different miniaturized micro-extraction devices in terms of their working principle and enhancement mechanism.

5 What are the major process parameters that can affect the extraction efficiency in a continuous extraction process?

6 Classify the active and passive techniques for intensification of heat and mass transfer process.

7 Classify the types of flow patterns in a continuous multiphase flow system. Also, discuss their importance for heat and mass transfer operations.

8 What is secondary flow and how does it affect the micro-flow extraction process?

9 Define the concept of flow inversion. Why does a device with flow inversion provide better extraction efficiency over a straight tube?

10 Discuss the effect of the aqueous phase volume fraction on the extraction efficiencies for toluene–water–acetone and butyl acetate–water–acetone systems.

Bibliography

Abas, F.O. (2008). Vanadium oxide recovery from spent catalysts by chemical leaching. *Engineering and Technology* 26: 1–24.

Abbas, M., Kaddour, S., and Trari, M. (2014). Kinetic and equilibrium studies of cobalt adsorption on apricot stone activated carbon. *Journal of Industrial and Engineering Chemistry* 20: 745–751.

Abbott, A.P., Capper, G., Davies, D.L. et al. (2003). Novel solvent properties of choline chloride/urea mixtures. *Chemical Communications* (1): 70–71.

Abbott, A.P., Boothby, D., Capper, G. et al. (2004). Deep eutectic solvents formed between choline chloride and carboxylic acids: versatile alternatives to ionic liquids. *Journal of the American Chemical Society* 126: 9142–9147.

Abdel-Aal, E.A. (2000). Kinetics of sulfuric acid leaching of low-grade zinc silicate ore. *Hydrometallurgy* 55: 247–254.

Abdel-Aal, E.A. and Rashad, M.M. (2004). Kinetic study on the leaching of spent nickel oxide catalyst with sulfuric acid. *Hydrometallurgy* 74: 189–194.

Abe, M., Hansupalak, N., Ookawara, S. et al. (2016). Bulk polymerization in "CFI" microreactor. Presented at *The Flow Chemistry Society, at IMRET14*, Beijing, China (12–14 September 2016).

Abhilash and Pandey, B.D. (2011). Role of ferric ions in bioleaching of uranium from low tenor Indian ore. *Canadian Metallurgical Quarterly* 50 (2): 102–112.

Abolhasani, M., Coley, C.W., Xie, L. et al. (2015). Oscillatory microprocessor for growth and *in situ* characterization of semiconductor nanocrystals. *Chemistry of Materials* 27: 6131–6138.

Adamczuk, A. and Kołodyńska, D. (2015). Equilibrium, thermodynamic and kinetic studies on removal of chromium, copper, zinc and arsenic from aqueous solutions onto fly ash coated by chitosan. *Chemical Engineering Journal* 274: 200–212.

Adamo, A., Beingessner, R.L., Behnam, M. et al. (2016). On-demand continuous flow production of pharmaceuticals in a compact, reconfigurable system. *Science* 352: 61–67.

Adamson, A.W. (1954). A proposed approach to the chelate effect. *Journal of the American Chemical Society* 76: 1578–1579.

Agricultural Waste Management Systems (2008). Agricultural waste management field handbook. http://irrigationtoolbox.com/NEH/Part651_AWMFH/awmfh-chap9.pdf (accessed 10 December 2018).

Ajmal, M., Rao, R.A.K., Ahmad, R., and Ahmad, J. (2000). Adsorption studies on *Citrus reticulata* (fruit peel of orange): removal and recovery of Ni(II) from electroplating wastewater. *Journal of Hazardous Materials* B79: 117–131.

Akcil, A., Erust, C., Gahan, C.S. et al. (2015). Precious metal recovery from waste printed circuit boards using cyanide and non-cyanide lixiviants – A review. *Waste Management* 45: 258–271.

Akpomie, K.G. and Dawodu, F.A. (2014). Efficient abstraction of nickel(II) and manganese(II) ions from solution onto an alkaline-modified montmorillonite. *Journal of Taibah University for Science* 8 (4): 343–356.

Akpomie, K.G. and Dawodu, F.A. (2015). Physicochemical analysis of automobile effluent before and after treatment with an alkaline-activated montmorillonite. *Journal of Taibah University for Science* 9: 465–476.

Akpomie, K.G. and Dawodu, F.A. (2016). Acid-modified montmorillonite for sorption of heavy metals from automobile effluent. *Journal of Basic and Applied Sciences* 5: 1–12.

Ali, A.A.H. and Hunaidi, T.A. (2004). Breakthrough curves and column design parameters for sorption of lead ions by natural zeolite. *Environmental Technology* 25: 1009–1019.

Alinnor, I.J. (2007). Adsorption of heavy metal ions from aqueous solution by fly ash. *Fuel* 86 (5–6): 853–857.

Allain, E. and Gaballah, I. (1994). Recycling of strategic metals from industrial slag by hydro and pyrometallurgical processes. *Resources, Conservation and Recycling* 10: 75–85.

Al-Mansi, N.M. and Abdel Monem, N.M. (2002). Recovery of nickel oxide from spent catalyst. *Waste Management* 22: 85–90.

Alpat, S.K., Ozbayrak, O., Alpat, S., and Akcay, H. (2008). The adsorption kinetics and removal of cationic dye, Toluidine Blue O, from aqueous solution with Turkish zeolite. *Journal of Hazardous Materials* 151: 213–220.

Altansukh, B., Haga, K., Ariunbolor, N., and Kawamura, S. (2016). Leaching and adsorption of gold from waste printed circuit boards using iodine-iodide solution and activated carbon. *Engineering Journal* 20: 29–40.

Amana, T., Kazi, A.A., Sabri, M.U., and Banoa, Q. (2008). Potato peels assolid waste for the removal of heavy metal copper(II) from wastewater/industrial effluent. *Colloids and Surfaces B: Biointerfaces* 63: 116–121.

Amer, H., El-Gendy, A., and El-Haggar, S. (2017). Removal of lead (II) from aqueous solutions using rice straw. *Water Science and Technology* 76 (5): 1011–1021.

Amiri, F., Yaghmaei, S., Mousavi, S.M., and Sheibani, S. (2011). Recovery of metals from spent refinery hydrocracking catalyst using adapted *Aspergillus niger*. *Hydrometallurgy* 109: 65–71.

Amiri, F., Mousavi, S.M., Yaghmaei, S., and Barati, M. (2012). Bioleaching kinetics of a spent refinery catalyst using *Aspergillus niger* at optimal conditions. *Biochemical Engineering Journal* 67: 208–217.

Andrea, M. and Keith, S. (2002). Leaching and electrochemical recovery of copper, lead and tin from scrap printed circuit boards. *Journal of Chemical Technology and Biotechnology* 77: 449–457.

Andreatta, A.E., Charnley, M.P., and Brennecke, J.F. (2015). Using ionic liquids to break the ethanol–ethyl acetate azeotrope. *ACS Sustainable Chemistry & Engineering* 3: 3435–3444.

Angelini, G., Chiappe, C., Maria, P.D. et al. (2005). Determination of the polarities of some ionic liquids using 2-nitrocyclohexanone as the probe. *Journal of Organic Chemistry* 70: 8193–8196.

Anlauf, H. (2007). Recent developments in centrifuge technology. *Separation and Purification Technology* 58: 242–246.

Appleby, D., Hussey, C.L., Seddon, K.R., and Turp, J.E. (1986). Room-temperature ionic liquids as solvents for electronic absorption spectroscopy of halide complexes. *Nature* 323: 614–616.

Aras, G. and Crowther, D. (2009). Corporate sustainability reporting: a study in disingenuity. *Journal of Business Ethics* 87 (Suppl 1): 279–288.

Armand, M., Endres, F., MacFarlane, D.R. et al. (2009). Ionic-liquid materials for the electrochemical challenges of the future. *Nature Materials* 8: 621–629.

Arshadi, M. and Mousavi, S.M. (2015). Multi-objective optimization of heavy metals bioleaching from discarded mobile phone PCBs: simultaneous Cu and Ni recovery using *Acidithiobacillus ferrooxidans*. *Separation and Purification Technology* 147: 210–219.

Askari, M.A., Hiroyoshi, N., Tsunekawa, M. et al. (2005). Rhenium extraction in bioleaching of Sarcheshmeh molybdenite concentrate. *Hydrometallurgy* 80: 23–31.

Arterburn, R.A. (1976). *The Sizing of Hydrocyclones*. Menlo Park, CA: Krebs Engineers.

Assmann, N., Kaiser, S., and Rudolf Von Rohr, P. (2012). Supercritical extraction of vanillin in a microfluidic device. *Journal of Supercritical Fluids* 67: 149–154.

Assmann, N., Ładosz, A., and Rudolf von Rohr, P. (2013). Continuous micro liquid–liquid extraction. *Chemical Engineering and Technology* 36: 921–936.

ASSOCHAM-cKinetics study (2016). Associated Chambers of Commerce and Industry of India (ASSOCHAM)–cKinetics joint study. 'India's WEEE growing at 30% per annum. http://www.assocham.org/newsdetail.php?id=5725 (accessed 21 September 2018).

Atalay, T.S., Kılıçarslan, A., and Sarıdede, M.N. (2015). *Energy Technology 2015: Carbon Dioxide Management and Other Technologies* (eds. A. Jha, C. Wang, N.R. Neelameggham, et al.), 201–207. TMS (The Minerals, Metals & Materials Society).

Aung, K.M. and Ting, Y.P. (2005). Bioleaching of spent fluid catalytic cracking catalyst using *Aspergillus niger*. *Journal of Biotechnology* 116 (2): 159–170.

Awasthi, A.K. and Li, J. (2017). An overview of the potential of eco-friendly hybrid strategy for metal recycling from WEEE. *Resources, Conservation and Recycling* 126: 228–239.

Awasthi, A.K., Zlamparet, G.I., Zeng, X., and Li, J. (2017). Evaluating waste printed circuit boards recycling: opportunities and challenges, a mini review. *Waste Management and Research* 35: 346–356.

Aylmore, M.G. and Muir, D.M. (2001). Thiosulfate leaching of gold—A review. *Minerals Engineering* 14: 135–174.

Ayuso, E.A., Sanchez, A.G., and Querol, X. (2003). Purification of metal electroplating waste waters using zeolites. *Water Research* 37: 4855–4862.

Babai, A. and Mudering, A.-V. (2006). Crystal engineering in ionic liquids. The crystal structures of [Mppyr]$_3$[NdI$_6$] and [Bmpyr]$_4$[NdI$_6$][Tf$_2$N]. *Inorganic Chemistry* 45: 4874–4876.

Babay, P.A., Emilio, C.A., Ferreyra, R.E. et al. (2001). Kinetics and mechanisms of EDTA photocatalytic degradation with TiO$_2$. *Water Science and Technology* 44: 179–185.

Babel, S. and Kurniawan, T.A. (2003). Low-cost adsorbents for heavy metals uptake from contaminated water: a review. *Journal of Hazardous Materials* B97: 219–243.

Baghalha, M. and Ebrahimpour, O. (2007). Structural changes and surface activities of ethylbenzene dehydrogenation catalysts during deactivation. *Applied Catalysis A: General* 326 (2): 143–151.

Baldé, C.P., Wang, F., Kuehr, R., and Huisman, J. (2015). *The Global E-Waste Monitor – 2014*. Bonn, Germany: United Nations University. IAS – SCYCLE.

Banerjee, S.B. (2002). Organisational strategies for sustainable development: developing a research agenda for the new millennium. *Australian Journal of Management* 27: 105–119.

Bansode, R.R., Losso, J.N., Marshall, W.E. et al. (2003). Adsorption of metal ions by pecan shell-based granular activated carbons. *Bioresource Technology* 89: 115–119.

Barakat, M.A. (2008). Adsorption of heavy metals from aqueous solutions on synthetic zeolite. *Research Journal of Environmental Sciences* 2 (1): 13–22.

Barrera-Cortés, J., Manilla-Pérez, E., and Poggi-Varaldo, H.M. (2006). Oxygen transfer to slurries treated in a rotating drum operated at atmospheric pressure. *Bioprocess and Biosystems Engineering* 29 (5–6): 391–398.

Basaldella, E.I., Vázquez, P.G., Iucolano, F., and Caputo, D. (2007). Chromium removal from water using LTA zeolites: effect of pH. *Journal of Colloid and Interface Science* 313: 574–578.

Bautista-Martinez, J.A., Tang, L., Belieres, J.P. et al. (2009). Hydrogen redox in protic ionic liquids and a direct measurement of proton thermodynamics. *Journal of Physical Chemistry C* 113: 12586–12593.

Begum, Z.A., Rahman, I.M.M., Sawai, H. et al. (2013). Effect of extraction variables on the biodegradable chelant-assisted removal of toxic metals from artificially contaminated European reference soils. *Water, Air, & Soil Pollution* 224: 1381–1402.

Behera, S.K. and Sukla, L.B. (2012). Microbial extraction of nickel from chromite overburdens in the presence of surfactant. *The Transactions of Nonferrous Metals Society of China* 22: 2840–2845.

Benz, K., Jackel, K.P., Regenauer, K.J. et al. (2001). Utilization of micromixers for extraction processes. *Chemical Engineering and Technology* 24: 11–17.

Beolchini, F., Fonti, V., Ferella, F., and Vegliò, F. (2010). Metal recovery from spent refinery catalysts by means of biotechnological strategies. *Journal of Hazarduous Material* 178: 529–534.

Berthod, A., Ruiz-Angel, M.J., and Carda-Broch, S. (2008). Ionic liquids in separation techniques. *Journal of Chromatography A* 1184: 6–18.

Bharadwaj, A. and Ting, Y.P. (2013). Bioleaching of spent hydrotreating catalyst by thermophilic and mesophilic acidophiles: effect of decoking. *Advances in Materials Research* 825: 280–283.

Bhatt, A.I., Bond, A.M., MacFarlane, D.R. et al. (2006). A critical assessment of electrochemistry in a distillable room temperature ionic liquid, DIMCARB. *Green Chemistry* 8: 161–171.

Bibri, M. (2008). Corporate sustainability/CSR communications and value creation: a marketing approach. Master's thesis. Retrieved from WorldCat Dissertations (OCLC: 747412678).

Birloaga, I., Michelis, I.D., Ferella, F. et al. (2013). Study on the influence of various factors in the hydrometallurgical processing of waste printed circuit boards for copper and gold recovery. *Waste Management* 33: 935–941.

Birloaga, I., Coman, V., Kopacek, B., and Vegliò, F. (2014). An advanced study on the hydrometallurgical processing of waste computer printed circuit boards to extract their valuable content of metals. *Waste Management* 34: 2581–2586.

Bittner, B., Wrobel, R.J., and Milchert, E. (2012). Physical properties of pyridinium ionic liquids. *Journal of Chemical Thermodynamics* 55: 159–165.

Blais, J.F., Tyagi, R.D., and Auclair, J.C. (1993). Metals removal from sewage sludge by indigenous iron-oxidizing bacteria. *Journal of Environmental Science and Health, Part A: Environmental Science and Engineering and Toxicology* 28 (2): 443–467.

Blanchard, G., Maunaye, M., and Martin, G. (1984). Removal of heavy metals from waters by means of natural zeolites. *Water Research* 18: 1501–1507.

Blaylock, M.J., Salt, D.E., Dushenkov, S. et al. (1997). Enhanced accumulation of Pb in Indian mustard by soil-applied chelating agents. *Environmental Science and Technology* 31: 860–865.

Bloom, H. (1961). Eleventh Spiers Memorial Lecture. Structural models for molten salts and their mixtures. *Discussions of the Faraday Society* 32: 7–13.

Bonhote, P., Dias, A.P., Papageorgiou, N. et al. (1996). Hydrophobic, highly conductive ambient-temperature molten salts. *Inorganic Chemistry* 35: 1168–1178.

Boon, J.A., Levisky, J.A., Pflug, J.L., and Wilkes, J.S. (1986). Friedel–Crafts reactions in ambient-temperature molten salts. *Journal of Organic Chemistry* 51: 480–483.

Borowiec, M., Huculak, M., Hoffmann, K., and Hoffmann, J. (2009). Biodegradation of selected substances used in liquid fertilizers as an element of Life Cycle Assessment. *Polish Journal of Chemical Technology* 11: 1–3.

Bosco, S.M.D., Jimenez, R.S., and Carvalho, W.A. (2005). Removal of toxic metals from wastewater by Brazilian natural scolecite. *Journal of Colloid and Interface Science* 281: 424–431.

Bosecker, K. (1997). Bioleaching: metal solubilization by microorganisms. *FEMS Microbiology Reviews* 20: 591–604.

Bosso, S.T. and Enzweiler, J. (2002). Evaluation of heavy metal removal from aqueous solution onto scolecite. *Water Research* 36: 4795–4800.

Bowman-James, K. (2005). Alfred Werner revisited: the coordination chemistry of anions. *Accounts of Chemical Research* 38: 671–678.

Brandl, H., Bosshard, R., and Wegmann, M. (2001). Computer-munching microbes: metal leaching from electronic scrap by bacteria and fungi. *Hydrometallurgy* 59: 319–326.

Brandl, H., Lrandl, H., Lehmann, S. et al. (2008). Biomobilization of silver, gold, and platinum from solid waste materials by HCN-forming microorganisms. *Hydrometallurgy* 94 (1–4): 14.

Breuer, P.L. and Jeffrey, M.I. (2000). Thiosulfate leaching kinetics of gold in the presence of copper and ammonia. *Minerals Engineering* 13: 1071–1081.

Breysse, M., Mariadassou, G.D., Geantet, G. et al. (2003). Deep desulfurization reactions, catalysts and technological challenges. *Catalysis Today* 84: 129–138.

Brusselaers, J., Hageluken, C., Mark, F. et al. (2005). An eco-efficient solution for plastics-metals-mixtures from electronic waste: the integrated metals smelter. *5th Identiplast 2005, the Biennial Conference on the Recycling and Recovery of Plastics Identifying the Opportunities for Plastics Recovery*, Brussels, Belgium.

Bryant, R.D., McGroarty, K.M., Costerton, J.W., and Laishley, E.J. (1983). Isolation and characterization of a new acidophilic *Thiobacillus* species (*T. albertis*). *Canadian Journal of Microbiology* 29 (9): 1159–1170.

Burgstaller, W. and Schinner, F. (1993). Leaching of metals with fungi. *Journal of Biotechnology* 27: 91–116.

Byerley, J.J., Fouda, S.A., and Rempel, G.L. (1973a). Kinetics and mechanism of the oxidation of thiosulphate ions by copper (II) ions in aqueous ammonia solution. *Journal of the Chemical Society, Dalton Transactions* 8: 889–893.

Byerley, J.J., Fouda, S.A., and Rempel, G.L. (1973b). The oxidation of thiosulfate in aqueous ammonia by copper (II) oxygen complexes. *Inorganic and Nuclear Chemistry Letters* 9: 879–883.

Cadena, L.E.S., Arroyo, Z.G., Lara, M.A.G., and Quiroz, Q.D. (2015). Cell-phone recycling by solvolysis for recovery of metals. *Journal of Materials Science and Chemical Engineering* 3: 52–57.

Canongia Lopes, J.N. and Rebelo, L.P.N. (2010). Ionic liquids and reactive azeotropes: the continuity of the aprotic and protic classes. *Physical Chemistry Chemical Physics* 12: 1948–1952.

Cao, Z., Wu, Z., and Sundén, B. (2018). Dimensionless analysis on liquid–liquid flow patterns and scaling law on slug hydrodynamics in cross-junction microchannels. *Chemical Engineering Journal* 344: 604–615.

Carlin, R.T. and Fuller, J. (1997). Ionic liquid-polymer gel catalytic membrane. *Chemical Communications* (15): 1345–1346.

Carlin, R.T. and Osteryoung, R.A. (1990). Complexation of Cp_2MCl_2 in a chloroaluminate molten salt: relevance to homogeneous Ziegler-Natta catalysis. *Journal of Molecular Catalysis* 63: 125–129.

Carmichael, A.J. and Seddon, K.R. (2000). Polarity study of some 1-alkyl-3-methylimidazolium ambient-temperature ionic liquids with the solvatochromic dye, Nile Red. *Journal of Physical Organic Chemistry* 13: 591–595.

Carroll, A.B. (1991). The pyramid of corporate social responsibility: toward the moral management of organizational stakeholders. *Business Horizons* 34: 39–48.

Cassity, C.G., Mirjafari, A., Mobarrez, N. et al. (2013). Ionic liquids of superior thermal stability. *Chemical Communications* 49: 7590–7592.

Castro, L.A. and Martins, A.H. (2009). Recovery of tin and copper by recycling of printed circuit boards from obsolete computers. *Brazilian Journal of Chemical Engineering* 26: 649–657.

Cayumil, R., Khanna, R., Rajarao, R. et al. (2016). Concentration of precious metals during their recovery from electronic waste. *Waste Management* 57: 121–130.

Cervera-Padrell, A.E., Morthensen, S.T., Lewandowski, D.J. et al. (2012). Continuous hydrolysis and liquid–liquid phase separation of an active pharmaceutical ingredient intermediate using a mini-scale hydrophobic membrane separator. *Organic Process Research and Development* 16: 888–900.

Cevasco, G. and Chiappe, C. (2014). Are ionic liquids a proper solution to current environmental challenges? *Green Chemistry* 16: 2375–2385.

Chalkley, M.E., Lakshmanan, V.I., Conard, B.R., and Wheeland, K.G. (eds.) (1989). *Tailings and Effluent Management*. Pergamon Press.

Chancerel, P., Meskers, C.E., Hagelüken, C., and Rotter, V.S. (2009). Assessment of precious metal flows during preprocessing of waste electrical and electronic equipment. *Journal of Industrial Ecology* 13: 791–810.

Chaudhary, A.J., Donaldson, J.D., Boddington, S.C., and Grimes, S.M. (1993). Heavy metals in the environment. II. A hydrochloric acid leaching process for the recovery of nickel value from a spent catalyst. *Hydrometallurgy* 34: 137–150.

Chauhan, G., Pant, K.K., and Nigam, K.D.P. (2012). Extraction of nickel from spent catalyst using biodegradable chelating agent EDDS. *Industrial and Engineering Chemistry Research* 51: 10354–10363.

Chauhan, G., Pant, K.K., and Nigam, K.D.P. (2013a). Metal recovery from hydroprocessing spent catalyst: a green chemical engineering approach. *Industrial and Engineering Chemistry Research* 52: 16724–16736.

Chauhan, G., Pant, K.K., and Nigam, K.D.P. (2013b). Development of green technology for extraction of nickel from spent catalyst and its optimization using response surface methodology. *Green Processing and Synthesis* 2: 259–271.

Chauhan, G., Pant, K.K., and Nigam, K.D.P. (2015a). Chelation technology: a promising green approach for resource management and waste minimization. *Environmental Science: Processes & Impacts* 17: 12–40.

Chauhan, G., Pant, K.K., and Nigam, K.D.P. (2015b). Conceptual mechanism and kinetics studies of chelating agent assisted metal extraction from spent catalyst. *Journal of Industrial and Engineering Chemistry* 27: 373–383.

Chauhan, G., Stein, M., Seidel-Morgenstern, A. et al. (2015c). The thermodynamics and biodegradability of chelating agents upon metal extraction. *Chemical Engineering Science* 137: 768–785.

Chauhan, G., Jadhao, P., Pant, K.K., and Nigam, K.D.P. (2018). Novel technologies and conventional processes for recovery of metals from waste electrical and electronic equipment: challenges & opportunities – A review. *Journal of Environmental Chemical Engineering* 6: 1288–1304.

Chauvin, Y., Gilbert, B., and Guibard, I. (1990). Catalytic dimerization of alkenes by nickel complexes in organochloroaluminate molten salts. *Journal of the Chemical Society, Chemical Communications* (23): 1715–1716.

Chen, Y., Feng, Q., Shao, Y. et al. (2006). Research on recycling of valuable metals in spent Al_2O_3 based catalyst. *Minerals Engineering* 19: 94–97.

Chen, D., Bi, X., Zhao, J. et al. (2009). Pollution characterization and diurnal variation of PBDEs in the atmosphere of an WEEE dismantling region. *Environmental Pollution* 157: 1051–1057.

Chen, A., Dietrich, K.N., Huo, X., and Ho, S.M. (2010). Developmental neurotoxicants in e-waste: an emerging health concern. *Environmental Health Perspectives* 119: 431–438.

Chen, Y., Chen, Z., and Wang, H. (2012). Enhanced fluorescence detection using liquid–liquid extraction in a microfluidic droplet system. *Lab on a Chip* 12: 4569–4575.

Cheremisinoff, P.E. (1989). *Library of Environmental Pollution Control Technology*, 557–591. Texas: Gulf Publishing Co.

Cheung, C.W., Porter, J.F., and McKay, G. (2001). Sorption kinetic analysis for the removal of cadmium ions from effluents using bone char. *Water Research* 35: 605–621.

Chi, D., Lee, J.C., Pandey, B.D. et al. (2010). Bacterial cyanide generation in the presence of metal ions (Na^+, Mg^{2+}, Fe^{2+}, Pb^{2+}) and gold bioleaching from waste PCBs. *Journal of Chemical Engineering of Japan* 44: 692–700.

Chiu, K.K., Ye, Z.H., and Wong, M.H. (2005). Enhanced uptake of As, Zn, and Cu by *Vetiveria zizanioides* and *Zea mays* using chelating agents. *Chemosphere* 60: 1365–1375.

Chojnacka, K. (2010). Biosorption and bioaccumulation--the prospects for practical applications. *Environment International* 36: 299–307.

Christmann, K. (2010). *Adsorption. Lecture Series 2010/2011. Modern Methods in Heterogeneous Catalysis Research*. Berlin: Institut für Chemie und Biochemie, Freie Universität.

Chum, H.L., Koch, V.R., Miller, L.L., and Osteryoung, R.A. (1975). Iron complexes and hexamethylbenzene in a room temperature molten salt. *Journal of the American Chemical Society* 97: 3264–3265.

Chung, S.H., Lopato, R., Greenbaum, S.G. et al. (2007). Nuclear magnetic resonance study of the dynamics of imidazolium ionic liquids with $-CH_2Si(CH_3)_3$ vs $-CH_2C(CH_3)_3$ substituents. *Journal of Physical Chemistry B* 111: 4885–4893.

Ciceri, D., Perera, J.M., and Stevens, G.W. (2014). The use of microfluidic devices in solvent extraction. *Journal of Chemical Technology and Biotechnology* 89: 771–786.

Cincotti, A., Mameli, A., Locci, A.M. et al. (2006). Heavy metals uptake by Sardinian natural zeolites: experiment and modeling. *Industrial and Engineering Chemistry Research* 45: 1074–1084.

Coleman, D. and Gathergood, N. (2010). Biodegradation studies of ionic liquids. *Chemical Society Reviews* 39: 600–637.

Collins, C., Steg, L., and Koning, M. (2007). Customers' values, beliefs on sustainable corporate performance, and buying behavior. *Psychology and Marketing* 24: 555–577.

Couillard, D. and Mercier, G. (1991). Optimum residence time (in CSTR and Airlift reactor) for bacterial leaching of metals from sewage sludge. *Water Research* 25: 211–218.

Couling, D.J., Bernot, R.J., Docherty, K.M. et al. (2006). Assessing the factors responsible for ionic liquid toxicity to aquatic organisms via quantitative structure-property relationship modeling. *Green Chemistry* 8: 82–90.

CRC (1992). *Handbook of Chemistry and Physics*, 73e (ed. D.R. Lide). Boca Raton, FL: CRC Press.

Cubas, A.L.V., Machado, M.D.M., Gross, F. et al. (2014). Inertization of heavy metals present in galvanic sludge by DC thermal plasma. *Environmental Science and Technology* 48: 2853–2861.

Cui, J. and Zhang, L. (2008). Metallurgical recovery of metals from electronic waste: a review. *Journal of Hazardous Materials* 158: 228–256.

Culbertson, C.T., Jacobson, S.C., and Ramsey, J.M. (2000). Microchip devices for high efficiency separations. *Analytical Chemistry* 72: 5814–5819.

Dąbrowski, A. (2001). Adsorption—from theory to practice. *Advances in Colloid and Interface Science* 93: 135–224.

Dai, S., Ju, Y.H., and Barnes, C.E. (1999). Solvent extraction of strontium nitrate by a crown ether using room-temperature ionic liquids. *Journal of the Chemical Society, Dalton Transactions* 8: 1201–1202.

Dai, Y., Sun, Q., Wang, W. et al. (2018). Utilizations of agricultural waste as adsorbent for the removal of contaminants: a review. *Chemosphere* 211: 235–253.

Das, A.P. and Mishra, S. (2010). Biodegradation of the metallic carcinogen hexavalent chromium Cr(VI) by an indigenously isolated bacterial strain. *Journal of Carcinogenesis* 9 (1): 6.

Das, T., Ayyappan, S., and Chaudhury, G.R. (1999). Factors affecting bioleaching kinetics of sulfide ores using acidophilic micro-organisms. *Biometals* 12 (1): 1–10.

Davis, J.H. Jr., (2004). Task-specific ionic liquids. *Chemistry Letters* 33: 1072–1077.

Davis, J.H. Jr., and Forrester, K.J. (1999). Thiazolium-ion based organic ionic liquids (OILs). Novel OILs which promote the benzoin condensation. *Tetrahedron Letters* 40: 1621–1622.

Davis, A.P. and Singh, I. (1995). Washing of zinc(II) from contaminated soil column. *Journal of Environmental Engineering* 121: 174–185.

Day, F.G. (1984). Recovery of platinum group metals, gold and silver from scrap. US Patent 4, 427, 442.

De Groot, R.S. (1992). *Functions of Nature: Evaluation of Nature in Environmental Planning, Management and Decision Making*. Groningen: Wolters-Noordhoff.

Deepatana, A. and Valix, M. (2008). Steric hindrance effect on adsorption of metal–organic complexes onto aminophosphonate chelating resin. *Desalination* 218: 297–303.

Dehchenari, M.A., Hosseinpoor, S., Aali, R. et al. (2017). Simple method for extracting gold from electrical and electronic wastes using hydrometallurgical process. *Environmental Health Engineering and Management Journal* 4: 55–58.

Deng, J., Guo, J., Zhou, X. et al. (2014). Hazardous substances in indoor dust emitted from waste TV recycling facility. *Environmental Science and Pollution Research International* 21: 7656–7667.

Department of Fertilizers (2013). *Annual Report 2012–13*. Government of India, Ministry of Chemicals and Fertilizers http://fert.nic.in/sites/default/files/Annual_Report2012-13.pdf.

Devi, S., Gupta, C., Jat, S.L., and Parmar, M.S. (2017). Crop residue recycling for economic and environmental sustainability: the case of India. *Open Agriculture* 2: 486–494.

Dietz, M.L. and Dzielawa, J.A. (2001). Ion-exchange as a mode of cation transfer into room-temperature ionic liquids containing crown ethers: implications for the 'greenness' of ionic liquids as diluents in liquid–liquid extraction. *Chemical Communications* (20): 2124–2125.

Dobson, R.S. and Burgess, J.E. (2007). Biological treatment of previous metal refinery wastewater: a review. *Minerals Engineering* 20: 519–532.

Docherty, K.M., Dixon, J.K., and Kulpa, C.F. (2007). Biodegradability of imidazolium and pyridinium ionic liquids by an activated sludge microbial community. *Biodegradation* 18: 481–493.

Domańska, U. and Rękawek, A. (2009). Extraction of metal ions from aqueous solutions using imidazolium based ionic liquids. *Journal of Solution Chemistry* 38: 739–751.

Dong, T., Hua, Y., Zhang, O., and Zhou, D. (2009). Leaching of chalcopyrite with Brønsted acidic ionic liquid. *Hydrometallurgy* 99: 33–38.

Ducker, W. and West, J.R. (1971). *The Manufacture of Sulphuric Acid*. Huntington, NY: Robert E. Krieger Publishing.

Dufresne, P. (2007). Hydroprocessing catalysts regeneration and recycling. *Applied Catalysis A: General* 322: 67–75.

Dunn, J., Wendell, E., Carda, D.D., and Storbeck, T.A. (1991). Chlorination process for recovering gold values from gold alloys. US Patent US5004500.

Dzulkefly, K., Hraon, M.J., Lim, W.H., and Woon, C.C. (2002). Recovery of nickel from spent hydrogenation catalyst using chelating resin. *Journal of Oleo Science* 51: 749–751.

Earle, M.J., Esperança, J.M., Gilea, M.A. et al. (2006). The distillation and volatility of ionic liquids. *Nature* 439: 831–834.

Edwards, K.J. and Rutenberg, A.D. (2001). Microbial response to surface microtopography: the role of metabolism in localized mineral dissolution. *Chemical Geology* 180 (1–4): 19–32.

Egorov, V.M., Djigailo, D.I., Momotenko, D.S. et al. (2010). Task-specific ionic liquid trioctylmethylammonium salicylate as extraction solvent for transition metal ions. *Talanta* 80: 1177–1182.

Elaiwai, A., Hitchcock, P.B., Seddon, K.R. et al. (1995). Hydrogen bonding in imidazolium salts and its implications for ambient-temperature halogenoaluminate(III) ionic liquids. *Journal of the Chemical Society, Dalton Transactions* 21: 3467–3472.

Elkington, J. (2005). Enter the triple bottom line. In: *The Triple Bottom Line. Does It All Add up?* (eds. A. Henriques and J. Richardson). London: Earthscan.

Elliott, H.A. and Brown, G.A. (1989). Comparative evaluation of NTA and EDTA for extractive decontamination of Pb-polluted soils. *Water, Air, and Soil Pollution* 45: 361–369.

Ellis, W.D., Fogg, T.C., and Tafuri, A.N. (1986). *Proceedings of the 12th Annual Research Symposium*. Cincinnati, OH: USEPA Hazardous Waste Engineering Research Laboratory.

El-Sayed, G., El-Sheikh, R., and Farag, N. (2015). Maize stalks as a cheap biosorbent for removal of Fe (II) from aqueous solution. *International Research Journal of Pure and Applied Chemistry* 6: 66–76.

EPA (2001). Electronics: a new opportunity for waste prevention, reuse and recycling. http://www.epa.gov/epr (accessed 9 January 2018).

EPA (2008). Hazardous waste listings. http://www.epa.gov/osw/hazard/wastetypes/pdfs/listing-ref.pdf (8 June 2018).

Erdem, E., Karapinar, N., and Donat, R. (2004). The removal of heavy metal cations by natural zeolites. *Journal of Colloid and Interface Science* 280: 309–314.

European Commission, Directive 2002/96/EC of the European Parliament and of the Council of 27 January 2003 on Waste Electrical and Electronic Equipment, Official Journal L 37, 2003.

European Council (2016). The producer responsibility principle of the WEEE directive. http://ec.europa.eu/environment/waste/weee/pdf/final_rep_okopol.pdf (accessed 15 April 2018).

European Environmental Agency (2012). Annual report 2012 and environmental statement 2013. https://www.eea.europa.eu/publications/eea-annual-report-2012/download (accessed 03 May 2017).

Evans, D.F., Chen, S.H., Schriver, G.W., and Arnett, E.M. (1981). Thermodynamics of solution of nonpolar gases in a fused salt. "Hydrophobic bonding" behavior in a nonaqueous system. *Journal of the American Chemical Society* 103: 481–482.

Fabry, F., Rehmet, C., Rohani, V., and Fulcheri, L. (2013). Waste gasification by thermal plasma: a review. *Waste and Biomass Valorization* 4: 421–439.

Fanali, C., Micalizzi, G., Dugo, P., and Mondello, L. (2017). Ionic liquids as stationary phases for fatty acid analysis by gas chromatography. *Analyst* 142: 4601–4612.

Fangueiro, D., Bermond, A., Santos, E. et al. (2002). Heavy metal mobility assessment in sediments based on a kinetic approach of the EDTA extraction: search for optimal experimental conditions. *Analytica Chimica Acta* 459: 245–256.

Fannin, A.A., Floreani, D.A., King, L.A. et al. (1984). Properties of 1,3-dialkylimidazolium chloride-aluminum chloride ionic liquids. 2. Phase transitions, densities, electrical conductivities, and viscosities. *Journal of Physical Chemistry* 88: 2614–2621.

Farooq, U., Khan, M., Athar, M., and Kozinski, J.A. (2011). Effect of modification of environmentally friendly biosorbent wheat (*Triticum aestivum*) on the biosorptive removal of cadmium(II) ions from aqueous solution. *Chemical Engineering Journal* 171 (2): 400–410.

Feng, N.C. and Guo, X.Y. (2012). Characterization of adsorptive capacity and mechanisms on adsorption of copper, lead and zinc by modified orange peel. *Transactions of the Nonferrous Metals Society of China* 22 (5): 1224–1231.

Feng, D., Deventer, J.S.J.V., and Aldrich, C. (2004). Removal of pollutants from acid mine wastewater using metallurgical by product slags. *Separation and Purification Technology* 40 (1): 61–67.

Feng, R., Zhao, D., and Guo, Y. (2010). Revisiting characteristics of ionic liquids: a review for further application development. *Journal of Environmental Protection* 1: 95–104.

Feng, N., Guo, X., Liang, S. et al. (2011). Biosorption of heavy metals from aqueous solutions by chemically modified orange peel. *Journal of Hazardous Materials* 185 (1): 49–54.

Fischer, K. and Bipp, H.P. (2002). Removal of heavy metals from soil components and soils by natural chelating agents. Part II. Soil extraction by sugar acids. *Water, Air, & Soil Pollution* 138: 271–288.

Fries, D.M., Voitl, T., and von Rohr, P.R. (2008). Liquid extraction of vanillin in rectangular microreactors. *Chemical Engineering and Technology* 31: 1182–1187.

Fry, S.E. and Pienta, N.J. (1985). Effects of molten salts on reactions. Nucleophilic aromatic substitution by halide ions in molten dodecyltributylphosphonium salts. *Journal of the American Chemical Society* 107: 6399–6400.

Gaballah, I. and Djona, M. (1995). Recovery of Co, Ni, Mo, V from unroasted spent hydrorefining catalysts by selective chlorination. *Metallurgical and Materials Transactions B: Process Metallurgy and Materials Processing Science* 26B: 41–50.

Gaballah, I., Allain, E., and Djona, M. (1994). Chlorination kinetics of refractory metal oxides (MoO_3, Nb_2O_5, Ta_2O_5 and V_2O_5). In: *Light Metals, Proc. of the technical sessions. TMS Light Metals Committee, 123rd TMS Annual Meeting San Francisco, Feb. 27-March 3 1994* (ed. U. Mannweiler), 1153–1161. TMS.

Gabriel, S. and Weiner, J. (1888). Ueber einige Abkömmlinge des Propylamins (About some derivatives of propylamine). *Berichte der Deutschen Chemischen Gesellschaft* 21: 2669–2679.

Gadd, G.M. (1999). Fungal production of citric and oxalic acid: importance in metal speciation, physiology and biogeochemical processes. *Advances in Microbial Physiology* 41: 47–92.

Gaillochet, C. and Chalmin, P. (2009). From waste to resources. *World Waste Survey 2009*.

Gandhi, N.M.S., Selladurai, V., and Santhi, P. (2006). Unsustainable development to sustainable development: a conceptual model. *Management of Environment Quality: An International Journal* 17: 654–672.

Gao, S., Luo, X., NieEr, Z. et al. (2010). Role of acidithiobacillus ferrooxidans in bioleaching of copper. *Chinese Journal of Environment Engineering* 4 (3): 677–682.

Gao, N.N., Wang, J.X., Shao, L., and Chen, J.F. (2011). Removal of carbon dioxide by absorption in microporous tube-in-tube microchannel reactor. *Industrial and Engineering Chemistry Research* 50: 6369–6374.

García-Mendieta, A., TeresaOlguín, M., and Solache-Ríos, M. (2012). Biosorption properties of green tomato husk (*Physalis philadelphica* Lam) for iron, manganese and iron–manganese from aqueous systems. *Desalination* 284: 167–174.

Gathergood, N. and Scammells, P.J. (2002). Design and preparation of room-temperature ionic liquids containing biodegradable side chains. *Australian Journal of Chemistry* 55: 557–560.

Gathergood, N., Garcia, M.T., and Scammells, P.J. (2004). Biodegradable ionic liquids: Part I. Concept, preliminary targets and evaluation. *Green Chemistry* 6: 166–175.

Gathergood, N., Scammells, P.J., and Teresa Garcia, M. (2006). Biodegradable ionic liquids: Part III. The first readily biodegradable ionic liquids. *Green Chemistry* 8: 156–160.

Gecol, H., Miakatsindila, P., Ergican, E., and Sage, R.H. (2006). Biopolymer coated clay particles for the adsorption of tungsten from water. *Desalination* 197 (1–3): 165–178.

Gedik, K. and Imamoglu, I. (2008). Affinity of clinoptilolite-based zeolites towards removal of Cd from aqueous solutions. *Separation Science and Technology* 43: 1191–1207.

Gerayeli, F., Ghojavand, F., Mousavi, S.M. et al. (2013). Screening and optimization of effective parameters in biological extraction of heavy metals from refinery spent catalysts using a thermophilic bacterium. *Separation and Purification Technology* 118: 151–161.

Gericke, M., Neale, J.W., and Stade, P.V. (2009). A Mintek perspective of the past 25 years in minerals bioleaching. *Southern African Institute of Mining and Metallurgy* 109 (10): 567–585.

Gerike, P. and Fischer, W.K. (1979). A correlation study of biodegradability determinations with various chemicals in various tests. *Ecotoxicology and Environmental Safety* 3: 159–173.

Ghosh, S. and Das, A.P. (2017). Bioleaching of manganese from mining waste residues using *Acinetobacter* sp. *Geology, Ecology, and Landscapes* 1 (2): 77–83.

Glauser, J., Blagoev, M., and Kishi, A. (2010). Chelating agent. www.sriconsulting.com/CEH/Public/Reports/515.5000/chelating%2520agents.gif (accessed December 2012).

Goel, S. and Gautam, A. (2010). Effect of chelating agents on mobilization of metal from waste catalyst. *Hydrometallurgy* 101: 120–125.

Goel, S., Pant, K.K., and Nigam, K.D.P. (2009). Extraction of nickel from spent catalyst using fresh and recovered EDTA. *Journal of Hazardous Materials* 171: 253–261.

Gontard, N., Sonesson, U., Birkved, M. et al. (2018). A research challenge vision regarding management of agricultural waste in a circular bio-based economy. *Critical Reviews in Environmental Science and Technology* 48: 614–654.

Gothard, N.A., Mara, M.W., Huang, J. et al. (2012). Strong steric hindrance effect on excited state structural dynamics of Cu(I) diimine complexes. *Journal of Physical Chemistry A* 116: 1984–1992.

Goyal, P. and Srivastava, S. (2009). Characterization of novel Zea mays based biomaterial designed for toxic metals biosorption. *Journal of Hazardous Materials* 172: 1206–1211.

Grant, K., Goldizen, F.C., Sly, P.D. et al. (2013). Health consequences of exposure to e-waste: a systematic review. *The Lancet Global Health* 1: 350–361.

Greaves, T.L. and Drummond, C.J. (2008). Ionic liquids as amphiphile self-assembly media. *Chemical Society Reviews* 37: 1709–1726.

Gunther, A. and Jensen, K.F. (2006). Multiphase microfluidics: from flow characteristics to chemical and materials synthesis. *Lab on a Chip* 6: 1487–1503.

Guo, X., Liu, J., Qin, H. et al. (2015). Recovery of metal values from waste printed circuit boards using an alkali fusion–leaching–separation process. *Hydrometallurgy* 156: 199–205.

Gupta, V.K., Jain, C.K., Ali, I. et al. (2003). Removal of cadmium and nickel from wastewater using bagasse fly ash—a sugar industry waste. *Water Research* 37 (16): 4038–4044.

Gürsel, I.V., Aldiansyah, F., Wang, Q. et al. (2015). Continuous metal scavenging and coupling to one-pot copper-catalyzed azide–alkyne cycloaddition click reaction in flow. *Chemical Engineering Journal* 270: 468–475.

Gürsel, I.V., Kurt, K.S., Aalders, J.A. et al. (2016). Utilization of milli-scale coiled flow inverter in combination with phase separator for continuous flow liquid–liquid extraction processes. *Chemical Engineering Journal* 283: 855–868.

Gurung, M., Adhikari, B.B., Kawakita, H. et al. (2013). Recovery of gold and silver from spent mobile phones by means of acidothiourea leaching followed by adsorption using biosorbent prepared from persimmon tannin. *Hydrometallurgy* 133: 84–93.

Gutnikov, G. (1971). Method of recovering metals from spent hydrotreating catalysts. US Patent 3, 567, 433.

Ha, V.H., Lee, J.C., Jeong, J. et al. (2010). Thiosulfate leaching of gold from waste mobile phones. *Journal of Hazardous Materials* 178: 1115–1119.

Ha, V.H., Lee, J.C., Huynh, T.H. et al. (2014). Optimizing the thiosulfate leaching of gold from printed circuit boards of discarded mobile phone. *Hydrometallurgy* 149: 118–126.

Halasz, I., Kim, S., and Marcus, B. (2002). Hydrophilic and hydrophobic adsorption on Y zeolites. *Molecular Physics* 100: 3123–3132.

Han, D. and Row, K.H. (2010). Recent applications of ionic liquids in separation technology. *Molecules* 15: 2405–2426.

Hancock, R.D. and Marsicano, F. (1978). Parametric correlation of formation constants in aqueous solution. 1. Ligands with small donor atoms. *Inorganic Chemistry* 17: 560–564.

Hart, S.L. and Milsten, M.B. (2003). Creating sustainable value. *Academy of Management Executive* 17: 56–69.

Hartman, R.L., Sahoo, H.R., Yen, B.C., and Jensen, K.F. (2009). Distillation in microchemical systems using capillary forces and segmented flow. *Lab on a Chip* 9: 1843–1849.

Harvey, T.J., Merwe, W.V.D., and Afewu, K. (2002). The application of the GeoBiotics GEOCOAT biooxidation technology for the treatment of sphalerite at Kumba resources' Rosh Pinah mine. *Minerals Engineering* 15: 823–829.

Havlik, T., Orac, D., Petranikova, M. et al. (2010). Leaching of copper and tin from used printed circuit boards after thermal treatment. *Journal of Hazardous Materials* 183: 866–873.

He, Y. and Xu, Z. (2015). Recycling gold and copper from waste printed circuit boards using chlorination process. *RSC Advances* 5: 8957–8964.

Heintz, W. (1862). Uber dem amoniaktypus angehorige suren. *Analytical Chemistry and Pharmacy* 122: 257–294.

Herrera, M.N., Escobar, B., Parra, N., and Vargas, T. (1998). Bioleaching of refractory gold concentrates at high pulp densities in a nonconventional rotating-drum reactor. *Minerals and Metallurgical Processing* 15 (2): 15–19.

Herreros, O., Quiroz, R., and Viñals, J. (1999). Dissolution kinetics of copper, white metal and natural chalcocite in Cl_2/Cl^- media. *Hydrometallurgy* 51: 345–357.

Heukelem, A., Reuter, M., Huisman, J. et al. (2004). Eco efficient optimization of pre-processing and metal smelting. In: *Electronics Goes Green 2004: Driving*

Forces for Future, Electronics (eds. H. Reichl, H. Griese and H. Pötter), 657–661. Germany.

Heydarian, A., Mousavi, S.M., Vakilchap, F., and Baniasadi, M. (2018). Application of a mixed culture of adapted acidophilic bacteria in two-step bioleaching of spent lithium–ion laptop batteries. *Journal of Power Sources* 378: 19–30.

Himmler, S., Hörmann, S., van Hal, R. et al. (2006). Transesterification of methylsulfate and ethylsulfate ionic liquids—an environmentally benign way to synthesize long-chain and functionalized alkylsulfate ionic liquids. *Green Chemistry* 8: 887–894.

Hirayama, H., Takai, K., Inagaki, F. et al. (2005). Thiobacter subterraneus gen. nov., sp. nov., an obligately chemolithoautotrophic, thermophilic, sulfur-oxidizing bacterium from a subsurface hot aquifer. *International Journal of Systematic and Evolutionary Microbiology* 55: 467–472.

Holbrey, J.D. and Seddon, K.R. (1999). The phase behaviour of 1-alkyl-3-methylimidazolium tetrafluoroborates; ionic liquids and ionic liquid crystals. *Journal of the Chemical Society, Dalton Transactions* 13: 2133–2140.

Holbrey, J. and Welton, T. (2018). Obituary: Kenneth R. Seddon: 1950–2018. *Green Chemistry* 20: 776–776.

Hong, Y. and Valix, M. (2014). Bioleaching of electronic waste using acidophilic sulfur oxidizing bacteria. *Journal of Cleaner Production* 65: 465–472.

Hong, P.K., Li, C., Banerjii, S.K., and Wang, Y. (2002). Feasibility of metal recovery from soil using DTPA and its biostability. *Journal of Hazardous Materials* 94: 253–272.

Hong, P.K., Cai, X., and Cha, Z. (2008). Pressure- assisted chelation extraction of lead from contaminated soil. *Environmental Pollution* 153: 14–21.

Hoornweg, D. and Bhada-Tata, P. (2012). *What a Waste: A Global Review of Solid Waste Management*, Urban Development Series; Knowledge Papers No. 15. Washington, DC: World Bank. https://openknowledge.worldbank.org/handle/10986/17388.License: CCBY3.0IGO.

Horeh, N.B., Mousavi, S.M., and Shojaosadati, S.A. (2016). Bioleaching of valuable metals from spent lithium-ion mobile phone batteries using *Aspergillus niger*. *Journal of Power Sources* 320: 257–266.

Horn, R.G., Evans, D.F., and Ninhamt, B.W. (1988). Double-layer and solvation forces measured in a molten salt and its mixtures with water. *Journal of Physical Chemistry* 92: 3531–3537.

Hoshino, T. (2013). Preliminary studies of lithium recovery technology from seawater by electrodialysis using ionic liquid membrane. *Desalination* 317: 11–16.

Howard, J.L. and Shu, J. (1996). Sequential extraction analysis of heavy metals using a chelating agent (NTA) to counteract resorption. *Environmental Pollution* 91: 89–96.

Hsu, E., Barmak, K., West, A.C., and Park, H.A. (2019). Advancements in the treatment and processing of electronic waste with sustainability: a review of metal extraction and recovery technologies. *Green Chemistry* 21: 919–936.

Hu, X., Zhao, M., Song, G., and Huang, H. (2011). Modification of pineapple peel fibre with succinic anhydride for Cu^{2+}, Cd^{2+} and Pb^{2+} removal from aqueous solutions. *Environmental Technology* 32 (7–8): 739–746.

Huang, J.W. and Cunningham, S.D. (1996). Lead phytoextraction: species variation in lead uptake and translocation. *New Phytologist* 134: 75–84.

Huang, J., Chen, J., Berti, W., and Cunningham, S.D. (1997). Phytoremediation of lead-contaminated soils: role of synthetic chelates in lead phytoextraction. *Environmental Science and Technology* 31: 800–805.

Huang, P.M., Li, Y., and Sumner, M.E. (2011). *Handbook of Soil Sciences: Properties and Processes*, 2e, 12–45. CRC Press.

Huang, J., Chen, M., Chen, H. et al. (2014). Leaching behavior of copper from waste printed circuit boards with Brønsted acidic ionic liquid. *Waste Management* 34: 483–488.

Huh, Y.S., Jeong, C., Chang, H.N. et al. (2010). Rapid separation of bacteriorhodopsin using a laminar-flow extraction system in a microfluidic device. *Biomicrofluidics* 4: 014103.

Hurley, F.H. and Wier, T.P. (1951). Electrodeposition of metals from fused quaternary ammonium salts. *Journal of the Electrochemical Society* 98: 203–206.

Idris, J., Musa, M., Yin, C.-Y., and Hamid, K.H.K. (2010). Recovery of nickel from spent catalyst from palm oil hydrogenation process using acidic solutions. *Journal of Industrial and Engineering Chemistry* 16: 251–255.

Ilyas, S., Anwar, M.A., Niazi, S.B., and Ghauri, M.A. (2007). Bioleaching of metals from electronic scrap by moderately thermophilic acidophilic bacteria. *Hydrometallurgy* 88 (1–4): 180–188.

Ilyas, S., Ryun, C.H., Bhatti, H.N. et al. (2010). Column bioleaching of metals from electronic scrap. *Hydrometallurgy* 101 (3–4): 135.

Implementation of E-Waste Rules (2011). *Central Pollution Control Board*. Delhi: Ministry for Envrionment and Forest. http://cpcb.nic.in/cpcbold/ImplimentationE-Waste.pdf.

In, G., Kim, Y.-S., and Choi, J.-M. (2008). Study on solvent extraction using salen(NEt$_2$)$_2$ as a chelating agent for determination of trace Cu(II), Mn(II), and Zn(II) in water samples. *Bulletin of the Korean Chemical Society* 29: 969–973.

Isildar, A., Vossenberg, V.D.J., Rene, E.R. et al. (2015). Two-step bioleaching of copper and gold from discarded printed circuit boards (PCB). *Waste Management* https://doi.org/10.1016/j.wasman.2015.11.033.

Ivascanu, S.T. and Roman, O. (1975). Nickel recovery from spent catalysts: I. Solvation process. *Buletinul Institutului Politehnic din Iaşi Sectia II* 2: 47–51.

Jackson, W.G., McKeon, J.A., and Cortez, S. (2004). Alfred Werner's inorganic counterparts of racemic and mesomeric tartaric acid: a milestone revisited. *Inorganic Chemistry* 43: 6249–6254.

Jadhao, P., Chauhan, G., Pant, K.K., and Nigam, K.D.P. (2016). Greener approach for the extraction of copper metal from electronic waste. *Waste Management* 57: 102–112.

Jadhav, U. and Hocheng, H. (2015). Hydrometallurgical recovery of metals from large printed circuit board pieces. *Scientific Reports* 5: 14574–14582.

Jadhav, U., Su, C., and Hocheng, H. (2016). Leaching of metals from printed circuit board powder by an *Aspergillus niger* culture supernatant and hydrogen peroxide. *RSC Advances* 6: 43442–43452.

Jastorff, B., Störmann, R., Ranke, J. et al. (2003). How hazardous are ionic liquids? Structure-activity relationships and biological testing as important elements for sustainability evaluation. *Green Chemistry* 5: 136–142.

Jemere, A.B., Martinez, D., Finot, M., and Harrison, D.J. (2009). Capillary electrochromatography with packed bead beds in microfluidic devices. *Electrophoresis* 30: 4237–4244.

Jena, P.K. and Brocchi, E.A. (1997). Metal extraction through chlorine metallurgy. *Mineral Processing and Extractive Metallurgy Review* 16: 211–237.

Jenkin, G.R.T., Al-Bassam, A.Z.M., Harris, R.C. et al. (2016). The application of deep eutectic solvent ionic liquids for environmentally-friendly dissolution and recovery of precious metals. *Minerals Engineering* 87: 18–24.

Jensen, W.B. (2005). The origin of the 18-electron rule. *Journal of Chemical Education* 82: 28–29.

Jensen, M.P., Dzielawa, J.A., Rickert, P., and Dietz, M.L. (2002). EXAFS investigations of the mechanism of facilitated ion transfer into a room-temperature ionic liquid. *Journal of the American Chemical Society* 124: 10664–10665.

Jensen, K.F., Reizman, B.J., and Newman, S.G. (2014). Tools for chemical synthesis in microsystems. *Lab on a Chip* 14: 3206–3212.

Jessop, P.G. (2011). Searching for green solvents. *Green Chemistry* 13: 1391–1398.

Jiang, W., Tao, T., and Liao, Z. (2011). Removal of heavy metal from contaminated soil with chelating agents. *Open Journal of Soil Science* 1: 70–76.

Jiménez-Cedillo, M.J., Olguín, M.T., Fall, C., and Colín, A. (2011). Adsorption capacity of iron- or iron–manganese-modified zeolite-rich tuffs for As(III) and As(V) water pollutants. *Applied Clay Science* 54 (3–4): 206–216.

Jin, J., Liu, G.L., Shaoyuan, S., and Cong, W. (2010). Studies on the performance of a rotating drum bioreactor for bioleaching processes — oxygen transfer, solids distribution and power consumption. *Hydrometallurgy* 103 (1): 30–34.

Jin, J., Shi, S.-Y., Liu, G.-L. et al. (2012). Numerical simulation of flow and particle collision in a rotating-drum bioreactor. *Chemical Engineering and Technology* 35 (2): 287–293.

Jing-ying, L., Xiu-li, X., and Wen-quan, L. (2012). Thiourea leaching gold and silver from the printed circuit boards of waste mobile phones. *Waste Management* 32: 1209–1212.

Jordan, C.H., Sanhueza, A., Gautier, V. et al. (2006). Electrochemical study of the catalytic influence of *Sulfolobus metallicus* in the bioleaching of chalcopyrite at 70 °C. *Hydrometallurgy* 83: 55–62.

Juang, R.S. and Shiau, R.C. (2000). Metal removal from aqueous solutions using Chitosan-enhanced membrane filtration. *Journal of Membrane Science* 165: 159–167.

Jujun, R., Jie, Z., Jian, H., and Zhang, J. (2015). Novel designed bioreactor for recovering precious metals from waste printed circuit boards. *Scientific Reports* 5: 13481.

Junior, A.C.G., Meneghel, A.P., Rubio, F. et al. (2013). Applicability of *Moringa oleifera* Lam. pie as an adsorbent for removal of heavy metals from waters. *Revista Brasileira de Engenharia Agrícola e Ambiental* 17: 94–99.

Kahhat, R. and Williams, E. (2009). Product or waste? Importation and end-of-life processing of computers in Peru. *Environmental Science and Technology* 43: 6010–6016.

Kamberovic, Z., Korac, M., and Ranitovic, M. (2011). Hydrometallurgical process for extraction of metals from electronic waste-part II: development of the processes for the recovery of copper from printed circuit boards (PCB). *Association of Metallurgical Engineers of Serbia AMES* 17: 139–149.

Kamio, E., Seike, Y., Yoshizawa, H. et al. (2011). Microfluidic extraction of docosahexaenoic acid ethyl ester: comparison between slug flow and emulsion. *Industrial and Engineering Chemistry Research* 50: 6915–6924.

Kar, B.B., Datta, P., and Misra, V.N. (2004). Spent catalyst: secondary source for molybdenum recovery. *Hydrometallurgy* 72: 87–92.

Kar, B.B., Murthy, B.V.R., and Misra, V.N. (2005). Extraction of molybdenum from spent catalyst by salt roasting. *International Journal of Mineral Processing* 76: 143–147.

Kashid, M.N., Harshe, Y.M., and Agar, D.W. (2007). Liquid–liquid slug flow in a capillary: an alternative to suspended drop or film contactors. *Industrial and Engineering Chemistry Research* 46: 8420–8430.

Kashid, M.N., Gupta, A., Renken, A., and Kiwi-Minsker, L. (2010). Numbering-up and mass transfer studies of liquid–liquid two-phase micro-structured reactors. *Chemical Engineering Journal* 158: 233–240.

Kashid, M.N., Renken, A., and Kiwi-Minsker, L. (2011). Influence of flow regime on mass transfer in different types of microchannels. *Industrial and Engineering Chemistry Research* 50: 6906–6914.

Kateja, K., Agarwal, H., Saraswat, A. et al. (2016). Continuous precipitation of process related impurities from clarified cell culture supernatant using a novel coiled flow inversion reactor (CFIR). *Biotechnology Journal* 11: 1320–1331.

Kauffmann, G.B. (1983). Nikolaï semenovich kurnakov, the reaction (1893) and the man (1860–1941) a ninety-year retrospective view. *Polyhedron* 2: 855–863.

Kaufmann, T.G., Kaldor, A., Stuntz, G.F. et al. (2000). Catalysis science and technology for cleaner transportation fuels. *Catalysis Today* 62: 77–90.

Kaur, P.J., Kardam, V., Pant, K.K. et al. (2016). Characterization of commercially important Asian bamboo species. *European Journal of Wood and Wood Products* 74 (1): 137–139.

Kawai, A., Hidemori, T., and Shibuya, K. (2004). Polarity of room-temperature ionic liquid as examined by EPR spectroscopy. *Chemistry Letters* 33: 1464–1465.

Kaya, M. (2016). Recovery of metals and nonmetals from electronic waste by physical and chemical recycling processes. *Waste Management* 57: 64–90.

Kaza, S., Yao, L.C., Bhada-Tata, P., and Frank, V.W. (2018). *What a Waste 2.0: A Global Snapshot of Solid Waste Management to 2050*, Urban Development. Washington, DC: World Bank. https://openknowledge.worldbank.org/handle/10986/30317.License:CCBY3.0IGO.

Kenig, E.Y., Lautenschleger, Y., Su, A. et al. (2013). Micro-separation of fluid systems: a state-of-the-art review. *Separation and Purification Technology* 120: 245–264.

Kestenbaum, H., Oliveira, A.L.D., Schmidt, W., and Schuth, F. (2002). Silver-catalyzed oxidation of ethylene to ethylene oxide in a microreaction system. *Industrial & Engineering Chemsitry Research* 41: 710–719.

Ketai, H., Li, L., and Wenying, D. (2008). Research on recovery logistics network of waste electronic and electrical equipment in China. *2008 3rd IEEE Conference on Industrial Electronics and Applications, ICIEA 2008*, p. 1797.

Khalifa, E.B., Rzig, B., Chakroun, R., and Nouagui, H. (2019). Application of response surface methodology for chromium removal by adsorption on low-cost biosorbent. *Chemometrics and Intelligent Laboratory Systems* 189: 18–26.

Khaliq, A., Rhamdhani, M.A., Brooks, G., and Masood, S. (2014). Metal extraction processes for electronic waste and existing industrial routes: a review and Australian perspective. *Resources* 14: 152–179.

Khan, Z.U.H., Kong, D., Chen, Y. et al. (2015). Ionic liquids based fluorination of organic compounds using electrochemical method. *Journal of Industrial and Engineering Chemistry* 31: 26–38.

Kılıçarslan, A. and Sarıdede, M.N. (2015). *Energy Technology 2015: Carbon Dioxide Management and other Technologies* (eds. A. Jha, C. Wang, N.R. Neelameggham, et al.), 208–215. TMS (The Minerals, Metals & Materials Society).

Kilicarslan, A., Saridede, M.N., Stopic, S., and Friedrich, B. (2014). Use of ionic liquid in leaching process of brass wastes for copper and zinc recovery. *International Journal of Minerals, Metallurgy, and Materials* 21: 138–143.

Kim, K. and Cho, J.W. (1997). Selective recovery of metals from spent desulfurization catalyst. *Korean Journal of Chemical Engineering* 14: 162–167.

Kim, H. and Park, D. (2004). Characteristics of fly ash/sludge slags vitrified by thermal plasma. *Journal of Industrial and Engineering Chemistry* 10: 234–238.

Kim, C.H., Woo, S.I., and Jeon, S.H. (2000). Recovery of platinum-group metals from recycled automotive catalytic converters by carbochlorination. *Industrial and Engineering Chemistry Research* 39: 1185–1192.

Kim, C., Lee, Y., and Ong, S.K. (2003). Factors affecting EDTA extraction of lead from lead-contaminated soils. *Chemosphere* 51: 845–853.

Kim, E.Y., Kim, M.S., and Lee, J.C. (2010). Leaching behaviour of copper using electro-generated chlorine in hydrochloric acid solution. *Hydrometallurgy* 100: 95–102.

Kinnunen, P.H.M., Robertson, W.J., Plumb, J.J. et al. (2003). The isolation and use of iron-oxidizing, moderately thermophilic acidophiles from the Collie coal mine for the generation of ferric iron leaching solution. *Applied Microbiology and Biotechnology* 60: 748–753.

Kletzin, A., Urich, T., Muller, F. et al. (2004). Dissimilatory oxidation and reduction of elemental sulfur in thermophilic archaea. *Journal of Bioenergetics and Biomembranes* 36: 77–91.

Kluner, T., Hempel, D.C., and Nortemann, B. (1998). Metabolism of EDTA and its metal chelates by whole cells and cell-free extracts of strain BNC1. *Applied Microbiology and Biotechnology* 49: 194–201.

Klutz, S., Kurt, S.K., Lobedann, M., and Kockmann, N. (2015a). Narrow residence time distribution in tubular reactor concept for Reynolds number range of 10–100. *Chemical Engineering Research and Design* 95: 22–33.

Klutz, S., Magnus, J., Lobedann, M. et al. (2015b). Developing the biofacility of the future based on continuous processing and single-use technology. *Journal of Biotechnology* 213: 120–130.

Klutz, S., Lobedann, M., Bramsiepe, C., and Schembecker, G. (2016). Continuous viral inactivation at low pH value in antibody manufacturing. *Chemical Engineering and Processing: Process Intensification* 102: 88–101.

Kociałkowski, W.Z., Diatta, J.B., and Grzebisz, W. (1999). Evaluation of chelating agents as heavy metals extractants in agricultural soils under threat of contamination. *Polish Journal of Environmental Studies* 8: 149–154.

Köddermann, T., Wertz, C., Heintz, A., and Ludwig, R. (2006). The association of water in ionic liquids: a reliable measure of polarity. *Angewandte Chemie International Edition* 45: 3697–3702.

Kolar, E., Catthoor, R.P.R., Kriel, F.H. et al. (2016). Microfluidic solvent extraction of rare earth elements from a mixed oxide concentrate leach solution using Cyanex® 572. *Chemical Engineering Science* 148: 212–218.

Kolencık, M., Urık, M., Cernansky, S. et al. (2013). Leaching of zinc, cadmium, lead and copper from electronic scrap using organic acids and the *Aspergillus niger* strain. *Fresenius Environmental Bulletin* 22: 3673–3679.

Konishi, Y., Nishibura, H., and Asai, S. (1998). Bioleaching of sphalerite by the acidophilic thermophile *Acidianus brierleyi*. *Hydrometallurgy* 47: 339–352.

Kralj, J.G., Sahoo, H.R., and Jensen, K.F. (2007). Integrated continuous microfluidic liquid–liquid extraction. *Lab on a Chip* 7: 256–263.

Kubička, D. and Horáček, J. (2011). Deactivation of HDS catalysts in deoxygenation of vegetable oils. *Applied Catalysis A: General* 394 (1–2): 9–17.

Kukawka, R., Pawlowska-Zygarowicz, A., Dutkiewicz, M. et al. (2016). New approach to hydrosilylation reaction in ionic liquids as solvent in microreactor system. *RSC Advances* 6: 61860–61868.

Kumar, V. and Nigam, K.D.P. (2005). Numerical simulation of steady flow fields in coiled flow inverter. *International Journal of Heat and Mass Transfer* 48: 4811–4828.

Kumar, S., Ruth, W., Sprenger, B., and Kragl, U. (2006). On the biodegradation of ionic liquid 1-butyl-3-methylimidazolium tetrafluoroborate. *Chimica Oggi* 24: 24–26.

Kumar, V., Mridha, M., Gupta, A.K., and Nigam, K.D.P. (2007). Coiled flow inverter as a heat exchanger. *Chemical Engineering Science* 62: 2380–2396.

Kumar, A.C., Perumal, R., Narayanan, L., and Kumar, A.J. (2011). Use of corn cob as low cost adsorbent for the removal of nickel (II) from aqueous solution. *International Journal of Advanced Biotechnology and Research (IJBR)* 5 (3): 325–330.

Kumar, P.S., Ramalingam, S., Sathyaselvabala, V. et al. (2012). Removal of cadmium(II) from aqueous solution by agricultural waste cashew nut shell. *Korean Journal of Chemical Engineering* 29: 756–768.

Kurt, S.K., Gelhausen, M.G., and Kockmann, N. (2015). Axial dispersion and heat transfer in a milli/microstructured coiled flow inverter for narrow residence time distribution at laminar flow. *Chemical Engineering and Technology* 7: 1122–1130.

Kurt, S.K., Gürsel, I.V., Hessel, V. et al. (2016a). Liquid–liquid extraction system with microstructured coiled flow inverter and other capillary for single-stage extraction applications. *Chemical Engineering Journal* 284: 764–777.

Kurt, S.K., Akhtar, M., Nigam, K.D.P., and Kockmann, N. (2016b). Modular concept of a smart scale helically coiled tubular reactor for continuous operation of multiphase reaction system. *Proceeding of the ASME 2016 Summer Heat Transfer Conference, Fluid Engineering Division Summer Meeting, and International*

Conference on Nanochannels, Microchannels, and Minichannels, Washington, DC, USA (10–14 July 2016).

Kyzas, G.Z. (2012). Commercial coffee wastes as materials for adsorption of heavy metals from aqueous solutions. *Materials* 5: 1826–1840.

Labunsk, I., Harrad, S., Santillo, D. et al. (2013). Levels and distribution of polybrominated diphenyl ethers in soil sediment and dust samples from various electronic waste recycling sites within Giuyu town, Southern China. *Environmental Science: Processes & Impacts* 15: 503–511.

Lagergren, S. (1898). Zur theorie der sogenannten adsorption gelˆster stoffe, Kungliga Svenska Vetenskapsakademiens. *Handlingar* 24 (4): 1–39.

Lai, H.Y. and Chen, Z.S. (2005). The EDTA effect on phytoextraction of single and combined metals-contaminated soils using rainbow pink (*Dianthus chinensis*). *Chemosphere* 60: 1062–1071.

Lam, K.F., Cao, E., Sorensen, E., and Gavriilidis, A. (2011). Development of multistage distillation in a microfluidic chip. *Lab on a Chip* 11: 1311–1317.

Langsch, R., Zalucky, J., Haase, S., and Lange, R. (2014). Investigation of a packed bed in a mini channel with a low channel-to-particle diameter ratio: flow regimes and mass transfer in gas–liquid operation. *Chemical Engineering and Processing: Process Intensification* 75: 8–18.

Lasheen, M.R., Ammar, N.S., and Ibrahim, H.S. (2012). Adsorption/desorption of $Cd(II)$, $Cu(II)$ and $Pb(II)$ using chemically modified orange peel: equilibrium and kinetic studies. *Solid State Sciences* 14 (2): 202–210.

Launiere, C.A. and Gelis, A.V. (2016). High precision droplet-based microfluidic determination of americium (III) and lanthanide (III) solvent extraction separation kinetics. *Industrial and Engineering Chemistry Research* 55: 2272–2276.

Lee, J.-M. (2012). Extraction of noble metal ions from aqueous solution by ionic liquids. *Fluid Phase Equilibria* 319: 30–36.

Lee, T., Park, T., and Lee, J.H. (2004). Waste green sands as reactive media for the removal of zinc from water. *Chemosphere* 56 (6): 571–581.

Lee, J.Y., Rao, S.V., Kumar, B.N. et al. (2010). Nickel recovery from spent Raneynickel catalyst through dilute sulfuric acid leaching and soda ash precipitation. *Journal of Hazardous Materials* 176: 1122–1125.

Lee, C.H., Tang, L.W., and Popuri, S.R. (2011). A study on the recycling of scrap integrated circuits by leaching. *Waste Management and Research* 29: 677–685.

Lester, J.N., Sterrit, R.M., and Kirk, P.W. (1983). Significance and behaviour of heavy metals in waste water treatment processes. II. Sludge treatment and disposal. *Science of the Total Environment* 30: 45–83.

Leung, A.O.W., Chan, J.Y., Xing, G. et al. (2010). Body burdens of polybrominated diphenyl ethers in childbearing-aged women at an intensive electronic-waste recycling site in China. *Environmental Science and Pollution Research International* 17: 1300–1313.

Levin, J. (1989). Observations on the Bond standard grindability test, and a proposal for a standard grindability test for fine materials. *Journal of the South African Institute of Mining and Metallurgy* 89: 13–21.

Lewandowski, A. and Swiderska-Mocek, A. (2009). Ionic liquids as electrolytes for Li–ion batteries-an overview of electrochemical studies. *Journal of Power Sources* 194: 601–609.

Li, J., Lu, H., Xu, Z., and Zhou, Y. (2008). Critical rotational speed model of the rotating roll electrode in corona electrostatic separation for recycling waste printed circuit boards. *Journal of Hazardous Materials* 154: 331–336.

Li, S.W., Xu, J.H., Wang, Y.J. et al. (2009). Low-temperature bonding of poly-(methyl methacrylate) microfluidic devices under an ultrasonic field. *Journal of Micromechanics and Microengineering* 19: 015035.

Li, B., Zhang, S., Wu, W., Liang, L., Jiang, S., Chen, L., and Li, Y. (2017). Imidazolium-based ionic liquid-catalyzed hydrosilylation ofimines and reductive amination of aldehydes using hydrosilane as the reductant. *RSC Advances* 7: 31795–31799.

Li, X., Li, Z., Orefice, M., and Binnemans, K. (2019). Metal recovery from spent samarium–cobalt magnets using a trichloride ionic liquid. *ACS Sustainable Chemistry & Engineering* 7 (2): 2578–2584.

Liao, S., Sackmann, J., Tollk, O. et al. (2015). Ultrasonic fabrication of micro nozzles from a stack of PVDF foils for generating and characterizing microfluidic dispersions. *Microsystem Technologies* 23: 695–702.

Lim, T.T., Tay, J.H., and Wang, J.Y. (2004). Chelating-agent enhanced heavy metal extraction from a contaminated acidic soil. *Journal of Environmental Engineering, ASCE* 130: 59–66.

Lim, T.-T., Chui, P.-C., and Goh, K.-H. (2005). Process evaluation for optimization of EDTA use and recovery for heavy metal removal from a contaminated soil. *Chemosphere* 58: 1031–1040.

Liskowitz, J.W., Chan, P.C., Trattner, R., and Shieh, M. (1980). Leachate treatment utilizing flyash-clay mixture. *Proceedings of National Conference on Hazardous and Toxic Waste Management*, New Jersey, Volume 2, pp. 515–544.

Liu, Y., Louie, T.M., Payne, J. et al. (2001). Identification, purification, and characterization of iminodiacetate oxidase from the EDTA-degrading bacterium BNC1. *Applied and Environmental Microbiology* 67: 696–701.

Liu, H.L., Chiu, C.W., and Cheng, Y.C. (2003). The effects of metabolites from indigenous *Acidithiobacillus thiooxidans* and temperature on the bioleaching of cadmium from soil. *Biotechnology and Bioengineering* 83: 638–645.

Liu, Z.C., Meng, X.H., Zhang, R., and Xu, C.M. (2009). Friedel-crafts acylation of aromatic compounds in ionic liquids. *Petroleum Science and Technology* 27: 226–237.

Liu, A., Ren, F., Lin, W.Y., and Wang, J.Y. (2015). A review of municipal solid waste environmental standards with a focus on incinerator residues. *International Journal of Sustainable Built Environment* 4: 165–188.

Lobedann, M., Klutz, S., and Kurt, S.K. (2015). Device and method for continuous virus inactivation. W.O. Patent No. 2015135844 A1.

de los Ríos, A.P., Hernández-Fernández, F.J., Lozano, L.J. et al. (2010). Removal of metal ions from aqueous solutions by extraction with ionic liquids. *Journal of Chemical and Engineering Data* 55: 605–608.

Luo, C., Shen, Z., and Li, X. (2005). Enhanced phytoextraction of Cu, Pb, Zn and Cd with EDTA and EDDS. *Chemosphere* 59: 1–11.

Maase, M. and Massonne, K. (2005). Biphasic acid scavenging utilizing ionic liquids: the first commercial process with ionic liquids. In: *Ionic Liquids IIIB: Fundamentals, Progress, Challenges, and Opportunities*, ACS Symposium Series, vol. 902 (eds. R.D. Rogers and K.R. Seddon), 126–132.

MacFarlane, D.R., Pringle, J.M., Johnsson, K.M. et al. (2006). Lewis base ionic liquids. *Chemical Communications* (18): 1905–1917.

Maduabuchi, M.N. (2018). Agricultural waste materials as a potential adsorbent for removal of heavy metals in wastewater. *Open Access Journal of Waste Management & Xenobiotics* 1 (1): 000104.

Magnuson, D.K., Bodley, J.W., and Evans, D.F. (1984). The activity and stability of alkaline phosphatase in solutions of water and the fused salt Ethylammonium nitrate. *Journal of Solution Chemistry* 13: 583–587.

Maguyon, M.C.C., Alfafara, C.G., Migo, V.P. et al. (2012). Recovery of copper from spent solid printed-circuit-board (PCB) wastes of a PCB manufacturing facility by two-step sequential acid extraction and electrochemical deposition. *Journal of Environmental Science and Management* 15: 17–27.

Makinen, J., Bacher, J., Kaartinen, T. et al. (2015). The effect of flotation and parameters for bioleaching of printed circuit boards. *Minerals Engineering* 75: 26–31.

Mandal, P.K. and Samnta, A. (2005). Fluorescence studies in a pyrrolidinium ionic liquid: polarity of the medium and solvation dynamics. *Journal of Physical Chemistry B* 109: 15172–15177.

Mandal, M.M., Kumar, V., and Nigam, K.D.P. (2010). Augmentation of heat transfer performance in coiled flow inverter vis-à-vis conventional heat exchanger. *Chemical Engineering Science* 65: 999–1007.

Mandal, M.M., Serra, C., Hoarau, Y., and Nigam, K.D.P. (2011a). Numerical modeling of polystyrene synthesis in coiled flow inverter. *Microfluidics and Nanofluidics* 10: 415–423.

Mandal, M.M., Aggarwal, P., and Nigam, K.D.P. (2011b). Liquid–Liquid mixing in coiled flow inverter. *Industrial and Engineering Chemistry Research* 50: 13230–13235.

Maneesuwannarat, S., Vangnai, A.S., Yamashita, M., and Thiravetyan, P. (2016). Bioleaching of gallium from gallium arsenide by *Cellulosimicrobium funkei* and its application to semiconductor/electronic wastes. *Process Saftey and Environmental Protection* 99: 80–87.

Manouchehri, N., Besancon, S., and Bermond, A. (2006). Major and trace metal extraction from soil by EDTA: equilibrium and kinetic studies. *Analytica Chimica Acta* 559: 105–112.

Manzoor, Q., Nadeem, R., Iqbal, M. et al. (2013). Organic acids pretreatment effect on *Rosa bourbonia* phyto-biomass for removal of Pb(II) and Cu(II) from aqueous media. *Bioresource Technology* 132: 446–452.

Marafi, M. and Furimsky, E. (2005). Selection of organic agents for reclamation of metals from spent hydroprocessing catalysts. *Erdol Erdgas Kohle* 121: 93–96.

Marafi, M. and Stanislaus, A. (2008). Solid catalyst waste management: a review, Part I – developments in hydroprocessing catalyst waste reduction and use. *Resources, Conservation and Recycling* 52: 859–873.

Marcus, A. and Fremeth, A. (2009). Green management matters regardless. *Academy of Management Perspectives* 23: 17–26.

Markosyan, G.E. (1972). A new iron-oxidizing bacterium: *Leptospirillum ferrooxidans* gen. nov. sp. *Biology Journal of Armenia* 25: 26–29 (in Russian).

Márquez, M., Gaspar, J., Bessler, K.E., and Magela, G. (2006). Process mineralogy of bacterial oxidized gold ore in São Bento Mine (Brasil). *Hydrometallurgy* 83: 114–123.

Marre, S., Adamo, A., Basak, S. et al. (2010). Design and packaging of microreactors for high pressure and high temperature applications. *Industrial and Engineering Chemistry Research* 49: 11310–11320.

Marsden, J.O. and House, C.I. (2006). *The Chemistry of Gold Extraction*, 2e, 651. Littleton, CO: Society for Mining, Metallurgy and Exploration Inc.

Marszałkowska, B., Regel-Rosocka, M., Nowak, L., and Wisniewski, M. (2010). Quaternary phosphonium salts as effective extractants of zinc (II) and iron (III) ions from acid pickling solutions. *Polish Journal of Chemical Technology* 12: 1–5.

Master, K.L. (2010). Hydrophilic Vs Hydrophobic Absorbents: Don't Be Confused by Terminology. Article Source: http://EzineArticles.com/5604670 (last assessed in December 2018).

Matsumoto, M., Shimizu, S., Sotoike, R. et al. (2017). Exceptionally high electric double layer capacitances of oligomeric ionic liquids. *Journal of the American Chemical Society* 139: 16072–16075.

Mayor of London (2010). The Mayor's Draft Municipal Waste Management Strategy London's Wasted Resource. https://www.london.gov.uk/sites/default/files/gla_migrate_files_destination/draft-mun-waste-strategy-jan2010.pdf (last assessed in December 2018).

McCluskey, A., Lawrance, G.A., Leitch, S.K. et al. (2002). Ionic liquids industrial applications for green chemistry. *Journal of the American Chemical Society* 818: 199–212.

McKay, G., Allen, S.J., McConvey, I.F., and Otterburn, M.S. (1981). Transport processes in the sorption of colored ions by wood particles. *Journal of Colloid and Interface Science* 80: 323–339.

Meers, E., Ruttens, A., Hopgood, M.J. et al. (2005). Comparison of EDTA and EDDS as potential soil amendments for enhanced phytoextraction of heavy metals. *Chemosphere* 58: 1011–1022.

Meshram, P., Abhilash, Pandey, B.D., and Mankhand, T.R. (2016). Acid baking of spent lithium ion batteries for selective recovery of major metals: a two-step process. *Journal of Industrial and Engineering Chemistry* 43: 117–126.

Messadi, A., Mohamadou, A., Boudesocque, S. et al. (2013). Task-specific ionic liquid with coordinating anion for heavy metal ion extraction: cation exchange versus ion-pair extraction. *Separation and Purification Technology* 107: 172–178.

Millsap, W. and Reisler, N. (1978). Cotters's new plant diets on spent catalysts. *Engineering and Mining Journal* 179: 105–107.

Ministry of the Environment, Government of Japan (2008). The world in transition, and Japan's efforts to establish a sound material cycle society. Available at: http://www.env.go.jp/en/recycle/smcs/arep/2008gs_full.pdf. Accessed on 18.03.2018.

Ministry of Environment, Office of Sound Material-cycle Society Waste Management and Recycling Department (2008). The world in transition, and Japan's efforts to establish a sound material-cycle society. https://www.env.go.jp/en/recycle/smcs/a-rep/2008gs_full.pdf (accessed 12 December 2018).

Miserere, S., Mottet, G., Taniga, V. et al. (2012). Fabrication of thermoplastics chips through lamination based techniques. *Lab on a Chip* 12: 1849–1856.

Mishra, P.C. and Patel, R.K. (2009). Removal of lead and zinc ions from water by low cost adsorbents. *Journal of Hazardous Materials* 168 (1): 319–325.

Mishra, D., Kim, D.J., Ahn, J.G., and Rhee, Y.H. (2005). Bioleaching: a microbial process of metal recovery; a review. *Metals and Materials International* 11 (3): 249–256.

Mishra, D., Chaudhary, G.R., Kim, D.J., and Ahn, J.G. (2010). Recovery of metal values from spent catalyst using leaching-solvent extraction technique. *Hydrometallurgy* 101: 35–40.

Monter, M.M.L., Olguin, M.T., and Rios, M.J.S. (2007). Lead sorption by a Mexican, clinoptilolite-rich tuff. *Environmental Science and Pollution Research International* 14: 397–403.

Montero, R., Guevara, A., and De la Torre, E. (2012). *International Mineral Processing Congress (IMPC)*, September, New Delhi, India, p. 3513.

Moreno, M., Simonetti, E., Appetecchi, G.B. et al. (2017). Ionic liquid electrolytes for safer lithium batteries. *Journal of the Electrochemical Society* 164: A6026–A6031.

Morgan, G.T. and Drew, H.D.K. (1920). CLXII.-Researches on residual affinity and co-ordination. Part II. Acetylacetones of selenium and tellurium. *Journal of the Chemical Society, Transactions* 117: 1456–1465.

Mousavi, S.M., Yaghmaei, S., Salimi, F., and Jafari, A. (2006). Influence of process variables on bio-oxidation of ferrous sulfate by an indigenous *Acidithiobacillus ferrooxidans*, Part I, Flask experiments. *Fuel* 85 (17–18): 2555–2560.

Mozes, N., Marchal, F., Hermesse, M.P. et al. (1987). Immobilization of microorganisms by adhesion: interplay of electrostatic and nonelectrostatic interactions. *Biotechnology and Bioengineering* 30: 439–450.

Mridha, M. and Nigam, K.D.P. (2008). Coiled flow inverter as an inline mixer. *Chemical Engineering Science* 63: 1724–1732.

Mu, X., Liang, Q., Hu, P. et al. (2010). Selectively modified microfluidic chip for solvent extraction of Radix Salvia Miltiorrhiza using three-phase laminar flow to provide double liquid–liquid interface area. *Microfluidics and Nanofluidics* 9: 365–373.

Mulak, W., Miazga, B., and Szymczycha, A. (2005). Kinetics of nickel leaching from spent catalyst in sulphuric acid solution. *International Journal of Mineral Processing* 77: 231–235.

Mulak, W., Szymczycha, A., Lesniewicz, A., and Zyrnick, W. (2006). Preliminary results of metals leaching from a spent hydrodesulphurization (HDS) catalyst. *Physicochemical Problems of Mineral Processing* 40: 69–76.

Mulligan, C.N. and Cloutier, R.G. (2003). Bioremediation of metal contamination. *Environmental Monitoring and Assessment* 84 (1–2): 45–60.

Mulligan, C.N., Kamali, M., and Gibbs, B.F. (2004). Bioleaching of heavy metals from a low-grade mining ore using *Aspergillus niger*. *Journal of Hazardous Materials* 110: 77–84.

Murthy, D.S.R. and Prasad, P.M. (1996). Leaching of gold and silver from Miller Process dross through non-cyanide leachants. *Hydrometallurgy* 42: 27–33.

Murthy, D.S.R., Kumar, V., and Rao, K.V. (2003). Extraction of gold from an Indian low-grade refractory gold ore through physical beneficiation and thiourea leaching. *Hydrometallurgy* 68: 125–130.

Mutch, M.L. and Wilkes, J.S. (1998). Thermal analysis of 1-ethyl-3-methylimidazolium tetrafluoroborate molten salt. *Proceedings – Electrochemical Society PV* 11: 254–260.

Myerson, A.S., Brooklyn, N.Y., Ernst, W.R., and Roswell, Ga. (1987). United States Patent Number: 4698321. Regeneration of HDS Catalysts. filed 8 August 1985 and issued 6 October 1987.

Myres, R.T. (1978). Thermodynamics of chelation. *Inorganic Chemistry* 17: 953–958.

Nah, W., Hwang, K.Y., Jeon, C., and Choi, H.B. (2006). Removal of Pb ion from water by magnetically modified zeolite. *Minerals Engineering* 19 (14): 1452–1455.

Najam, R. and Muzaffar, S.A. (2016). Adsorption capability of sawdust of *Populus alba* for Pb(II), Zn(II) and Cd(II) ions from aqueous solution. *Desalination and Water Treatment* 57 (59): 1–17.

Namasivayam, C. and Ranganathan, K. (1993). Waste Fe (III)/Cr (III) hydroxide as adsorbent for the removal of Cr(VI) from aqueous solution and chromium plating industry wastewater. *Environmental Pollution* 82 (3): 255–261.

Nan, H. and Anderson, J.L. (2018). Ionic liquid stationary phases for multidimensional gas chromatography. *TrAC Trends in Analytical Chemistry* 105: 367–379.

Neduein, R.M., Duriae, E.K.M., Boskoviae, G. et al. (1998). Deactivation of HDS catalysts. *Studies in Surface Science and Catalysis* 113: 399–404.

Netzahuatl-Muñoz, A.R., Cristiani-Urbina, M.C., and Cristiani-Urbina, E. (2015). Chromium biosorption from Cr(VI) aqueous solutions by *Cupressus lusitanica* bark: kinetics, Equilibrium and Thermodynamic Studies. *Plos One* 10 (9): e0137086.

Nguyen, T.A.H., Ngo, H.H., Guo, W.S. et al. (2013). Applicability of agricultural waste and by-products for adsorptive removal of heavy metals from wastewater. *Bioresource Technology* 148: 574–585.

Nie, H.C.Y., Zhu, N., Wu, P. et al. (2015). Isolation of *Acidithiobacillus ferrooxidans* strain Z1 and its bioleaching mechanism of copper from waste printed circuit boards. *Journal of Chemical Technology and Bioechnology* 90: 714–721.

Nigam, K.D.P. (2008). Baffle and tube for a heat exchanger. US Patent US007337835B2, 4 March 2008.

Niu, H., Pan, L., Su, H., and Wang, S. (2009). Effects of design and operating parameters on CO_2 absorption in microchannel contactors. *Industrial and Engineering Chemistry Research* 48: 8629–8634.

Niu, Z., Huang, Q., Xin, B. et al. (2016). Optimization of bioleaching conditions for metal removal from spent zinc-manganese batteries using response surface methodology. *Journal of Chemical Technology and Bioechnology* 91: 608–617.

Njoku, V.O., Oguzie, E.E., Obi, C. et al. (2011). Adsorption of copper(II) and lead(II) from aqueous solutions onto a nigerian natural clay. *Austrailian Journal of Basic Applied Science* 5: 346–353.

Nockemann, P., Pellens, M., Van Hecke, K. et al. (2010). Cobalt(II) complexes of nitrile-functionalized ionic liquids. *Chemistry A European Journal* 16: 1849–1858.

Noël, T., Kuhn, S., Musacchio, A.J. et al. (2011). Suzuki–Miyaura cross-coupling reactions in flow: multistep synthesis enabled by a microfluidic extraction. *Angewandte Chemie International Edition* 50: 5943–5946.

Norris, P.R. (1997). Thermophiles and bioleaching. In: *Biomining*, Biotechnology Intelligence Unit (ed. D.E. Rawlings). Berlin, Heidelberg: Springer-Verlag, 247–258.

Norris, P.R., Barr, D.W., and Hinson, D. (1988). Iron and mineral oxidation by acidophilic bacteria: affinities for iron and attachment to pyrite. In:

Biohydrometallurgy, Proceedings of the International Symposium 1987 (eds. P.R. Norris and D.P. Kelly), 43–60. Kew, Surrey: Science and Technology Letters.

Nörtemann, B. (1999). Biodegradation of EDTA. *Applied Microbiology and Biotechnology* 51: 751–759.

Nortemann, B. (2005). Biodegradation of chelating agents: EDTA, DTPA, PDTA, NTA, and EDDS. In: *Biogeochemistry of Chelating Agents* (eds. B. Nowack and J.M. VanBriesen), 150–170. Washington, DC: Americal Chemical Society.

Novak, U., Pohar, A., Plazl, I., and Žnidaršič, P. (2012). Ionic liquid-based aqueous two-phase extraction within a microchannel system. *Separation and Purification Technology* 97: 172–178.

Nuengmatcha, P., Mahachai, R., and Chanthai, S. (2016). Adsorption capacity of the as-synthetic graphene oxide for the removal of alizarin red dye from aqueous solution. *Oriental Journal of Chemistry* 32: 1399–1410.

Nuthakki, B., Greaves, T.L., Krodkiewska, I. et al. (2007). Protic ionic liquids and iconicity. *Australian Journal of Chemistry* 60: 21–28.

O'Brien, M., Koos, P., Browne, D.L., and Ley, S.V. (2012). A prototype continuous-flow liquid–liquid extraction system using open-source technology. *Organic & Biomolecular Chemistry* 10: 7031–7036.

OECD (2001). *Extended Producer Responsibility: A Guidance Manual for Governments*. Paris: OECD.

OECD Guidelines (1992). *OECD Guidelines for the Testing of Chemicals, Section 3*, Test No. 301: Ready Biodegradability. OECD Publishing. https://doi.org/10.1787/9789264070349-en.

Ofomaja, A.E. (2008). Sorptive removal of methylene blue from aqueous solution using palm kernel fibre: effect of fibre dose. *Biochemical Engineering Journal* 40: 8–18.

Ofomaja, A.E. and Ho, Y.S. (2006). Equilibrium sorption of anionic dye from aqueous solution by palm kernel fiber as sorbent. *Dyes and Pigments* 74 (1): 60–66.

Ogonczyk, D., Wegrzyn, J., Jankowski, P. et al. (2010). Bonding of microfluidic devices fabricated in polycarbonate. *Lab on a Chip* 10: 1324–1327.

Ohashi, A., Sugaya, M., and Kim, H. (2011). Development of a microfluidic device for measurement of distribution behavior between supercritical carbon dioxide and water. *Analytical Sciences* 27: 567–569.

Ojeda, M.W., Perino, E., and Ruiz, M.d.C. (2009). Gold extraction by chlorination using a pyrometallurgical process. *Minerals Engineering* 22: 409–411.

Okubo, Y., Toma, M., Ueda, H. et al. (2014). Flow inversion: an effective means to scale-up controlled radical polymerization tubular microreactors. *Macromolecular Reaction Engineering* 8: 597–603.

Olivier-Bourbigou, H., Magna, L., and Morvan, D. (2010). Ionic liquids and catalysis: recent progress from knowledge to applications. *Applied Catalysis A: General* 373: 1–56.

Olson, G.J. and Clark, T.R. (2008). Bioleaching of molybdenite. *Hydrometallurgy* 93: 10–15.

Ostrowski, M. and Sklodowska, A. (1993). Bacterial and chemical leaching pattern on copper ores of sandstone and limestone type. *World Journal of Microbiological Biotechnology* 9: 328–333.

Oter, O. and Akcay, H. (2007). Use of natural clinoptilolite to improve, water quality: sorption and selectivity studies of lead(II), copper(II), zinc(II), and nickel(II). *Water Environment Research* 79: 329–335.

Pacholec, F., Butler, H.T., and Poole, C.F. (1982). Molten organic salt phase for gas–liquid chromatography. *Analytical Chemistry* 54: 1938–1941.

Painter, H.A., Miura, K., Tosima, Y. et al. (1993). OECD Guidelines for Testing of Chemicals, Paris, France, p. 1.

Palma, L.D., Gonzini, O., and Mecozzi, R. (2011). Use of different chelating agents for heavy metal extraction from contaminated harbour sediment. *Chemistry and Ecology* 27: 97–106.

Panday, A.K. (2008). Kinetic study of ion exchange column operation for ultrapure water application. Master thesis. West Bengal, India: National Institute of Technology Durgapur.

Panday, K.K., Prasad, G., and Singh, V.N. (1985). Copper(II) removal from aqueous solutions by fly ash. *Water Research* 19 (7): 869–873.

Pandey, P., Sambi, S.S., Sharma, S.K., and Singh, S. (2009). *Proceedings of the World Congress on Engineering and Computer Science*. San Francisco, CA: Newswood Limited (20–22 October 2009).

Parida, D., Serra, C.A., Garg, D.K. et al. (2014). Coil flow inversion as a route to control polymerization in microreactors. *Macromolecules* 47: 3282–3287.

Park, Y.J. and Fray, D.J. (2009). Recovery of high purity precious metals from printed circuit boards. *Journal of Hazardous Materials* 164: 1152–1158.

Park, K.H., Reddy, B.R., Mohapatra, D., and Nam, C.W. (2006). Hydrometallurgical processing and recovery of molybdenum trioxide from spent catalyst. *International Journal of Mineral Processing* 80: 261–265.

Park, D., Yun, Y.-S., and Park, J.M. (2010). The past, present, and future trends of biosorption. *Biotechnology and Bioprocess Engineering* 15: 86–102.

Parker, S.P. (1980). *Encyclopaedia of Environmental Sciences*, 2e. New York: McGraw Hill.

Pastre, J.C., Génisson, Y., Saffon, N. et al. (2010). Synthesis of novel room temperature chiral ionic liquids. application as reaction media for the heck arylation of aza-endocyclic acrylates. *Journal of the Brazilian Chemical Society* 21: 821–836.

Pathak, A., Dastidar, M.G., and Sreekrishnan, T.R. (2009). Bioleaching of heavy metals from sewage sludge: a review. *Journal of Environmental Management* 90 (8): 2343–2353.

Paulechka, Y.U., Zaitsau, D.H., Kabo, G.J., and Strechan, A.A. (2005). Vapor pressure and thermal stability of ionic liquid 1-butyl-3-methylimidazolium bis(trifluoromethylsulfonyl)amide. *Thermochimica Acta* 439: 158–160.

Payne, J.W., Bolton, H.J., Campbell, J.A., and Xun, L. (1998). Purification and characterization of EDTA monooxygenase from the EDTA-degrading bacterium BNC1. *Journal of Bacteriology* 180: 3823–3827.

Peng, A.A., Hong-Chang, L., Zhen-Yuan, N., and Jin-Lan, X. (2012). Effect of surfactant Tween-80 on sulfur oxidation and expression of sulfur metabolism relevant genes of *Acidithiobacillus ferrooxidans*. *Transactions of Nonferrous Metals Society of China* 22: 3147–3155.

Pereiro, A.B., Araújo, J.M.M., Esperança, J.M.S.S. et al. (2011). Ionic liquids in separations of azeotropic systems – a review. *Journal of Chemical Thermodynamics* 46: 2–28.

Petter, P.M.H., Veit, H.M., and Bernardes, A.M. (2014). Evaluation of gold and silver leaching from printed circuit board of cellphones. *Waste Management* 34: 475–482.

Petter, P.M.H., Veit, H.M., and Bernardes, A.M. (2015). Leaching of gold and silver from printed circuit board of mobile phones. *REM: Revista Escola de Minas Ouro Preto* 68: 61–68.

Pfeiffer, W. and Offermann, H. (1942). Werner. *Journal für Praktische Chemie* 159: 313.

Porro, S., Ramirez, S., Reche, C. et al. (1997). Bacterial attachment: its role in bioleaching processes. *Process Biochemistry* 32: 573–578.

Pradhan, J.K. and Kumar, S. (2012). Metals bioleaching from electronic waste by *Chromobacterium violaceum* and *Pseudomonads* sp. *Waste Management and Research* 30: 1151–1159.

Pradhan, D., Ahn, J.-G., Kim, D.-J., and Lee, S.-W. (2009). Effect of Ni^{2+}, V^{4+} and Mo^{6+} concentration on iron oxidation by *Acidithiobacillus ferrooxidans*. *Korean Journal of Chemical Engineering* 26 (3): 736–741.

Priya, A. and Hait, S. (2018). Extraction of metals from high grade waste printed circuit board byconventional and hybrid bioleaching using *Acidithiobacillus ferrooxidans*. *Hydrometallurgy* 177: 132–139.

Puckett, J. and Smith, T. (2002). *Exporting Harm: The High-tech Trashing of Asia, The Basel Action Network*. Seattle, WA: Silicon Valley Toxics Coalition.

Qi-Ming, F., Yan-Hai, S., Guo-Fan, Z. et al. (2009). Kinetics of nickel leaching from roasting-dissolving residue of spent catalyst with sulfuric acid. *Journal of Central South University of Technology* 16: 410–415.

Quartacci, M.F., Baker, A.J.M., and Navari-Izzo, F. (2005). Nitrilotriacetate- and citric acid-assisted phytoextraction of cadmium by Indian mustard (*Brassica juncea* (L.) Czernj, *Brassicaceae*). *Chemosphere* 59: 1249–1255.

Quijada-Maldonado, E., Sánchez, F., Pérez, B. et al. (2018). Task-specific ionic liquids as extractants for the solvent extraction of molybdenum(VI) from aqueous solution using different commercial ionic liquids as diluents. *Industrial and Engineering Chemistry Research* 57: 1621–1629.

Quinet, P., Proost, J., and Lierde, A.V. (2005). Recovery of precious metals from electronic scrap by hydrometallurgical processing routes. *Minerals and Metallurgical Processing* 22 (1): 17–22.

Rabah, M.A. (1998). Combined hydro-pyrometallurgical method for the recovery of high lead/tin/bronze alloy from industrial scrap. *Hydrometallurgy* 47: 281–295.

Ramachandran, K. and Kikukawa, N. (2002). Thermal plasma in-flight treatment of electroplating sludge. *IEEE Transactions on Plasma Science* 30: 310–317.

Ramnial, T., Ino, D.D., and Clyburne, J.A.C. (2005). Phosphonium ionic liquids as reaction media for strong bases. *Chemical Communications* (3): 325–327.

Rana, M.S., Samano, V., Ancheyta, J., and Diaz, J.A.I. (2007). A review of recent advances on process technologies for upgrading of heavy oils and residua. *Fuel* 86: 1216–1231.

Rapaport, D. (2000). Are spent hydroprocessing catalysts listed as hazardous wastes? *Hydrocarbon Processing* 79: 11–22.

Rasoulnia, P. and Mousavi, S.M. (2016). V and Ni recovery from a vanadium-rich power plant residual ash using acid producing fungi: *Aspergillus niger* and *Penicillium simplicissimum*. *RSC Advances* 6: 9139–9151.

Rathore, A. and Nigam, K.D.P. (2015). Indian Patent entitled 185/DEL/2015 (Indian Institute of Technology Delhi).

Reddy, P.N., Padmaja, P., Reddy, B.V.S., and Rambabu, G. (2015). Ionic liquid/water mixture promoted organic transformations. *RSC Advances* 5: 51035–51054.

Regel-Rosocka, M., Rzelewska, M., Baczynska, M. et al. (2015). Removal of palladium(ii) from aqueous chloride solutions with cyphos phosphonium ionic liquids as metal ion carriers for liquid–liquid extraction and transport across polymer inclusion membranes. *Physicochemical Problems in Mineral Processing* 51: 621–631.

Ren, W.X., Li, P.J., Geng, Y., and Li, X.J. (2009). Biological leaching of heavy metals from a contaminated soil by *Aspergillus niger*. *Journal of Hazardous Materials* 167 (1–3): 164–169.

Repo, E., Warchol, J.K., Kurniawan, T.A., and Sillanpaa, M.E.T. (2010). Adsorption of Co(II) and Ni(II) by EDTA- and/or DTPA-modified chitosan: kinetic and equilibrium modeling. *Chemical Engineering Journal* 161: 73–82.

Reyna-González, J.M., Torriero, A.A., Siriwardana, A.I. et al. (2010). Extraction of copper(II) ions from aqueous solutions with a methimazole-based ionic liquid. *Analytical Chemistry* 82: 7691–7698.

Reyna-González, J.M., Galicia-Pérez, R., Reyes-López, J.C., and Aguilar-Martínez, M. (2012). Extraction of copper(II) from aqueous solutions with the ionic liquid 3-butylpyridinium bis(trifluoromethanesulfonyl)imide. *Separation and Purification Technology* 89: 320–328.

Ribot, J.C., Guerrero-Sanchez, C., Greaves, T.L. et al. (2012). Amphiphilic oligoether-based ionic liquids as functional materials for thermos-responsive ion gels with tunable properties via aqueous gelation. *Soft Matter* 8: 1025–1032.

Rickert, P.G., Antonio, M.R., Firestone, M.A. et al. (2007). Tetraalkylphosphonium polyoxometalate ionic liquids: novel, organic–inorganic hybrid materials. *Journal of Physical Chemistry B* 111: 4685–4692.

Rijkens, H.C. (2000). Membrane developments for natural gas conditionings. *Proceedings of the Symposium: Process Intensification: a challenge for the process Industry*, Rotterdam.

Ritcey, G.M. (1989). *Tailings Management*. Amsterdam: Elsevier.

Robinson, J. and Osteryoung, R.A. (1979). An electrochemical and spectroscopic study of some aromatic hydrocarbons in the room temperature molten salt system aluminum chloride-n-butylpyridinium chloride. *Journal of the American Chemical Society* 101: 323–327.

Rodrigues, M.L.M., Leao, V.A., Gomes, O. et al. (2015). Copper extraction from coarsely ground printed circuit boards using moderate thermophilic bacteria in a rotating-drum reactor. *Waste Management* 41: 148–158.

Rogers, K. and Hudson, B. (2011). The triple bottom line: the synergies of transformative perceptions and practices for sustainability. *OD Practitioner* 43: 3–9.

Rossi, G. (2001). The design of bioreactors. *Hydrometallurgy* 59: 217–231.

Ruamsap, N., Akaracharanya, A., and Dahl, C. (2003). Pyritic sulfur removal from lignite by *Thiobacilus ferrooxidans*: strain improvement. *Journal of Scientific Research Chulalongkorn University* 28 (1): 45–55.

Ruan, J.J., Zhu, X.J., Qian, Y.M., and Hu, J. (2014). A new strain for recovering precious metals from waste printed circuit boards. *Waste Management* 34: 901–907.

Ruiz, M., Sastre, A.M., and Guibal, E. (2000). Palladium sorption onglutaraldehyde-crosslinked chitosan. *Reactive and Functional Polymers* 45: 155–173.

Ruthven, D.M. (1984). *Principle of Adsorption and Adsorption Processes*, Chapter 2–3. New York: Wiley.

Sackmann, J., Burlage, K., Gerhardy, C. et al. (2015). Review on ultrasonic fabrication of polymer micro devices. *Ultrasonics* 56: 189–200.

Sahin, M., Akcil, A., Erust, C. et al. (2015). A potential alternative for precious metal recovery from WEEE: iodine leaching. *Separation Science and Technology* 50: 2587–2595.

Sahin, A.K., Voßenkaul, D., Stoltz, N. et al. (2017). Selectivity potential of ionic liquids for metal extraction from slags containing rare earth elements. *Hydrometallurgy* 169: 59–67.

Sahoo, H.R., Kralj, J.G., and Jensen, K.F. (2007). Multistep continuous-flow microchemical synthesis involving multiple reactions and separations. *Angewandte Chemie International Edition* 46: 5704–5708.

Sajidu, S.M.I., Persson, I., Masamba, W.R.L. et al. (2006). Removal of Cd^{2+}, Cr^{3+}, Cu^{2+}, Hg^{2+}, Pb^{2+} and Zn^{2+} cations and AsO_4^{3-} anions from aqueous solutions by mixed clay from Tundulu in Malawi and characterization of the clay. *Water SA* 32 (4): 519–526.

Salim, A., Fourar, M., Pironon, J., and Sausse, J. (2008). Oil-water two-phase flow in microchannels: flow patterns and pressure drop measurements. *Canadian Journal of Chemical Engineering* 86: 978–988.

Sand, W. and Gehrke, T. (2006). Extracellular polymeric substances mediate bioleaching/biocorrosion via interfacial processes involving iron(III) ions and acidophilic bacteria. *Research in Microbiology* 157: 49–56.

Sand, W., Gehrke, T., Hallmann, R., and Schippers, A. (1995). Sulfur chemistry, biofilm, and the (in) direct attack mechanism—a critical evaluation of bacterial leaching. *Applied Microbiology and Biotechnology* 43 (6): 961–966.

Sanders, J.R., Ward, E.H., and Hussey, C.L. (1986). Aluminum bromide-1-methyl-3-ethylimidazolium bromide ionic liquids I. Densities, viscosities, electrical conductivities, and phase transitions. *Journal of the Electrochemical Society* 133: 325–330.

Santhiya, D. and Ting, Y.P. (2005). Bioleaching of spent refinery processing catalyst using *Aspergillus niger* with high-yield oxalic acid. *Journal of Biotechnology* 116: 171–184.

Santhiya, D. and Ting, Y.P. (2006). Use of adapted *Aspergillus niger* in the bioleaching of spent refinery processing catalyst. *Journal of Biotechnology* 121: 62–74.

Sarioglan, S. (2013). Recovery of palladium from spent activated carbon-supported palladium catalysts. *Platinum Metals Review* 57: 289–296.

Sarvar, M., Salarirad, M.M., and Shabani, M.A. (2015). Characterization and mechanical separation of metals from computer Printed Circuit Boards (PCBs) based on mineral processing methods. *Waste Management* 45: 246–257.

Sasaki, Y., Sugo, Y., Suzuki, S., and Tachimori, S. (2001). The novel extractants di-glycolamides for the extraction of lanthanides and actinides in HNO_3–n-dodecane system. *Solvent Extraction and Ion Exchange* 19: 91–103.

Sawant, A.D., Rauta, D.G., Darvatkaraand, N.B., and Salunkhe, M.M. (2011). Recent developments of task-specific ionic liquids in organic synthesis. *Green Chemistry Letters and Reviews* 4: 41–54.

Saxena, A.K. and Nigam, K.D.P. (1984). Coiled configuration for flow inversion and its effect on residence time distribution. *AIChE Journal* 30: 363–368.

Scheffler, T.B., Hussey, C.L., Seddon, K.R. et al. (1983). Molybdenum chloro complexes in room-temperature chloroaluminate ionic liquids: stabilization of $[MoCl_4]^{2-}$ and $[MoCl_6]^{3}$. *Inorganic Chemistry* 22: 2099–2100.

Schippers, A., Rohwerder, T., and Sand, W. (1999). Intermediary sulfur compounds in pyrite oxidation: implications for bioleaching and biodepyritization of coal. *Applied Microbiology and Biotechnology* 52: 104–110.

Schneider, S., Hawkins, T., Rosander, M. et al. (2008). Ionic liquids as hypergolic fuels. *Energy and Fuels* 22: 2871–2872.

Schwarzenbach, G. (1952). The chelate effect. *Helvetica Chimica Acta* 35: 2344–2363.

Seddon, K.R. (1997). Ionic liquids for clean technology. *Journal of Chemical Technology and Biotechnology* 68: 351–356.

Seghezzo, L. (2009). The five dimensions of sustainability. *Environmental Politics* 18: 539–556.

Seki, H. and Yoshimoto, M. (2001). Deactivation of HDS catalyst in two-stage RDS process: II. Effect of crude oil and deactivation mechanism. *Fuel Processing Technology* 69 (3): 229–238.

Sen, G.S. and Bhattacharyya, K.G. (2011). Kinetics of adsorption of metal ions on inorganic materials: a review. *Advances in Colloid and Interface Science* 162 (1–2): 39–58.

Sepúlveda, A., Schluep, M., Fabrice, G.R. et al. (2010). A review of the environmental fate and effects of hazardous substances released from electrical and electronic equipments during recycling: examples from China and India. *Environmental Impact Assessment Review* 30: 28–41.

Serra Morenoa, J., Jeremias, S., Moretti, A. et al. (2015). Ionic liquid mixtures with tunable physicochemical properties. *Electrochimica Acta* 151: 599–608.

Shah, M.B., Tipre, D.R., Purohit, M.S., and Dave, S.R. (2015). Development of two-step process for enhanced biorecovery of Cu-Zn-Ni from computer printed circuit boards. *Journal of Bioscience and Bienegineering* 120: 167–173.

Shamim, A., Mursheda, A.K., and Rafiq, I. (2015). E-waste trading impact on public health and ecosystem services in developing countries. *International Journal of Waste Resources* 5: 188–199.

Shao, H.W., Lu, Y.C., Wang, K., and Luo, G.S. (2012). Liquid–liquid flow and mass transfer characteristics in micro-sieve array device with dual-sized pores. *Chemical Engineering Journal* 193–194: 96–101.

Shariat, M.H., Setoodeh, N., and Atash Dehghan, R. (2001). Optimizing conditions for hydrometallurgical production of purified molybdenum trioxide from roasted molybdenite of sarcheshme. *Minerals Engineering* 14: 815–820.

Sharma, A.K., Agarwal, H., Pathak, M. et al. (2016). Continuous refolding of a biotech therapeutic in a novel coiled flow inverter reactor. *Chemical Engineering Science* 140: 153–160.

Sharma, N., Chauhan, G., Kumar, A., and Sharma, S.K. (2017). Statistical optimization of heavy metals (Cu^{2+}/Co^{2+}) extraction from printed circuit board and mobile batteries using chelation technology. *Industrial and Engineering Chemistry Research* 56: 6805–6819.

Sharma, L., Nigam, K.D.P., and Roy, S. (2017a). Single phase mixing in coiled tubes and coiled flow inverters in different flow regimes. *Chemical Engineering Science* 160: 227–235.

Sharma, L., Nigam, K.D.P., and Roy, S. (2017b). Investigation of two-phase (oil-water) flow in coiled geometries using "Radioactive Particle Tracking-Time of Flight (RPT-TOF)" and "Radioactive Particle Tracking-Volume Fraction (RPT-VOF)" measurements. *Chemical Engineering Science* 170: 422–436.

Shen, S., Pan, T., Liu, X. et al. (2010). Adsorption of Rh(III) complexes from chloride solutions obtained by leaching chlorinated spent automotive catalysts on ion-exchange resin Diaion WA21J. *Journal of Hazardous Materials* 179: 104–112.

Sheng, P.P. and Etsell, T.H. (2007). Recovery of gold from computer circuit board scrap using aqua regia. *Waste Management and Research* 25: 380–383.

Shiflett, M.B. and Scurto, A.M. (2017). *Ionic Liquids: Current State and Future Directions*, ACS Symposium Series. Washington, DC: American Chemical Society.

Shimojo, K., Kurahashi, K., and Naganawa, H. (2008). Extraction behavior of lanthanides using a diglycolamide derivative TODGA in ionic liquids. *Dalton Transactions* 37: 5083–5088.

Shinde, S.S., Patil, S.N., Ghatge, A., and Kumar, P. (2015). Nucleophilic fluorination using imidazolium based ionic liquid bearing tert-alcohol moiety. *New Journal of Chemistry* 39: 4368–4374.

Singh, J. and Nigam, K.D.P. (2016). Pilot plant study for effective heat transfer area of coiled flow inverter. *Chemical Engineering and Processing: Process Intensification* 102: 219–228.

Singh, J., Verma, V., and Nigam, K.D.P. (2012). Flow characteristics of power-law fluids in coiled flow inverter. *Industrial and Engineering Chemistry Research* 52: 207–221.

Singh, J., Choudhary, N., and Nigam, K.D.P. (2014a). The thermal and transport characteristics of nanofluids in a novel three-dimensional device. *Canadian Journal of Chemical Engineering* 92: 2185–2201.

Singh, J., Kockmann, N., and Nigam, K.D.P. (2014b). Novel three-dimensional microfluidic device for process intensification. *Chemical Engineering and Processing* 86: 78–89.

Singh, J., Srivastava, V., and Nigam, K.D.P. (2016). Novel membrane module for permeate flux augmentation and process intensification. *Industrial and Engineering Chemistry Research* 55: 3861–3870.

Singh, H., Chauhan, G., Jain, A.K., and Sharma, S.K. (2017). Adsorptive potential of agricultural wastes for removal of dyes from aqueous solutions. *Journal of Environmental Chemical Engineering* 5: 122–135.

Singh, J., Montesinos-Castellanos, A., and Nigam, K.D.P. (2019a). Thermal and hydrodynamic performance of a novel passive mixer 'wavering coiled flow inverter'. *Chemical Engineering and Processing: Process Intensification* 141: 107536.

Singh, J., Montesinos-Castellanos, A., and Nigam, K.D.P. (2019b). Process intensification for compact and micro heat exchangers through innovative technologies: a review. *Industrial and Engineering Chemistry Research* https://doi.org/10.1021/acs.iecr.9b02082.

Sinha, R., Chauhan, G., Singh, A. et al. (2018). A novel eco-friendly hybrid approach for recovery and reuse of copper from electronic waste. *Journal of Environmental Chemical Engineering* 6: 1053–1061.

Sistani, K.R., Mays, D.A., Taylor, R.W., and Buford, C. (1995). Evaluation of four chemical extractants for metal determinations in wetland soils. *Communications in Soil Science and Plant Analysis* 26: 2167–2180.

Smirnova, S.V., Samarina, T.O., and Pletnev, I.V. (2015). Hydrophobic–hydrophilic ionic liquids for the extraction and determination of metal ions with water-soluble reagents. *Analytical Methods* 7: 9629–9635.

Smith, E.L., Abbott, A.P., and Ryder, K.S. (2014). Deep eutectic solvents (DESs) and their applications. *Chemical Reviews* 114: 11060–11082.

Snead, D.R. and Jamison, T.F. (2013). End-to-end continuous flow synthesis and purification of diphenhydramine hydrochloride featuring atom economy, inline separation, and flow of molten ammonium salts. *Chemical Science* 4: 2822–2827.

Soni, S., Sharma, L., Meena, P. et al. (2019). Compact coiled flow inverter for process intensification. *Chemical Engineering Science* 193: 312–324.

Sóvágó, I. and Gergely, A. (1976). Effects of steric factors on the equilibrium and thermodynamic conditions of mixed ligand complexes of the copper(II) ion with diamines. *Inorganica Chimica Acta* 20: 27–32.

Sreekrishnan, T.R. and Tyagi, R.D. (1995). Sensitivity of metal-bioleaching operation to process variables. *Process Biochemistry* 30 (1): 69–80.

Srichandan, H., Pathak, A., Singh, S. et al. (2014). Sequential leaching of metals fromspent refinery catalyst in bioleaching–bioleaching and bioleaching–chemical leaching reactor: comparative study. *Hydrometallurgy* 150: 130–143.

Steemson, M.L. (1999). The selection of a hydroxide precipitation/ammoniacal releach circuit for metal recovery from acid pressure leach liquors. In: *ALTA 1999 Nickel/Cobalt Pressure Leaching & Hydrometallurgy Forum*. Perth: ALTA Metallurgical Services.

Stegemann, H., Rhode, A., Reiche, A. et al. (1992). Room temperature molten polyiodides. *Electrochimica Acta* 37: 379–383.

Stephen, J.R. and Macnaughton, S.J. (1999). Developments in terrestrial bacterial remediation of metals. *Current Opinion in Biotechnology* 10 (3): 230–233.

Stojanovic, A., Kogelnig, D., Fischer, L. et al. (2010). Phosphonium and ammonium ionic liquids with aromatic anions: synthesis, properties, and platinum extraction. *Australian Journal of Chemistry* 63: 511–524.

Stolte, S., Abdulkarim, S., Arning, J. et al. (2008). Primary biodegradation of ionic liquid cations, identification of degradation products of 1-methyl-3-octylimidazolium chloride and electrochemical wastewater treatment of poorly biodegradable compounds. *Green Chemistry* 10: 214–242.

Sun, D.D., Tay, J.H., Cheong, H.K. et al. (2001). Recovery of heavy metals and stabilization of spent hydrotreating catalyst using a glass–ceramic matrix. *Journal of Hazardous Materials* B87: 213–223.

Sun, X.Q., Luo, H.M., and Dai, S. (2012). Ionic liquids-based extraction: a promising strategy for the advanced nuclear fuel cycle. *Chemical Reviews* 112: 2100–2128.

Sun, Z., Xiao, Y., Sietsma, J. et al. (2015). A cleaner process for selective recovery of valuable metals from electronic waste of complex mixtures of end-of-life electronic products. *Environmental Science and Technology* 49 (13): 7981–7998.

Sundberg, A., Uusi-Kyyny, P., and Alopaeus, V. (2009). Novel micro-distillation column for process development. *Chemical Engineering Research and Design* 87: 705–710.

Suzuki, Y., Wakatsuki, J., Tsubaki, M., and Sato, M. (2013). Imidazolium-based chiral ionic liquids: synthesis and application. *Tetrahedron* 69: 9690–9700.

Swain, C.G., Ohno, A., Roe, D.K. et al. (1967). Tetrahexylammonium benzoate, a liquid salt at 25 degree, a solvent for kinetics or electrochemistry. *Journal of the American Chemical Society* 89: 2648–2649.

Swatloski, R.P., Spear, S.K., Holbrey, J.D., and Rogers, R.D. (2002). Dissolution of cellose with ionic liquids. *Journal of the American Chemical Society* 124: 4974–4975.

Sykora, V., Pitter, P., Bitternova, I., and Lederer, T. (2001). Biodegradability of ethylenediamine-based complexing agents. *Water Research* 35: 2010–2016.

Taha, G.M., Arified, A.E., and El-Nahas, S. (2011). Removal efficiency of potato peels as a new biosorbent material for uptake of Pb(II), Cd(II) and Zn(II) from their aqueous solutions. *Journal of Solid Waste Technology and Management* 37 (2): 128–140.

Taha, A.A., Moustafa, A.H.E., Abdel-Rahman, H.H., and Abdul Al-Hameed, M.M.A. (2018). Comparative biosorption study of Hg (II) using raw and chemically activated almond shell. *Adsorption Science and Technology* 36: 521–548.

Tandy, S., Bossart, K., Mueller, R. et al. (2004). Extraction of heavy metals from soils using biodegradable chelating agents. *Environmental Science and Technology* 38: 937–944.

Tang, P.L., Lee, C.K., Low, K.S., and Zainal, Z. (2003). Sorption of Cr(VI) and Cu(II) in aqueous solution by ethylenediamine modified rice hull. *Environmental Technology* 24 (10): 1243–1251.

Tavakoli, H.Z., Abdollahy, M., Ahmadi, S.J., and Darban, A.K. (2017). Kinetics of uranium bioleaching in stirred and column reactors. *Minerals Engineering* 111: 36–46.

Telpoukhovskaia, A.M. and Orvig, C. (2013). Werner coordination chemistry and neurodegeneration. *Chemical Society Reviews* 42: 1836–1846.

Tempel, D.J., Henderson, P.B., Brzozowski, J.R. et al. (2008). High gas storage capacities for ionic liquids through chemical complexation. *Journal of the American Chemical Society* 130: 400–401.

Tetala, K.K.R., Swarts, J.W., Chen, B. et al. (2009). A three-phase microfluidic chip for rapid sample clean-up of alkaloids from plant extracts. *Lab on a Chip* 9: 2085–2092.

Tickner, J., Rajarao, R., Lovric, B. et al. (2016). Measurement of gold and other metals in electronic and automotive waste using gamma activation analysis. *Journal of Sustainable Metallurgy* 2: 296–303.

Tran, N.T., Ayed, I., Pallandre, A., and Taverna, M. (2010). Recent innovations in protein separation on microchips by electrophoretic methods: an update. *Electrophoresis* 31: 147–173.

Tripathi, A., Kumar, M., Sau, D.C. et al. (2012). Leaching of gold from the waste mobile phone printed circuit boards (PCBs) with ammonium thiosulphate. *International Journal of Metallurgical Engineering* 1: 17–21.

Tsydenova, O. and Bengtsson, M. (2011). Chemical hazards associated with treatment of waste electrical and electronic equipment. *Waste Management* 31: 45–58.

Tyagi, R.D. and Couillard, D. (1987). Bacterial leaching of metals from digested sewage sludge. *Process Biochemistry* 22 (4): 114–117.

Tyagi, R.D. and Couillard, D. (1989). Bacterial leaching of metals from sludge. In: *Encyclopedia of Environmental Control Technology: Wastewater Treatment Technology* (ed. P. Cheremisinoff), vol. 3, Gulf Professional Publishing, ISBN: 9780872012479.

Tyagi, R.D., Couillard, D., and Tran, F. (1988). Heavy metals removal from anaerobically digested sludge by chemical and microbiological methods. *Environmental Pollution* 50: 295–316.

U.S. Energy Information Administration (2018). Short-Term Energy Outlook January 2018.

Uetz, T. and Egli, T. (1992). Characterization of an inducible, membrane-bound iminodiacetate dehydrogenase from *Chelatobacter heintzii* ATCC 29600. *Biodegradation* 3: 423–434.

United Nations Department of Economic and Social Affairs (2018). Population division. https://www.un.org/development/desa/en/news/population/2018-revision-of-world-urbanization-prospects.html (accessed 12 April 2018).

United Nations University (2009). Set world standards for electronics recycling, reuse to curb e-waste exports to developing countries, experts urge, *Science Daily*. http://www.sciencedaily.com/releases/2009/09/090915140919.htm (accessed 24 June 2016).

United States Environmental Protection Agency (USEPA) (2003). Hazardous waste management system. *Federal Register 68* 202: 59935–59940.

US Environmental Protection Agency (2017). Criteria for the definition of solid waste and solid and hazardous waste exclusions. https://www.epa.gov/hw/criteria-definition-solid-waste-and-solid-and-hazardous-waste-exclusions (accessed 04 July 2017).

USEPA (2000). *Guidance on Cumulative Risk Assessment of Pesticide Chemicals That Have a Common Mechanism of Toxicity*. Washington, DC 20460: U.S. Environmental Protection Agency https://www.epa.gov/sites/production/files/2015-07/documents/guidance_on_common_mechanism.pdf.

Uysal, M. and Ar, I. (2007). Removal of Cr(VI) from industrial wastewaters by adsorption. Part I: Determination of optimum conditions. *Journal of Hazardous Materials* 149 (2): 482–491.

Valix, M., Usai, F., and Malik, R. (2001). Fungal bio-leaching of low grade laterite ores. *Minerals Engineering* 14: 197–203.

Vallet, V., Wahlgren, U., and Grenthe, I. (2003). Chelate effect and thermodynamics of metal complex formation in solution: a quantum chemical study. *Journal of the American Chemical Society* 125: 14941–14950.

Van Gerven, T. and Stankiewicz, A. (2009). Structure, energy, synergy, times - the fundamentals of process intensification. *Industrial & Engineering Chemsitry Research* 48: 2465–2474.

Vander Hoogerstraete, T., Onghena, B., and Binnemans, K. (2013). Homogeneous liquid–liquid extraction of rare earths with the betaine-betainium bis(trifluoromethylsulfonyl)imide ionic liquid system. *International Journal of Molecular Sciences* 14: 21353–21377.

Vandevivere, P.C., Saveyn, H., Verstraete, W. et al. (2001). Biodegradation of Metal–[S,S]-EDDS complexes. *Environmental Science and Technology* 35: 1765–1770.

Vashisth, S. and Nigam, K.D.P. (2007). Experimental investigation of two phase pressure drop in coiled flow inverter: an experimental investigation. *Industrial and Engineering Chemistry Research* 46: 5043–5050.

Vashisth, S. and Nigam, K.D.P. (2008a). Liquid-phase residence time distribution for two-phase flow in coiled flow inverter. *Industrial and Engineering Chemistry Research* 47: 3630–3638.

Vashisth, S. and Nigam, K.D.P. (2008b). Experimental investigation of void fraction and flow patterns in coiled flow inverter. *Chemical Engineering and Processing: Process Intensification* 47: 1281–1291.

Vashisth, S., Kumar, V., and Nigam, K.D.P. (2008). A review on the potential application of curved geometries in process industries. *Industrial and Engineering Chemistry Research* 47: 3291–3337.

Vassil, A.D., Kapulnik, Y., Raskin, I., and Salt, D.E. (1998). The role of EDTA in lead transport and accumulation by Indian mustard. *Plant Physiology* 117: 447–453.

Vdović, N., Jurina, I., Škapin, S.D., and Sondi, I. (2010). The surface properties of clay minerals modified by intensive dry milling – revisited. http://fulir.irb.hr/2195/2/Vdovic_Surface_properties_2010_appl%20clay%20sci_48_575m.pdf (accessed 01 June 2018).

Vegliò, F., Beolchini, F., Nardini, A., Toro, L. (2000). Bioleaching of a pyrrhotiteore by sulfooxidans strain: kinetic analysis, *Chemical Engineering Science* 55 (4): 783–795.

Veit, H.M., Bernardes, A.M., Ferreira, J.Z. et al. (2006). Recovery of copper from printed circuit boards scraps by mechanical processing and electrometallurgy. *Journal of Hazardous Materials* 137: 1704–1709.

Veldbuizen, H. and Sippel, B. (1994). Mining discarded electronics. *Industrial Environment* 17: 7–11.

Vendilo, A.G., Esipova, E.V., Kovaleva, N.E., and Chernikova, E.A. (2012). Effect of chelating agents on the selectivity of a hydrophobic ionic liquid membrane. *Russian Journal of Inorganic Chemistry* 57: 751–753.

Vergara, M.A.V., Lijanova, I.V., Likhanova, N.V. et al. (2014). The removal of heavy metal cations from an aqueous solution using ionic liquids. *Canadian Journal of Chemical Engineering* 92: 1875–1881.

Vestola, E.A., Kuusenaho, M.K., Närhi, H.M. et al. (2010). Acid bioleaching of solid waste materials from copper, steel and recycling industries. *Hydrometallurgy* 103: 74–79.

Vicent-Luna, J.M., Idígoras, J., Hamad, S. et al. (2014). Ion transport in electrolytes for dye-sensitized solar cells: a combined experimental and theoretical study. *The Journal of Physical Chemistry C* 118: 28448–28455.

Villanueva, M., Coronas, A., García, J., and Salgado, J. (2013). Thermal stability of ionic liquids for their application as new absorbents. *Industrial and Engineering Chemistry Research* 52: 15718–15727.

Villar-Garcia, I.J., Lovelock, K.R.J., Men, S., and Licence, P. (2014). Tuning the electronic environment of cations and anions using ionic liquid mixtures. *Chemical Science* 5: 2573–2579.

Villarreal, M.S., Kharisov, B.I., Torres-Martinez, L.M., and Elizondo, V.N. (1999). Recovery of vanadium and molybdenum from spent petroleum catalyst. *Industrial and Engineering Chemistry Research* 38: 4624–4628.

Visser, A.E., Swatloski, R.P., Reichert, W.M. et al. (2001a). Task-specific ionic liquids for the extraction of metal ions from aqueous solutions. *Chemical Communications* (1): 135–136.

Visser, A.E., Swatloski, R.P., Griffin, S.T. et al. (2001b). Liquid/liquid extraction of metal ions in room temperature ionic liquids. *Separation Science and Technology* 36: 785–804.

Vogelaar, B.M., Steiner, P., Zijden, T.F.V.D. et al. (2007). Catalyst deactivation during thiophene HDS: the role of structural sulfur. *Applied Catalysis A: General* 318: 28–36.

Vuyyuru, K.R., Pant, K.K., Krishnan, V.V., and Nigam, K.D.P. (2010). Recovery of nickel from spent industrial catalysts using chelating agents. *Industrial and Engineering Chemistry Research* 49: 2014–2024.

Wakai, C., Oleinikova, A., Ott, M. et al. (2005). How polar are ionic liquids? Determination of the static dielectric constant of an imidazolium-based ionic liquid by microwave dielectric spectroscopy. *Journal of Physical Chemistry B* 109: 17028–17030.

Walden, P. (1914). Molecular weights and electrical conductivity of several fused salts. *Bulletin de l'Académie impériale des sciences de St.-Pétersbourg* 8: 405–422.

Wan Ngah, W.S. and Hanafiah, M.A. (2008). Removal of heavy metal ions from wastewater by chemically modified plant wastes as adsorbents: a review. *Bioresource Technology* 99 (10): 3935–3948.

Wang, K. and Luo, G. (2017). Microflow extraction: a review of recent development. *Chemical Engineering Science* 169: 18–33.

Wang, Z.-L., Jin, L.-Y., and Wei, L.H. (2008). Bis(benzimidazolium) naphthalene-1,5-disulfonate trihydrate. *Acta Crystallographica Section E: Structure Reports Online* 64: 674.

Wang, J., Bai, J., Xu, J., and Liang, B. (2009). Bioleaching of metals from printed wire boards by *Acidithiobacillus ferrooxidans* and *Acidithiobacillus thiooxidans* and their mixture. *Journal of Hazarduous Materials* 172 (2–3): 1100.

Wang, L., Yang, L., Li, Y. et al. (2010a). Study on adsorption mechanism of Pb(II) and Cu(II) in aqueous solution using PS-EDTA resin. *Chemical Engineering Journal* 163: 364–372.

Wang, X., Wang, M., Shi, L. et al. (2010b). Recovery of vanadium during ammonium molybdate production using ion exchange. *Hydrometallurgy* 104: 317–321.

Wang, K., Lu, Y.C., Xu, J.H., and Luo, G.S. (2011). Droplet generation in micro-sieve dispersion device. *Microfluidics and Nanofluidics* 10: 1087–1095.

Wang, F., Zhao, Y., Zhang, T. et al. (2015). Mineralogical analysis of dust collected from typical recycling line of waste printed circuit boards. *Waste Management* 43: 434–441.

Wang, G., Song, H., Li, R. et al. (2018). Olefin oligomerization via new and efficient Brönsted acidic ionic liquid catalyst systems. *Chinese Journal of Catalysis* 39: 1110–1120.

Wasserscheid, P. and Keim, W. (2000). Ionic-liquids – new solutions for transition metal catalysis. *Angewandte Chemie International Edition* 39: 3772–3789.

Wasserscheid, P., Dordon, C.M., Hilgers, C. et al. (2001). Ionic liquids: polar, but weakly coordinating solvents for the first biphasic oligomerisation of ethene to higher α-olefins with cationic Ni complexes. *Chemical Communications* (13): 1186.

Watanabe, M., Thomas, M.L., Zhang, S. et al. (2017). Application of ionic liquids to energy storage and conversion materials and devices. *Chemical Reviews* 117: 7190–7239.

Watling, H.R. (2006). The bioleaching of sulphide minerals with emphasis on copper sulphides–a review. *Hydrometallurgy* 84 (1–2): 81–108.

WCED (1987). *Our Common Future*, 1e. Oxford: Oxford University Press.

Wells, A.S. and Coombe, V.T. (2006). On the freshwater ecotoxicity and biodegradation properties of some common ionic liquids. *Organic Process Research and Development* 10: 794–798.

Welton, T. (2015). Solvents and sustainable chemistry. *Proceedings of the Royal Society A: Mathematical, Physcial and Engineering Sciences* 471: 502–528.

Welton, T. (2018). Ionic liquids: a brief history. *Biophysical Reviews* 10: 691–706.

Whitehead, J.A., Lawrence, G.A., and McCluskey, A. (2004). Green leaching: recyclable and selective leaching of gold-bearing ore in an ionic liquid. *Green Chemistry* 6: 313–315.

Whitehead, J.A., Zhang, J., Pereira, N. et al. (2007). Application of 1-alkyl-3-methyl-imidazolium ionic liquids in the oxidative leaching of sulphidic copper, gold and silver ores. *Hydrometallurgy* 88: 109–120.

Wierzbicki, A. and Davis, J.H. Jr., (2000). Envisioning the second generation of ionic liquid technology: design and synthesis of Task-specific Ionic Liquids (TSILs). In: *Proceedings of the Symposium on Advances in Solvent Selection and Substitution for Extraction*, 14F. New York: AIChE.

Wilkes, J.S. and Zaworotko, M.J. (1992). Air and water stable 1-ethyl-3-methylimidazolium based ionic liquids. *Journal of the Chemical Society, Chemical Communications* (13): 965–967.

Wilkes, J.S., Levisky, J.A., Wilson, R.A., and Hussey, C.L. (1982). Dialkylimidazolium chloroaluminate melts: a new class of room temperature ionic liquids for electrochemistry, spectroscopy, and synthesis. *Inorganic Chemistry* 21: 1263–1264.

Williams, D.B., Stoll, M.E., Scott, B.L. et al. (2005). Coordination chemistry of the bis(trifluoromethylsulfonyl)imide anion: molecular interactions in room temperature ionic liquids. *Chemical Communications* (11): 1438–1440.

Willner, J. and Fornalczyk, A. (2013). Extraction of metals from electronic waste by bacterial leaching. *Environment Protection Engineering* 39 (1): 197–208.

Willner, J., Fornalczyk, A., and Saternus, M. (2015). Selective recovery of copper from solutions after bioleaching electronic waste. *Proceeding 3rd International Conference "Biotechnology and Metals - 2014"*, Košice, Slovak Republic (17–19 September 2014).

Witek-Krowiak, A. and Reddy, D.H.K. (2013). Removal of microelemental Cr(III) and Cu(II) by using soybean meal waste–unusual isotherms and insights of binding mechanism. *Bioresource Technology* 127: 350–357.

Woitalka, A., Kuhn, S., and Jensen, K.F. (2014). Scalability of mass transfer in liquid–liquid flow. *Chemical Engineering Science* 116: 1–8.

Wong, L. and Henry, J.G. (1984). Decontaminating biological sludge for agricultural use. *Water Science and Technology* 17: 575–586.

Wong, J.W.C., Xiang, L., Gu, X.Y., and Zhou, L.X. (2004). Bioleaching of heavy metals from anaerobically digested sewage sludge using FeS_2 as an energy source. *Chemosphere* 55: 101–107.

Wong, F.F., Lin, C.M., Chang, C.P. et al. (2006). Recovery and reduction of spent nickel oxide catalyst via plasma sintering technique. *Plasma Chemistry and Plasma Processing* 26: 585–595.

World Bank 2018 Report (2018). Global Waste to Grow by 70 Percent by 2050 Unless Urgent Action is Taken: World Bank Report. https://www.worldbank.org/en/news/press-release/2018/09/20/global-waste-to-grow-by-70-percent-by-2050-unless-urgent-action-is-taken-world-bank-report (accessed 20 April 2019).

World Health Organization (2006). *Guidelines for Drinking-Water Quality*, 3e, 54. Geneva: World Health Organization.

Wu, F., Zhu, N., Bai, Y. et al. (2016). Highly safe ionic liquid electrolytes for sodium-ion battery: wide electrochemical window and good thermal stability. *ACS Applied Materials & Interfaces* 8: 21381–21386.

Xia, L., Yin, C., Dai, S. et al. (2010). Bioleaching of chalcopyrite concentrate using *Leptospirillum ferriphilum, Acidithiobacillus ferrooxidans* and *Acidithiobacillus thiooxidans* in a continuous bubble column reactor. *Journal of Industrial Microbiology and Biotechnology* 37 (3): 289–229.

Xia, M.C., Wang, Y.P., Peng, T.J. et al. (2017). Recycling of metals from pretreated waste printed circuit boards effectively in stirred tank reactor by a moderately thermophilic culture. *Journal of Bioscience and Bioengineering* 123 (6): 714–721.

Xiao, Y. and Malhotra, S.V. (2005). Friedel–Crafts acylation reactions in pyridinium based ionic liquids. *Journal of Organometallic Chemistry* 690: 3609–3613.

Xiu, F., Qi, Y., and Zhang, F. (2013). Recovery of metals from waste printed circuit boards by supercritical water pre-treatment combined with acid leaching process. *Waste Management* 33: 1251–1257.

Xiu, F.R., Qi, Y., and Zhang, F.S. (2015). Leaching of Au, Ag, and Pd from waste printed circuit boards of mobile phone by iodide lixiviant after supercritical water pre-treatment. *Waste Management* 41: 134–141.

Xu, J.H., Luo, G.S., Chen, G.G., and Tan, B. (2005). Mass transfer performance and two-phase flow characteristic in membrane dispersion mini-extractor. *Journal of Membrane Science* 249: 75–81.

Yan, W., Chen, C., and Chang, W. (2009). An investigation into sustainable product constructualization using a design knowledge hierarchy and Hopfield network. *Computers & Industrial Engineering* 56: 617–626.

Yang, T., Xu, Z., Wen, J., and Yang, L. (2009). Factors influencing bioleaching copper from waste printed circuit boards by Acidithiobacillus ferrooxidans. *Hydrometallurgy* 97 (1–2): 29–32.

Yang, H., Liu, J., and Yang, J. (2011). Leaching copper from shredded particles of waste printed circuit boards. *Journal of Hazardous Materials* 187: 393–400.

Yang, F., Kubota, F., Baba, Y. et al. (2013). Selective extraction and recovery of rare earth metals from phosphor powders in waste fluorescent lamps using an ionic liquid system. *Journal of Hazardous Materials* 254–255: 79–88.

Ye, C., Chen, G., and Yuan, Q. (2012). Process characteristics of CO_2 absorption by aqueous monoethanolamine in a microchannel reactor. *Chinese Journal of Chemical Engineering* 20: 111–119.

Yeung, A.T. and Gu, Y.Y. (2011). A review on techniques to enhance electrochemical remediation of contaminated soils. *Journal of Hazardous Materials* 195: 11–29.

Yip, T.C.M., Tsang, D.C.W., Ng, K.T.W., and Lo, I.M.C. (2009). Kinetic interactions of EDDS with soils. 1. Metal resorption and competition under EDDS deficiency. *Environmental Science and Technology* 43: 831–836.

Yoo, J.M., Lee, J.C., Kim, B.S. et al. (2004). Leaching of nickel from a hydrodesulphurization spent catalyst with ammonium sulfate. *Journal of Chemical Engineering of Japan* 37: 1129–1134.

Yoo, J.M., Jeong, J., Yoo, K. et al. (2009). Enrichment of the metallic components from waste printed circuit boards by a mechanical separation process using a stamp mill. *Waste Management* 29: 1132–1137.

Yoshizawa, M., Xu, W., and Angell, C.A. (2003). Ionic liquids by proton transfer: vapor pressure, conductivity, and the relevance of DeltapK_a from aqueous solutions. *Journal of the American Chemical Society* 125: 15411–15419.

Yoshizawa-Fujita, M., MacFarlane, D.R., Howlett, P.C., and Forsyth, M. (2006). A new Lewis-base ionic liquid comprising a mono-charged diamine structure: a highly stable electrolyte for lithium electrochemistry. *Electrochemistry Communications* 8: 445–449.

Yu, Z.X., Zhou, B.N., and Lu, Z.H. (1999). Hydro-electro metallurgical process for recovering scrap from cuprous chloride solution. *Shanghai Nonferrous Metals* 2: 24–28.

Yu, J., Williams, E., and Ju, M. (2010). Analysis of material and energy consumption of mobile phones in China. *Energy Policy* 38: 4135–4141.

Yuan, Z. and VanBriesen, J. (2008). The formation of intermediates in EDTA and NTA biodegradation. *Environmental Engineering Science* 23: 533–544.

Yusof, S.M.M. and Yahya, W.Z.N. (2016). Binary ionic liquid electrolyte for dye-sensitized solar cells. *Procedia Engineering* 148: 100–105.

Zafar, S. (2019). *A Primer on Agricultural Residues*. Bioenergy Consult. https://www.bioenergyconsult.com/tag/what-are-agro-residues/.

Zaitsau, D.H., Kabo, G.J., Strechan, A.A. et al. (2006). Experimental vapor pressures of 1-alkyl-3-methylimidazolium bis(trifluoromethylsulfonyl)imides and a correlation scheme for estimation of vaporization enthalpies of ionic liquids. *Journal of Physical Chemistry A* 110: 7303–7306.

Zhang, Z. and Zhang, F.S. (2013). Synthesis of cuprous chloride and simultaneous recovery of Ag and Pd from waste printed circuit boards. *Journal of Hazardous Materials* 261: 398–404.

Zhang, Y., Zheng, R., Zhao, J. et al. (2014). Characterization of H_3PO_4-treated rice husk adsorbent and adsorption of copper(II) from aqueous solution. *BioMed Research International* 2014: 496878, 1–8.

Zhang, R., Zhang, X., Tang, S., and Huang, A. (2015). Ultrasound-assisted HCl-NaCl leaching of lead-rich and antimony-rich oxidizing slag. *Ultrasonics Sonochemistry* 27: 187–191.

Zhang, L., Hessel, V., Peng, J. et al. (2017). Co and Ni extraction and separation in segmented micro-flow using a coiled flow inverter. *Chemical Engineering Journal* 307: 1–8.

Zhang, Q., Liu, H., Zhao, S. et al. (2019). Hydrodynamics and mass transfer characteristics of liquid–liquid slug flow in microchannels: the effects of temperature, fluid properties and channel size. *Chemical Engineering Journal* 358: 794–805.

Zhao, C.X. and Middelberg, A. (2011). Two-phase microfluidic flows. *Chemical Engineering Science* 66: 1394–1411.

Zhao, Y., Chen, G., and Yuan, Q. (2006). Liquid–liquid two-phase flow patterns in a rectangular microchannel. *AIChE Journal* 52: 4052–4060.

Zhao, F., Lu, Y.C., Wang, K., and Luo, G.S. (2016). Kinetic study on selective extraction of HCl and H_3PO_4 in a microfluidic device. *Chinese Journal of Chemical Engineering* 24: 221–225.

Zhou, Y. and Qu, J. (2017). Ionic liquids as lubricant additives: a review. *ACS Applied Materials & Interfaces* 9: 3209–3222.

Zhou, Y., Boudesocque, S., Mohamadou, A., and Dupont, L. (2015). Extraction of metal ions with task specific ionic liquids: influence of a coordinating anion. *Separation Science and Technology* 50: 38–44.

Zhou, Q., Gao, J., Li, Y. et al. (2017). Bioleaching in batch tests for improving sludge dewaterability and metal removal using acidithiobacillus ferrooxidans and acidithiobacillus thiooxidans after cold acclimation. *Water Science and Technology* 76 (6): 1347–1359.

Zhu, P. and Gu, G.B. (2002). Recovery of gold and copper from waste printed circuits. *Chinese Journal of Rare Metals* 26: 214–216.

Zimring, C.A. and Rathje, W.L. (2012). *Encyclopedia of Consumption and Waste: The Social Science of Garbage*. Sage Publications.

Ziogas, A., Cominos, V., Kolb, G. et al. (2012). Development of a microrectification apparatus for analytical and preparative applications. *Chemical Engineering and Technology* 35: 58–71.

Index

a
Acidithiobacillus albertis 96
Acidithiobacillus ferrooxidans 95, 96
acid leaching 42, 43, 50, 53–55, 57, 99, 113, 144, 149, 178
adsorbate 72–74, 76, 77, 85, 88
adsorption
   Green Adsorption 74–75
   case study 89–90
   definition 71–72
   hydrophilic compounds 72
   hydrophobic compounds 72–73
   innovative applications of 88, 89
   kinetic models 77–78
   metal uptake mechanisms 78–79
   polymer matrix 73–74
adsorption capacity
   definition 74
   Green Adsorbents
      adsorbent dosage 76–77
      co-ions effect 77
      influence of pH 75–76
      initial solute concentration 76
      temperature 76
adsorptive 72, 74, 76
agricultural residue 12, 20, 74, 79, 80
agricultural waste 20, 21, 74, 75, 79–82
agro-residue waste 11, 17, 21, 22, 80
alkali fusion–leaching process 58
alkali leaching 50, 57–59, 61
aminopolycarboxylates 127, 128, 143, 145, 152
anthropocentrism 5
aprotic ionic liquids 169

aqueous phase volume fractions 223–224
artificial ore 32–33
atom efficiency 10, 11

b
Bamako Convention 27, 29
Basel Action Network (BAN) 18, 27
base metals operations (BMO) 46
batch reactors 45, 105, 106, 119
batch to continuous process
   conservative design based on experience 185
   consistent manufacture to high tolerance 185
   phased development 185
   pilot plant designing
      batch scheduling 203, 204
      comminution circuit development 193–197
      dechelation 202, 204
      filtration system 199–201
      material balance 191
      reactor sizing and agitator selection 197–199
      shell and tube heat exchanger 201–202
   process design fabrication 190
Baylis–Hillman reaction 11
bioleaching process
   advantages 94–95
   aqueous phase 105
   batch process 105–106
   challenges 119–121
   chemical treatments 94

Sustainable Metal Extraction from Waste Streams, First Edition.
Garima Chauhan, Perminder Jit Kaur, K.K. Pant, and K.D.P. Nigam.
© 2020 Wiley-VCH Verlag GmbH & Co. KGaA. Published 2020 by Wiley-VCH Verlag GmbH & Co. KGaA.

bioleaching process *(contd.)*
  commercial metal extraction plants 121
  definition 93
  gaseous phase 105
  metal dissolution reaction 102
  metal extraction from waste
    industrial waste 115–118
    mineral waste 118
    municipal sewage sludge 119
    WEEE 111–115, 117
  microorganisms 94
    chemolithoautotrophic bacteria 95
    heterotrophic microbes 97
    mesophiles 95–96
    and metal surfaces 97–98
    thermophiles 96–97
  physicochemical factors
    mineral substrate 99
    oxygen and carbon dioxide content 98–99
    pH value of solution 99
    surface chemistry of metals 99–100
    surface properties 98
    surfactant and organic extractants 100
    temperature 99
  process efficiency, metal surface 97
  reactor design 100–101
  recovery of metals 94–95
  solid phase 105
biomethanation 8
biopolymers 88
biosorption 40, 74, 79, 88, 114
bounding process 73
bulk solution transport 73

c

calcinations 12, 37, 40, 58, 190
carbon neutral programs 7
catalyst deactivation 25–27
Central Pollution Control Board (CPCB) 28
centrifugation 185
chelate assisted extraction process 141
chelate effect 132–133
chelation
  definition 124
  ecotoxicological concerns and biodegradability 151–155
  ligands classification 124–127
  medicine and household activities 143
  metal-chelate complex formation 130–132
  metal extraction from metal-contaminated soil
    hydrometallurgical route 144–145
    phyto-remediation 145–147
    WEEE 149–151
  metal extraction from waste
    competing ions in reaction zone 141–142
    framework 133–134
    molar concentration of 138–139
    pH reaction 135–138
    reaction temperature 140–141
  metal extraction from metal-contaminated soil
    industrial waste 147–149
    hydrometallurgical route 144–145
  metal ions for complexation 129–130
  metal–ligand complex 127–128, 142–143
  process 200
  technology 12
chemisorption 74
chemolithoautotrophic bacteria 95
chiral ILs 170
chlorination
  extraction and refinement, precious metals 66
  gold extraction, waste 67
  hydrometallurgical processes 66, 69
  recover copper, waste printed circuit boards 67
circular economy (CE) 9
clay 36, 72, 79, 84–87

coconut shell charcoal 81
coiled flow inverter (CFI)
   liquid–liquid extraction 215, 216
   liquid–liquid micro-flow extraction
      aqueous phase volume fractions 223–224
      extraction efficiency 222–223
      flow patterns 220–222
      metal extraction operations 215
      methodology 218–220
      parameters 218
compound annual growth rate (CAGR) 19
continuously stirred-tank reactor (CSTR) 105–109
continuous micro-flow extraction, CFI
   liquid–liquid extraction 215, 216, 220
   metal extraction operations 215
   methodology 218–220
   parameters 218
conventional technology
   hydrometallurgical treatment of waste 50–69
   pyrometallurgical operations 40–49
corporate sustainable management 7
crushed metal enrichment (CME) 58
cyanide lixiviants 63

d
dechelation 147, 148, 190, 191, 200–203
deep eutectic solvents (DESs) 171
denticity 124, 125
designer solvents 159, 161, 173
diffusive transport 73
dimensions of sustainability 2–5
dry chlorination methods 66
Dutch Building Materials Decree 9

e
economic pillar 4
18-electron rule 128
electrical waste 15
electronic waste (e-waste) 11, 15, 17, 18, 20, 27, 28, 30, 34, 39, 40, 47, 54–56, 69, 120, 135, 144, 151, 182

Environment (Protection) Act 1986, 28
environmental functions 5
Environmental Management (EM) Act 17
Environmental Protection Agency 19, 29, 30
environmental scrutiny 171–173
Environment Canada 32
Environment-factor (E-factor) 10
ethyl ammonium nitrate 161
European Environment Agency (2012) 27
exopolysaccharide (EPS) layer 102

f
five dimensions sustainability 5
formal recycling waste 31
formamidine disulfide (FDS) 63
formation–dissociation equilibrium 141
four force model 5–7
fuel oil-based plants generate carbon waste 31

g
gravity separation methods 36
Green Adsorbents
   adsorption capacity
      adsorbent dosage 76–77
      co-ions effect 77
      influence of pH 75–76
      initial solute concentration 76
      temperature 76
   agricultural/industrial waste 74
   agricultural resources 79–81
   biosorbents 74
   clay 84–85
   industrial waste 85–88
   modified biopolymers 88
   zeolites 81–84
Green Adsorption 12 *see also* adsorption
Green Chemistry
   atom efficiency 10, 11
   chemical process 10
   definition 1

Green Chemistry (*contd.*)
  E-factor 10
  effective mass yield 11
  environmental efficiency 10
Greening the waste 8–10
green/low-cost adsorbents 74
Green Technology Development 11
green value stream mapping (GVSM) 7–8

h
halide leaching 66
hard and soft acids and bases (HSAB) theory 129, 130
hazardous wastes 21
  incineration 30–31
  legislations and regulations for 27–28
  recycling 31
  secured landfilling 29–30
Hazardous Wastes (Management & Handling) Rules 1989 28
heterotrophic microbes 97
high-density polyethylene (HDPE) 29
hybrid process 89, 90, 110
hydrocyclones 36, 190, 191, 193, 196–197
hydrometallurgical processes 12
hydrometallurgical treatment, of waste
  acidic medium 50, 55
  alkali medium 57
  halide 66
  lixiviants 60
hydrophilic compounds 72
hydrophobic compounds 72
hydroxy-carboxylates chelating agents 127

i
incineration 30
industrial waste 21
  bioleaching process 115
  chelation 147
  Green Adsorbents 85, 87
  pyrometallurgical treatment 40
informal recycling waste 27, 28, 31

integrated solid waste management (ISWM) approach 8
International Organization for Standardization (ISO) 3
ionic liquids (ILs) 12
  characteristic properties of
    conductivity 167
    coordination ability 166–167
    density 164
    melting point 162
    nonflammability 162–163
    polarity 166
    solubility 167
    thermal stability of 163–164
    vapor pressure 162–163
    viscosity 164–165
  classification of
    aprotic ionic liquids 169
    chiral 170
    deep eutectic solvents 171
    protic ionic liquids 169
    task specific ionic liquids 169
  definition 158
  dispersion forces 161
  diversity 158
  environmental scrutiny 171–173
  evolution 158–161
  extraction of metals
    aqueous media 173–176
    industrial solid waste/ores 176–177
    WEEE 177–178
  fundamental properties of 158
  neoteric solvents 158, 161
  solvents for sustainable development 161

j
Jahn–Teller distortion 128

k
Kasese bioleach plant 108
kinetic models, adsorption 77–78

l
lag phase 94, 100, 106
landfilling 8, 19, 30, 31, 144
  secured 29–30

leaching
  acidic 50–57
  halide 66–69
  lixiviants 60–66
Le Chatelier's Principle 139
Leptospirillum ferrooxidans 94
Lewis acids 129, 130, 167
life cycle assessment (LCA) 3
ligand, classification of 124–127
ligand field theory (LFT) 128
liquid–liquid extraction 185
liquid–liquid micro-flow extraction
  efficiency 222–223
  flow patterns 220–222
lixiviants 50, 53, 60–66, 68
logarithmic growth phase 106

m

mechanical and biological treatment (MBT) 8
mesophiles 95–96, 113
metal complexation process 127
metal dissolution reaction
  direct vs. indirect mechanism 102–103
  polysulfide mechanism 103–104
metal extraction from waste
  adsorption 93
  artificial ore 32–33
  bioleaching process
    industrial waste 115–118
    mineral waste 118
    municipal sewage sludge 119
    WEEE 111–115
  chelation
    competing ions in reaction zone 141–142
    framework 133–134
    molar concentration of 138–139
    pH reaction 135–138
    reaction temperature 140–141
  human health and environment 31–32
  hydrometallurgical treatment of waste 50–69
  overview of 40–41
  pyrometallurgical treatment of
    challenges of 49
    industrial waste 40–45
    WEEE 45–49
  metal–ligand complexation 127–128, 132, 135, 141, 147, 149, 155
  metal sulfide dissolution 103–104
  micro-flow extraction
    Co and Ni
      extractant concentration 229
      pH effect 225
      residence time 225
    continuous 212–229
    miniaturized extraction devices
      application of 211–214
      intensification 209–211
microorganisms
  chemolithoautotrophic bacteria 95
  direct vs. indirect mechanism 102–103
  heterotrophic microbes 97
  mesophiles 95–96
  metal surfaces 97–98
  polysulfide mechanism 103–104
  thermophiles 96–97
micro-reactors 188
mineral waste, bioleaching process 118
miniaturized extraction devices
  application of 211–214
  intensification 209–211
Ministry of Environment and Forests (MoEF) 28
mono-charged diamine based ILs 170
monolayer adsorption 73
municipal sewage sludge, extraction of metals 119
municipal waste (MSW) 15, 30, 34, 119

n

nano-magnetic polymers (NMPs) 73
National Environment policy 27
North Atlantic Treaty Organization (NATO) advanced research workshop 158
negligible vapor pressure 162, 163, 171
nitrilotriacetic acid (NTA) 128
non-cyanide lixiviants 60, 62, 64
nonhazardous industrial waste 21

o

Organization for Economic Cooperation and Development (OECD) 4

p

people factor, sustainability 4
persistent organic pollutants (POPs) 19, 30
phosphonates 127, 128
physisorption 74
phyto-remediation, chelation 145–147
planet factor, sustainability 5
plasma arc furnace 43, 45, 49
Pollution Control Committees (PCCs) 28
polydentate ligand 124, 132
polymer matrix 73–74
polyphosphates 126, 127
polysaccharides 73, 88, 171
primary valence 128
process intensification (PI)
  benefits of 183–184
  metal extraction processes
    centrifugation 185
    comminution 188–189
    drying 189
    liquid–liquid extraction 185–186
    mixing 186–188
    reactors 188
  micro-flow extraction devices 208
  principles of 184
profit factor, sustainability 4
protic ionic liquids (PILs) 163, 169
pyrometallurgical processes 12, 43, 45, 49, 178
pyrometallurgical treatment
  challenges of 49
  industrial waste 40–45
  WEEE 45–49

r

Raneynickel catalyst 50
recycling of wastes 8, 29, 31
reduce, reuse and recycle policy 8, 9, 20
refuse derived fuel (RDF) 9
Resource Conservation and Recovery Act (RCRA) 17, 21
Responsible Electronics Recycling Act (2011) 28
Rotterdam and Stockholm Conventions 28

s

scale-up technology
  batch to continuous process
    pilot plant designing 190
    process design fabrication 190
  tested pyrometallurgical methods 182
silicon valley toxics coalition (SVTC) 27, 29
social dimension of sustainability 4
solid adsorbents 71
solid waste management 15
spent catalyst 25, 37
  acid leaching 52
  salt-roasting 43
spent desulfurization catalyst 42
spent nickel oxide catalyst 45, 53
standard cyclone 196
State Pollution Control Boards (SPCBs) 28
Sulfobacillus thermosulfidooxidans 96
supported ionic liquid membranes (SILMs) 175
sustainability
  corporate sustainable management 7
  definition 2–3
  dimensions of 3–5
  financial advantages 2
  Four Force Model 5–7
  goal of 2
  TBL framework 2
  Type I (ISO 14024: Environment labels and declaration) 3
  Type II (ISO 14021: Self declared environmental claims) 3
  Type III (ISO 14025: Environment labels and declaration) 3
sustainable development (SD) 1, 5, 15, 17, 161, 183
switchable polarity solvents (SPS) 170

t

task specific ILs (TSILs)   161, 169
tested pyrometallurgical methods   182
2010 E-waste Management and
    Handling Rules   28
thermal plasmas   43, 45
thermophiles   96–97, 113, 115, 120
thiosulfate   50, 53, 60–66, 68, 69, 97,
    103, 104, 116
thiourea   50, 54, 60–66, 68, 69, 89
triple bottom line (TBL)   2

u

United Nations Environmental
    Programme (UNEP)   3

v

value stream mapping (VSM)   2, 7
volatile organic compounds (VOCs)
    158

w

waste
    agro-residue   20–24
    definition   17
    electrical and electronics   15
    Environmental Management (EM)
        Act   17
    generation and management   17
    hazardous   27–28
    hydrometallurgical treatment   50
    industrial   21
    metal recovery
        artificial ore   32–33
        human health and environment
            31–32
        wealth   33–34
    minimization   11
    municipal solid waste (MSW)   15
    pre-treatment
        calcination   37
        classification   36
        comminution   34–35
        "look and pick" disassembly
            operation   34
        screening/sieving   35–36
        segregation   36–37
    solid waste   15
Waste electrical and electronic
        equipment (WEEE)   17–20, 28,
        29, 32
    acid leaching   53
    bioleaching process   111–115
    chelating agents   149–151
    cyanide/non-cyanide lixiviant
        64–65
    ionic liquids   177–178
    pyrometallurgical treatment of
        45–49
waste management
    hierarchy   9
    projections of   16
Werner's theory   128
World Commission on Environment
        and Development (WCED)
        1, 5
World Health Organization's Children's
        Environmental Health   28
World Trade Organization
        (WTO)   4

z

zeolites   12, 71, 72, 79, 81–85